Current Topics in Neuroendocrinology Volume 7

Current Topics in Neuroendocrinology

Vol. 1: Sleep. Clinical and Experimental Aspects
1982. 47 figures. VII, 129 pages. DM 95,–. ISBN 3-540-11125-5
Neurobiology of REM Sleep. A Possible Role for Dopamine · Endocrine and Peptide Functions in the Sleep-Waking Cycle · Sleep Regulation: Circadian Rhythm and Homeostasis · Haemodynamic Changes During Sleep in Man

Vol. 2: Adrenal Actions on Brain
1982. 25 figures. V, 153 pages. DM 95,–. ISBN 3-540-11126-3
Glucocorticoids and Hippocampus: Receptors in Search of a Function · Glucocorticoids and the Developing Nervous System · Feedback Actions of Adrenal Steroid Hormones · Electrophysiologic and Clinical Aspects of Glucocorticoids on Certain Neural Systems · Adrenal Steroids and Behavioral Adaptation: Relationship to Brain Corticoid Receptors

Vol. 3: Central Cardiovascular Control. Basic and Clinical Aspects
1983. 71 figures. V, 192 pages. DM 106,–. ISBN 3-540-11350-9
Functional and Anatomic Aspects of Central Nervous Cardiovascular Regulation · Autonomic Nervous System and Blood Pressure Control in Normotensive and Hypertensive Conditions · Reflex Control of Circulation in Normotensive and Hypertensive Humans · Corticotropin-Releasing Factor: Central Nervous System Effects on the Sympathetic Nervous System and Cardiovascular Regulation · Neuropeptides and Central Blood Pressure Regulation · Centrally Acting Drugs as a Tool to Study Central Mechanisms of Blood Pressure Control · The Blood-Brain Barrier and its Role in the Control of Circulating Hormone Effects on the Brain

Vol. 4: Neurobiology of Vasopressin
1985. 53 figures. V, 203 pages. DM 98,–. ISBN 3-540-11351-7
Biosynthesis of Vasopressin · Electrophysiological Studies of the Magnocellular Neurons · Volume Regulation of Antidiuretic Hormone Secretion · Vasopressin, Cardiovascular Regulation and Hypertension · Neuroanatomical Pathways Related to Vasopressin

Vol. 5: Actions of Progesterone on the Brain
1985. 61 figures. V, 216 pages. DM 98,–. ISBN 3-540-13433-6
Antiprogestins · Progesterone Receptors in Brain and Hypophysis · Effects of Estrogen and Progesterone as Revealed by Neurophysiological Methods · Progesterone Receptors in CNS Correlated with Reproductive Behavior · Estradiol-Progesterone Interactions in the Reproductive Behavior of Female Rats · Behavioral Effects on Humans of Progesterone-Related Compounds During Development and in the Adult

Vol. 6: Neurobiology of Oxytocin
1986. 38 figures. V, 175 pages. DM 98,–. ISBN 3-540-15341-1
Biosynthesis of Oxytocin in the Brain and Peripheral Organs · Regulation of Oxytocin Release · Proteolytic Conversion of Oxytocin, Vasopressin, and Related Peptides in the Brain · Oxytocin and Behavior · Oxytocin as an Ovarian Hormone · Oxytocin and the Milk-Ejection Reflex

Morphology of Hypothalamus and Its Connections

Editors

D. Ganten and D. Pfaff

Contributors

Y. Arai, G. V. Childs, S. L. Cottingham, G. E. Hoffman
K. Inoue, H. Khachaturian, K. Kozasa
B. Krisch, K. Kurosumi, A. Matsumoto, Y. Nakai
M. Nishizuka, H. Ochiai, R. B. Page, M. Palkovits
D. Pfaff, C. J. Phelps, S. Shioda, J. R. Sladek, Jr.

With 94 Figures

Springer-Verlag
Berlin Heidelberg New York
London Paris Tokyo

Editors

Dr. DETLEV GANTEN, M.D., Ph.D.
Pharmakologisches Institut
Universität Heidelberg
Im Neuenheimer Feld 366
6900 Heidelberg/FRG

Dr. DONALD PFAFF, Ph.D.
Rockefeller University
York Avenue, and 66th Street
New York, NY 10021/USA

The picture on the cover has been taken from Nieuwenhuys R., Voogd J., van Huijzen Chr.: The Human Central Nervous System. 2nd Edition. Springer-Verlag Berlin Heidelberg New York 1981

ISBN-13: 978-3-642-71463-4 e-ISBN-13: 978-3-642-71461-0
DOI: 10.1007/ 978-3-642-71461-0

Library of Congress Cataloging in Publication Data. Morphology of hypothalamus and its connections. (Current topics in neuroendocrinology; v. 7) Includes bibliographies and index. 1. Hypothalamus–Anatomy. 2. Hypothalamo-hypophyseal system–Anatomy. 3. Hypothalamus–Innervation. I. Ganten, D. (Detlev), 1941– . II. Pfaff, Donald W., 1939– . III. Arai, Y. (Yasumasa) IV. Series. [DNLM: 1. Hypothalamus–anatomy & histology. W1 CU82Q v. 7/WL 312 M871] QM455.M693 1986 611'.81 86-20213

Softcover reprint of the hardcover 1st edition 1986

Typesetting, printing and bookbinding: Brühlsche Universitätsdruckerei, Giessen
2121/3130-543210

Contents

The Pituitary Portal System

R. B. PAGE

Contents

1 Introduction

By 1970, the picture seemed to be complete. Wislocki and King (1936) had first accurately described the portal system in 1936 after injecting monkeys, cats, and rabbits with India ink. They reported that superior hypophyseal arteries supplied the capillary bed of the median eminence. Its capillaries coalesced into long portal "venules" which passed between the median eminence and the pars distalis, where they then arborized into a secondary capillary bed. These capillaries in turn coalesced into lateral and inferior hypophyseal veins, which passed to the adjacent cavernous sinus. E. Scharrer and B. Scharrer (1940, 1944) had proposed the concept of neurosecretion, and ample evidence was forthcoming to support it as the

---

Milton S. Hershey Medical Center, The Pennsylvania State University, P. O. Box 850, Hershey, PA 17033, USA

Current Topics in Neuroendocrinology, Vol. 7

means by which the brain controls posterior pituitary (neural lobe) function (Verney 1947; Vigneaud et al. 1953; VanDyke 1955; Palay 1955, 1960; Sloper et al. 1960; Bargmann 1966).

Realizing that there were no nerve terminals in the pars distalis, Green and Harris (1947) proposed that the brain regulated anterior pituitary function by means of a "humoral relay through portal vessels" (i.e., they proposed to extend the concept of neurosecretion to explain how the brain regulates anterior pituitary function). In 1949 they supported this proposal with a report that blood flowed from the median eminence to the pars distalis through the portal vessels in the living rat. The isolation of thyrotropin-releasing hormone (TRH), a hormone which stimulates pituitary thyrotrophs to release thyroid-stimulating hormone (TSH), from the hypothalami of pigs and sheep by two independent research groups (Boler et al. 1969; Burgus et al. 1970) dispelled reasonable doubt.

Neural lobe function is the neurosecretion of oxytocin (OT) and vasopressin (AVP) into capillaries for transport by the systemic circulation to distant organs such as the uterus, breasts, and kidneys to regulate their function. Analogously, median eminence function is the neurosecretion of hypothalamic releasing and/or inhibiting hormones into capillaries for transport by restricted vascular routes to neighboring structures such as the pars tuberalis and pars distalis to regulate their function.

Central to the picture of the means by which the brain regulates the function of the pituitary gland is the relationship between neurosecretory axon terminals and capillaries and the anatomy and physiology of the pituitary portal system. This "final common path" (Scharrer 1965) between the brain and the pituitary gland is the subject of this review. It will consider the development of the portal system; the organization of the neurohypophysis and of the portal system; the pattern of blood flow in the pituitary gland; and the means by which the portal system regulates the amount and determines the distribution of blood flowing through the microcirculation of the neurohypophysis and thence into the anterior pituitary gland.

2 Development of the Portal System

2.1 Embryology

The pituitary gland is comprised of neural and glandular tissue. The neurohypophysis evaginates from the developing mammalian forebrain and is made up of the infundibulum (median eminence), infundibular stem, and infundibular process (neural lobe; Fig. 1). The adenohypophysis evaginates from the primitive stomedeum (Rathke's pouch) to form the pars tuberalis, pars intermedia, and pars distalis (Atwell 1926; Rioch et al. 1940; Wingstrand 1966). As the evaginations of the forebrain and stomodeum approach each other, they entrap a layer of mesenchyme between them. Although some hypothalamic projections to the neurohypophysis pass across this border to terminate in the pars tuberalis and pars intermedia, most of the hypothalamic projections to the neurohypophysis termi-

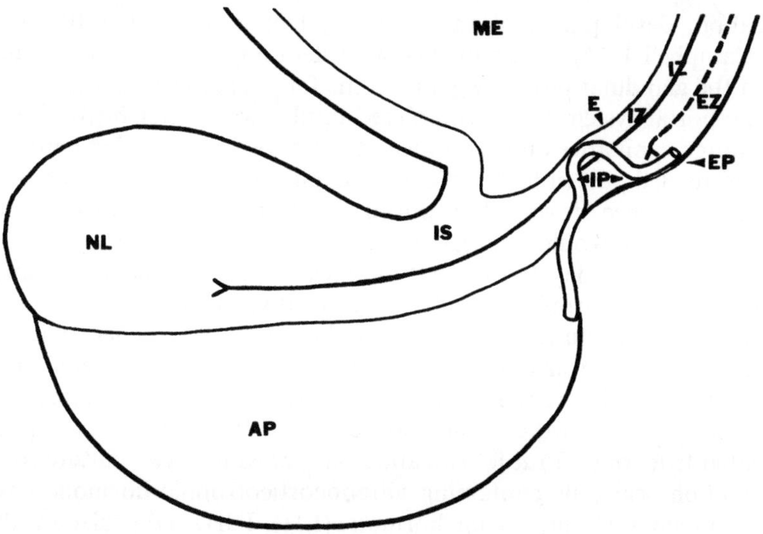

Fig. 1. Schematic representation of the rabbit pituitary gland in midsagittal section, stressing the neurohypophysis and its components: the median eminence (*ME*) or infundibulum, the infundibular stem (*IS*), and the neural lobe (*NL*). The median eminence is subdivided into an ependymal layer (*E*), an internal zone (*IZ*), and an external zone (*EZ*). The primary capillary plexus of the median eminence is subdivided into an external plexus (*EP*) on the surface of the median eminence and an internal plexus (*IP*) which penetrates into the substance of the median eminence. *AP*, anterior pituitary

nate there (J. D. Green 1948, 1951). The mesenchyme, entrapped between neuro- and adenohypophysis, gives rise to the mantle plexus of capillaries that invades both the developing neuro- and adenohypophysis and becomes the portal system (Wislocki 1937b; Enemar 1961; J. D. Green 1966). It is this vascular system, not a neuronal network, which conveys information between the brain and the pars distalis of the pituitary gland (Green and Harris 1947).

This account is based primarily upon the reports of Atwell (1926) and Wislocki (1937b). More recently, Takor and Pearse (1975) have proposed that the adenohypophysis may be of neuroectodermal and not of stomodeal origin. They note that some adenohypophyseal cells are capable of amine precursor uptake and decarboxylation (APUD) and thus can be considered a part of the APUD system. Cells in this system are believed to be of neuroectodermal origin (Pearse and Takor 1976). Pearse and Takor propose that the adenohypophysis arises from the ventral neural ridge, not from the invaginating stomodeum. Accounts of the development of the portal system have not been reevaluated in the light of their proposals.

2.2 Role in Adenohypophyseal Development

Although the role of the portal system in regulating the function of adenohypophyseal cells has been clear for many years (Harris 1955), its role in regulating the development of functional cell types within the adenohypophysis has not. In the rabbit, development of the portal system is completed 10 days before birth (Campbell 1966); hence the portal system might play a role in the development of the glandular pituitary gland in the fetus. However, in the rat development of the portal system is not completed until 1 week after birth (Glydon 1957). Although capillaries on the surface of the developing median eminence are present before birth, the (internal) plexus of capillary loops develops after the birth of the rat, and hence the role of the portal system in the regulation of adenohypophyseal development would not seem important in this species. Further, encephalectomy of fetal rats at day 16 of gestation does not prevent the development of melanotrophs, corticotrophs, gonadotrophs, or thyrotrophs, as determined by immunohistochemical study on day 21 (Chatelain et al. 1976; Begeot et al. 1981). Pituitary anlagen taken from day 12–15 fetal rats and grafted to the renal capsule show subsequent cytodifferentiation by light microscopic (LM) and transmission electron microscopic (TEM) criteria (Gash et al. 1977). When Rathke's pouches were isolated from day 12 rat fetuses and maintained in organ culture for 9 days adenohypophyseal cells containing adrenocorticotrophic hormone (ACTH), growth hormone (GH), luteinizing hormone (LH), TSH, and prolactin (Pro) were found by immunohistochemistry (Watanabe and Daikoku 1976). Clonal strains of anterior pituitary cells have been established from Rathke's pouch tissue of fetal rats. Radioimmunoassay of the medium demonstrates that some cell lines produce only Pro, some produce Pro and GH, and others produce Pro, Gh, and ACTH (Ishikawa et al. 1977). These studies suggest that the differentiation of anterior pituitary cells is programmed independently of the presence of hypothalamic hypophysiotropic factors carried to the anterior pituitary gland by portal vessels.

Shiino et al. (1978), however, reported that clonal cells derived from fetal Rathke's pouch epithelium produced more hormone in vitro and demonstrated more (secretory) granules on TEM examination if the culture medium was supplemented with fresh serum or median eminence extract. The development of fetal rat somatotrophs in vitro was aided if the culture medium was supplemented with insulin, cortisol, thyroxine, and glucagon. Growth hormone-releasing factor, when added to the medium, increased the size but not the number of somatotrophs (Hemming et al. 1984).

Nerves terminating in the perivascular space of median eminence capillaries in the mouse (Eurenius and Jarskar 1971) and rat (Halasz et al. 1972) can first be demonstrated by TEM on day 16 of gestation. At that time the internal plexus has not penetrated the median eminence, but the external plexus (mantle or supratuberal plexus) of median eminence capillaries has developed, as have its connections to the pars distalis (Halasz 1972). Gross and Baker (1977) reported that gonadotropin-releasing hormone (GnRH) first appears in the mouse median eminence on day 17 of gestation. Gonadotrophs (Gross and Baker 1979) also first appear in the pituitary on day 17 in the fetal mouse. They were found aggregated about capillaries in the ventral portion of the pars distalis. The appearance of gonadotrophs did not precede the differentiation of the external (mantle) plexus and its connections or the appearance of GnRH in the median eminence, but did precede the development of the internal plexus. While the hypothalamus and the portal system may not be necessary for programming the development of adenohypophyseal cells, the hypothalamus and a portal system uniting the rostral end of the neurohypophysis with the pars distalis appear to facilitate their development.

3 Organization of the Neurohypophysis

3.1 Neurohemal Contact

Although the neurohypophysis is a diverticulum of the brain, it is easily distinguished from the overlying hypothalamus by its lack of a blood–brain barrier. Following the injection of Trypan Blue into rabbits, cats, and monkeys, Wislocki and King (1936) demonstrated that the boundary separating the unstained hypothalamus from the blue-stained neurohypophysis lay at the level of the tuberoinfundibular sulcus. In the hypothalamus, passage of the trypan blue, which is bound to plasma albumin, from the capillaries into the neuropil is prohibited. The

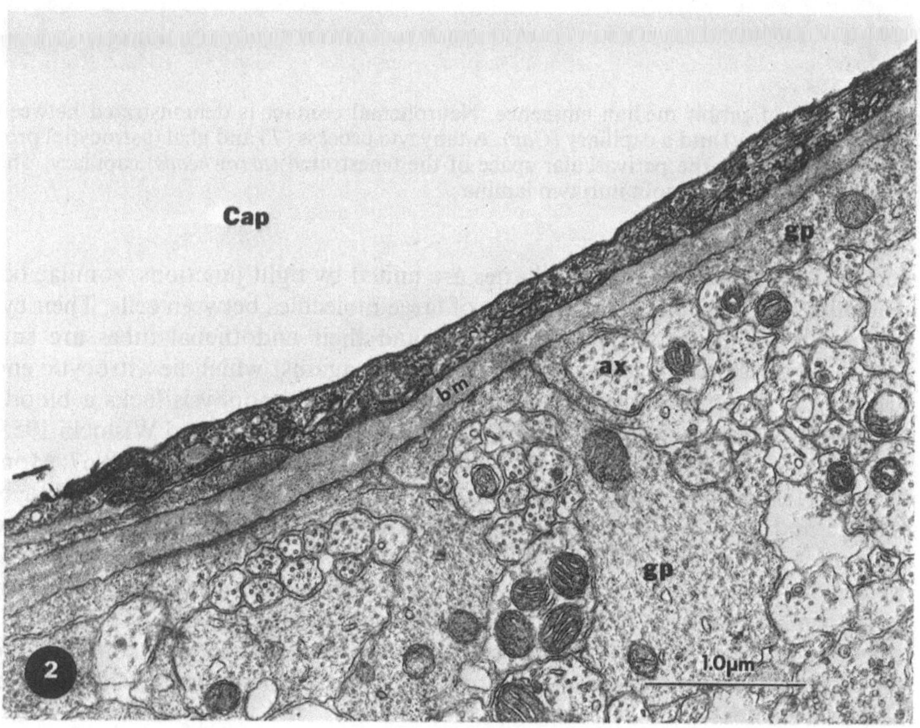

Fig. 2. TEM of rabbit arcuate nucleus. The morphological components of the blood–brain barrier are illustrated. Axons (*ax*) are separated from a hypothalamic capillary (*Cap*) by a thick single basement membrane (*bm*), and are enveloped by glial (astrocytic) processes (*gp*). The capillary endothelium is not fenestrated

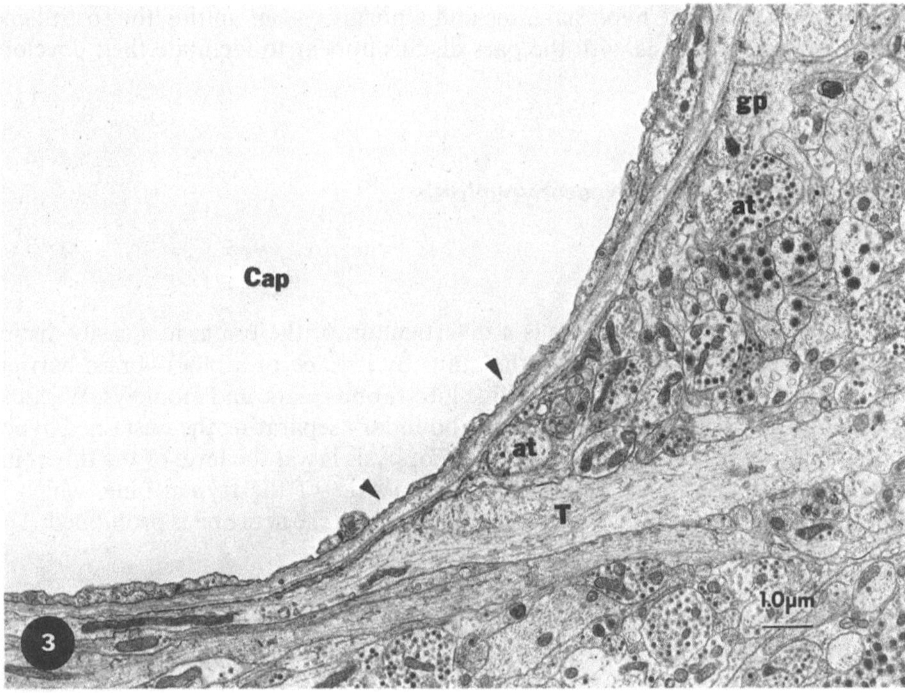

Fig. 3. TEM of rabbit median eminence. Neurohemal contact is demonstrated between axon terminals (*at*) and a capillary (*Cap*). A tanycyte process (*T*) and glial (astrocytic) process (*gp*) also lie in the perivascular space of the fenestrated (*arrowheads*) capillary. The basement membrane is split into two lamina

endothelial cells of cerebral capillaries are united by tight junctions, zonulae occludentes, which prohibit the passage of large molecules between cells. Their cytoplasm contains few transport vesicles, and their endothelial tubes are surrounded by a single dense basement membrane against which lie astrocytic end feet (Fig. 2; Brightman and Reese 1969). The neurohypophysis lacks a blood–brain barrier because its capillaries are fenestrated (Dempsey and Wislocki 1955; Lederis 1965; Kobayashi et al. 1966; Rinne 1966; Duffy and Menefee 1967; Monroe 1967; Akmayev 1969; Bergland and Torack 1969; Rodriguez 1969 a, b; Knigge and Scott 1970; Page and Dovey-Hartman 1984 a). These fenestrated capillaries are in turn surrounded by a double basement membrane. Axons, as well as glial end feet, terminate on the outer basement membrane (Fig. 3), an arrangement called neurohemal contact (Bern and Knowles 1966). The terminals which lie in the neurohypophysis are organized to discharge their contents into the perivascular space of fenestrated capillaries, not into a synaptic cleft between neuronal elements.

Transmission electron microscopic TEM examination reveals several types of terminals in the perivascular space of fenestrated capillaries when neurohypophyseal tissue is prepared with aldehyde fixative. For example, in the rabbit some terminals (those found predominantly in the neural lobe) contain large dense-core vesicles, about 250–350 nm in diameter. Other terminals, those found predominantly in the median eminence, contain large granular vesicles about 120 nm in diameter, while others do not contain any electron-dense vesicles. However, all three terminal types contain clear vesicles with an average diameter of about 50 nm (Duffy and Menefee 1965; Page and Dovey-Hartman 1984a). Similar observations have been made in the rat (Kobayashi et al. 1966; Rinne 1966; Monroe 1967), in the human (Lederis 1965; Bergland and Torack 1969), and in submammalian species, for example the toad (Rodriguez 1969a, b). However, the number of groups into which the terminals are classified, a process based on vesicle size, varies with author and species.

Axon terminals containing dense-core and lucent vesicles lie in the perivascular space of neurohypophyseal capillaries. They are the terminals of the supraopticohypophyseal tract, a magnocellular peptidergic neurosecretory system. Their cell bodies lie in the hypothalamic supraoptic and paraventricular nuclei (Sherlock et al. 1975). Hatton et al. (1976) presented evidence that there are two populations of magnocellular neurons in the paraventricular nucleus of the rat – one with an area of $121.4 \pm 5.4\ \mu m^2$ (mean ± SE), and a second with an area of $136.6 \pm 7.6\ \mu m^2$. On the other hand, Fisher et al. (1979) found a range of 120–290 μm^2, with a median area of 210 μm^2, in magnocellular perikarya in the supraoptic and paraventricular nuclei of the rat. Perhaps differences in tissue preparation (with resultant differences in tissue shrinkage) account for the discrepancies in measurement of the area of cell bodies in the magnocellular system. In any event, such cells are perceptibly larger than other (parvicellular) neurons in the supraoptic and paraventricular nuclei or cells in the suprachiasmatic nuclei, where the paracellular vasopressinergic cell bodies range in area from 40 to 100 μm^2 (Fisher et al. 1979).

OT and AVP, with their associated carrier proteins (neurophysin I and II, respectively), are synthesized in separate cells of this magnocellular system. The constituent amino acids of these peptide hormones are assembled on ribosomes, processed through the Golgi apparatus, and packaged in the large dense-core vesicles. The assembled and packaged peptide is a prohormone and is considerably larger than the biologically active AVP and OT and their associated neurophysins. The dense-core vesicles are transported down the axons to be stored in their terminals in the neural lobe. During transport, post-translational processing of the prohormones occurs. The peptide is enzymatically cleaved into smaller, biologically active components. With appropriate stimulation, the contents of the large dense-core vesicles are discharged from the axon terminals, and OT or AVP is released into the perivascular space of fenestrated capillaries (Palay 1945, 1960; Bargmann and Scharrer 1951; Sloper et al. 1960; Bargmann 1966; Norstrom et al. 1971; Zimmerman et al. 1974; Vandesande and Dierickx 1975; Flament-Durand et al. 1975; Zimmerman and Antunes 1976; Gainer et al. 1977; Antunes and Zimmerman 1978). Although it is generally accepted that the large dense-core vesicles stored in the axon terminals of magnocellular neurons contain peptide hormones (Leclerc and Pelletier 1974; Silverman and Zimmerman 1975) and release them in response to appropriate stimuli, the role of the small lucent vesicles has not been clarified. They resemble "synaptic vesicles," but they do not contain neurotransmitters, and recent work shows that they do not contain peptide hormones and are not associated with axonal endoplasmic reticulum, the Golgi apparatus, or microtubules. They may participate in the neurosecretory process, not by being involved with the packaging, transport, or exocytosis of neurosecretory material, but by endocytosis and possibly recycling of neurohypophyseal terminal membrane (Broadwell et al. 1984).

Axon terminals seen to contain large granular vesicles and small lucent vesicles on TEM examination also lie in the perivascular space of neurohypophyseal capillaries. These terminals are smaller than terminals of the supraopticohypophyseal tract (0.3–1.4 μm vs 1.5–7.0 μm in the rat; Monroe 1967). The average diameter of large granular vesicles contained within these parvicellular terminals is smaller on average than the diameter of dense-core vesicles in the axons and terminals of the supraopticohypophyseal tract – 90 μm vs 180 μm in the rat (Monroe 1967), and 120 μm vs 230 μm in the rabbit (Page and Dovey-Hartman 1984a). Axon terminals which contain large granular vesicles and small lucent vesicles may be aminergic or peptidergic (Bloom 1970; Ajika 1980; Schwab and Thoenen 1983). In terminals of the parvicellular peptidergic system, TEM immunohistochemical studies demonstrate peptide hormones localized within the large granular vesicles (Ajika 1979, 1980; Nakai et al. 1983; Shioda and Nakai 1983). Although Pelletier and Dube did localize a peptide, α-melanocyte-stimulating hormone (α-MSH), within small lucent vesicles 40–70 nm in diameter, Broadwell et al. (1984) propose that small lucent vesicles in terminals of paracellular pep-

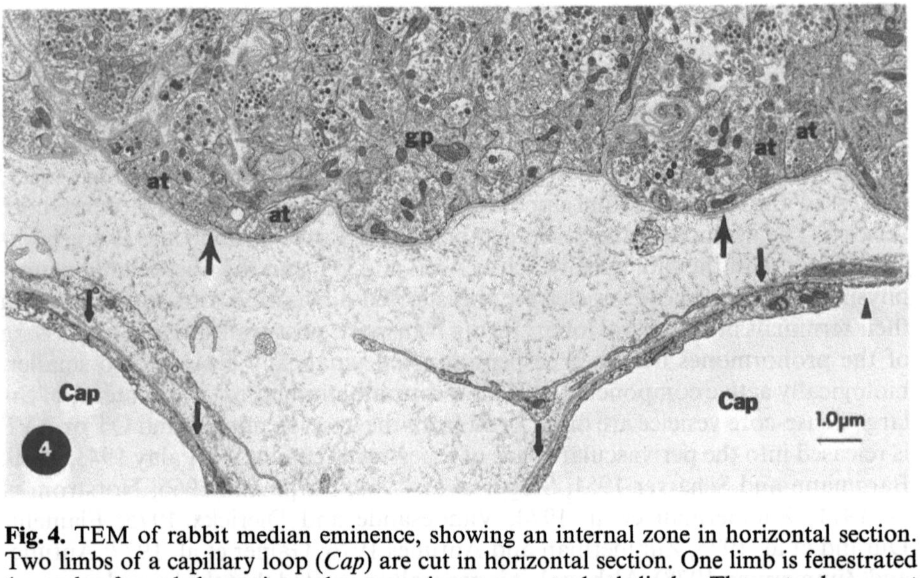

Fig. 4. TEM of rabbit median eminence, showing an internal zone in horizontal section. Two limbs of a capillary loop (*Cap*) are cut in horizontal section. One limb is fenestrated (*arrowhead*), and the perivascular space is common to both limbs. The outer basement membrane (basal lamina) is condensed at the margin of the contact zone common to both capillaries (*vertical large arrows*). Thickening of the plasmalemma of axon terminals (*at*) is occasionally seen. There is no evidence of synaptic specialization at the axon terminals in the perivascular space about this capillary formation. Note that the inner basement membrane is condensed about each individual capillary (*vertical small arrows*). *gp*, glial process. From Page and Dovey-Hartman (1984a)

tidergic neurons play a role in endocytosis and terminal membrane recycling as in the magnocellular neurosecretory system. In terminals of the aminergic systems, uptake of the false neurotransmitter 5-hydroxydopamine (5-OHDA) into small lucent and large granular vesicles occurs (Tranzer and Thoenen 1967; Bloom 1970; Richards and Tranzer 1970; Tranzer and Richards 1976; Page and Dovey-Hartman 1984a). Ajika (1979) employed immunohistochemical techniques and TEM to localize tyrosine hydroxylase preferentially in small lucent

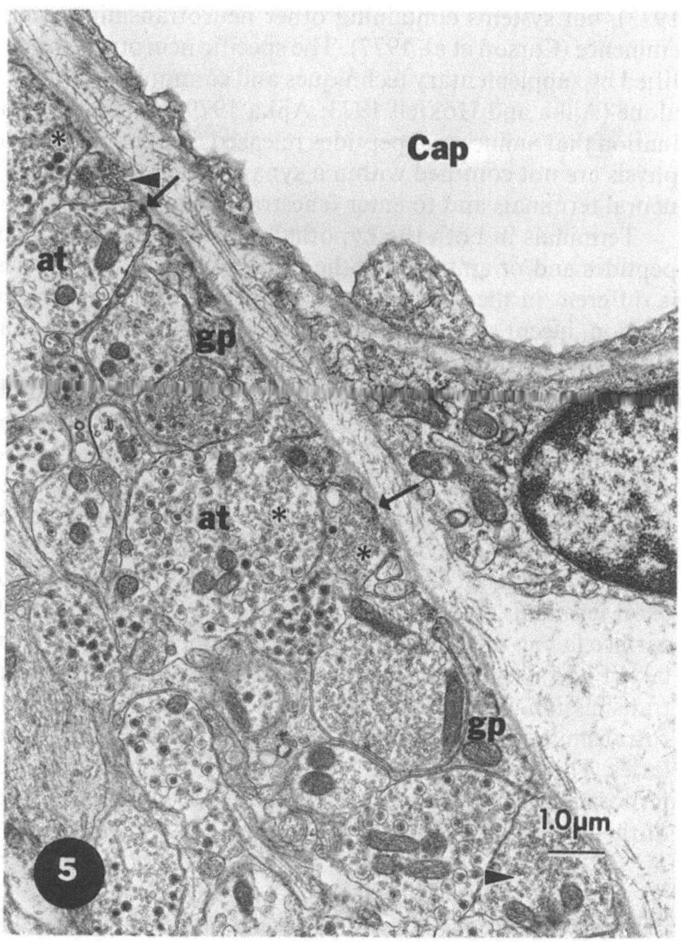

Fig. 5. TEM of rabbit median eminence: neurohemal contact with internal plexus capillary. Axons terminate in the perivascular zone about a capillary (*Cap*) with fenestrated endothelium. The contact zone is comprised of axon terminals containing large granular and small lucent vesicles (*), axon terminals containing small lucent vesicles (*arrowheads*), and glial processes (*gp*). The apparent extracellular location of lucent vesicles (at *upper arrowhead*) is infrequently seen or may be an artifact of sectioning. Note the lack of synaptic specializations between axon terminals, the occasional thickening of the plasmalemma of terminals abutting on the perivascular space (*small arrows*), and the glial processes, which are interposed between some axon terminals and the perivascular space. From Page and Dovey-Hartman (1984a)

vesicles within axon terminals in the rat. Schwab and Thoenen (1983) propose that large granular vesicles in aminergic terminals return used synaptic vesicle membrane and associated transmitter to the cell body. The contrary view, that exocytosis occurs from both large and small dense-core vesicles, is defended by Thureson-Klein (1983).

Axon terminals containing only small lucent vesicles are also seen on TEM examination of the neurohypophysis following aldehyde fixation (Page and Dovey-Hartman 1984a). Some of these axons are aminergic (Ajika and Hökfelt 1973), but systems containing other neurotransmitters terminate in the median eminence (Carson et al. 1977). The specific neurotransmitter in each must be identified by supplementary techniques and cannot be surmised by TEM examination alone (Ajika and Hökfelt 1973; Ajika 1979). One can surmise from TEM examination that amines and peptides released from nerve terminals in the neurohypophysis are not confined within a synaptic cleft, but are free to reach neighboring neural terminals and to enter fenestrated capillaries.

Terminals in both the hypothalamus and the neurohypophysis may contain peptides and/or amines, but the organization of the vesicles within the terminals is different in the two areas. In regions of the brain organized for neurotransmission, lucent synaptic vesicles are aggregated within aminergic terminals near a presynaptic thickening that borders a synaptic cleft. Terminals of peptidergic systems outside the neurohypophysis also synapse with other neurons. In presynaptic peptidergic terminals, the small lucent vesicles are aggregated near a presynaptic thickening, whereas large granular vesicles are randomly dispersed in the terminal (Johansson et al. 1980). This organizational pattern is seldom found in the neurohypophysis (Fig. 4). The intercellular spaces between axon terminals are remarkably free of synaptic specializations, and synapses are not often seen on TEM examination. Vesicles, both large and small, are randomly dispersed within axon terminals. Thickening of the axon terminal, when found, occurs at its free surface facing the outer layer of the basement membrane and a fenestrated capillary (Fig. 5). The peptidergic systems in the pituitary are organized not for neurotransmission, but for neurosecretion. Even its aminergic systems release their neurotransmitter (e.g., dopamine) into the perivascular space of fenestrated capillaries. The termination of a secretory neuron near a blood vessel without an interposed blood–brain barrier constitutes the fundamental organizational pattern of the neurohypophysis.

3.2 Regional Disposition of Neurosecretory Systems

Regional specialization of function within the neurohypophysis is achieved by segregation of neurosecretory systems according to the hormone produced in and secreted by its constituent neurons, and by the disposition of the vascular bed into which the hormone is secreted. At its caudal pole lies the neural lobe, in which aminergic and peptidergic fiber systems terminate. From these terminals secretions are carried by systemic vascular routes to regulate the function of target organs such as the kidney, heart, and uterus. At the rostral pole of the neurohypophysis lies the median eminence, or infundibulum (Fig. 6). In it terminate ami-

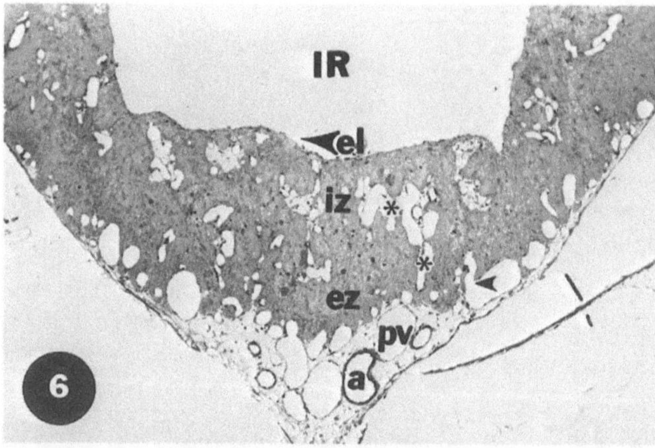

Fig. 6. LM of rabbit median eminence (coronal section). The dorsal boundary is delineated by the ependymal layer (*el*), which separates it from the infundibular recess (*IR*) of the third ventricle. The median eminence is comprised of the internal zone (*iz*) and external zone (*ez*). External plexus capillaries lie in the ventral surface of the external zone and drain to portal vessels (*black arrowheads*). Internal plexus capillaries arise from the external plexus (*) and penetrate the median eminence, some reaching the ependymal layer. The individual capillaries of a single unit in the internal plexus share a common extracellular space. Portal vessels (*pv*), along with an arteriole (*a*), lie in fibrous connective tissue beneath the median eminence. (From Page and Dovey-Hartman 1984a Methylene blue azure II)

nergic and peptidergic fiber systems descending from the hypothalamus and aminergic systems ascending from the brain stem. The secretions released from these terminals are carried by restricted vascular routes to the pars distalis of the adenohypophysis.

The ependymal layer of the median eminence forms the floor of the third ventricle. The epithelial lining that separates the neuropil of the median eminence from the ventricular fluid differs from the epithelium overlying the adjacent hypothalamus (Fig. 7a). Ependymal cells in the infundibular recess are not ciliated. Microvilli and apical blebs are present at their apical surface. Some ependymal cells are stretched between the ventricular surface and the perivascular space of capillaries in the median eminence; these elongated cells are called tanycytes. Adjacent ependymal cells overlying the median eminence are linked by tight junctions, which limit passive exchange of materials between the ventricle and the extracellular space of the median eminence. (For a review of this subject the reader is referred to Knigge et al. 1974.) The neuropil of the median eminence is thus protected from an uncontrolled leak of neurosecretions into the ventricular system.

The internal zone (Fig. 7a) lies beneath the ependymal layer of the vertebrate median eminence (Wingstrand 1966), and is subdivided into a hypependymal layer and a fiber layer (Kobayashi and Matsui, 1969). Within the hypependymal layer lie subependymal cells and pituicytes. In the hypependymal layer of the rat, axons of parvicellular peptidergic neurons projecting from the hypothalamus (Ajika 1980; Bugnon et al. 1982) and of aminergic neurons projecting from the

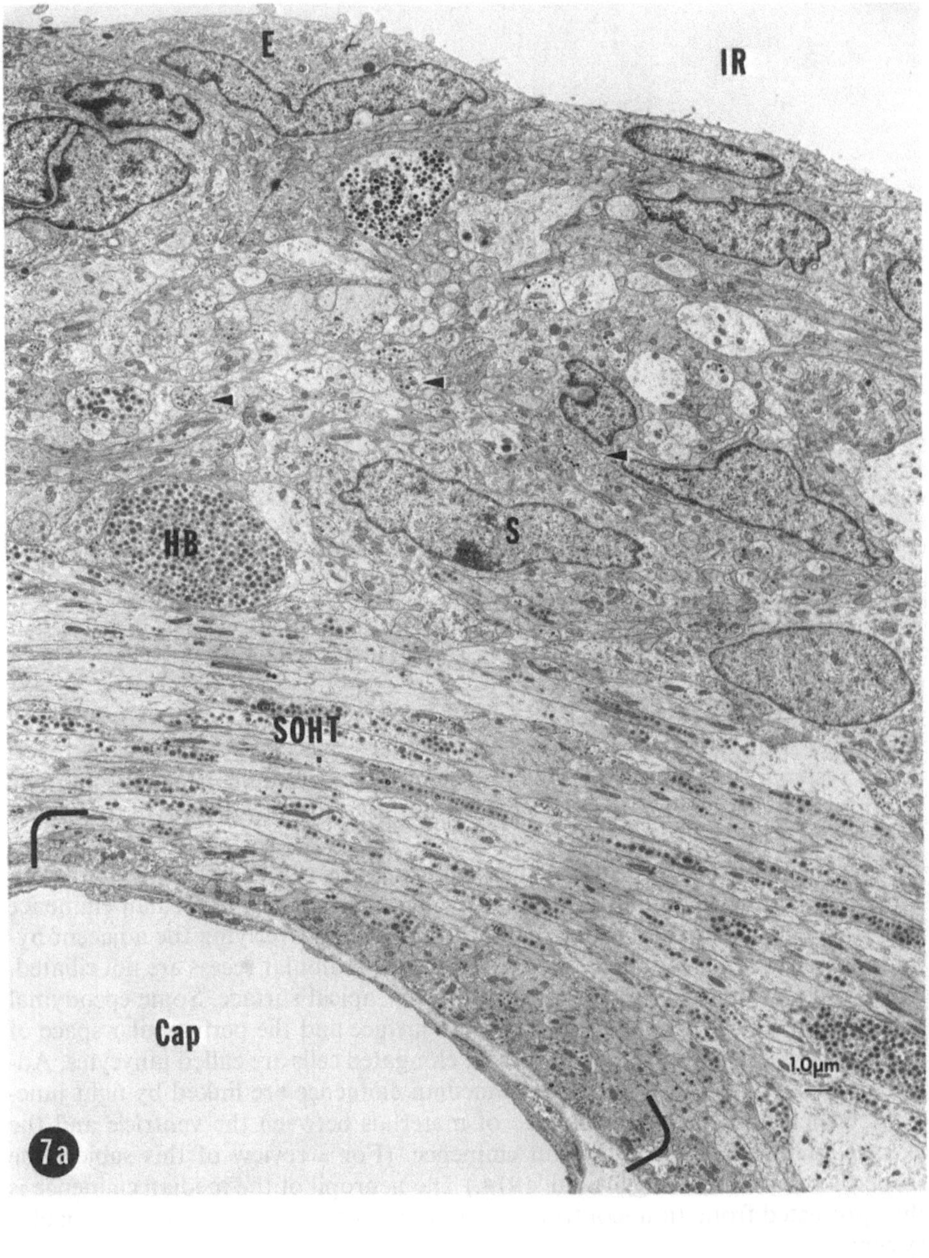
E
IR
HB
S
SOHT
Cap
1.0μm
7a

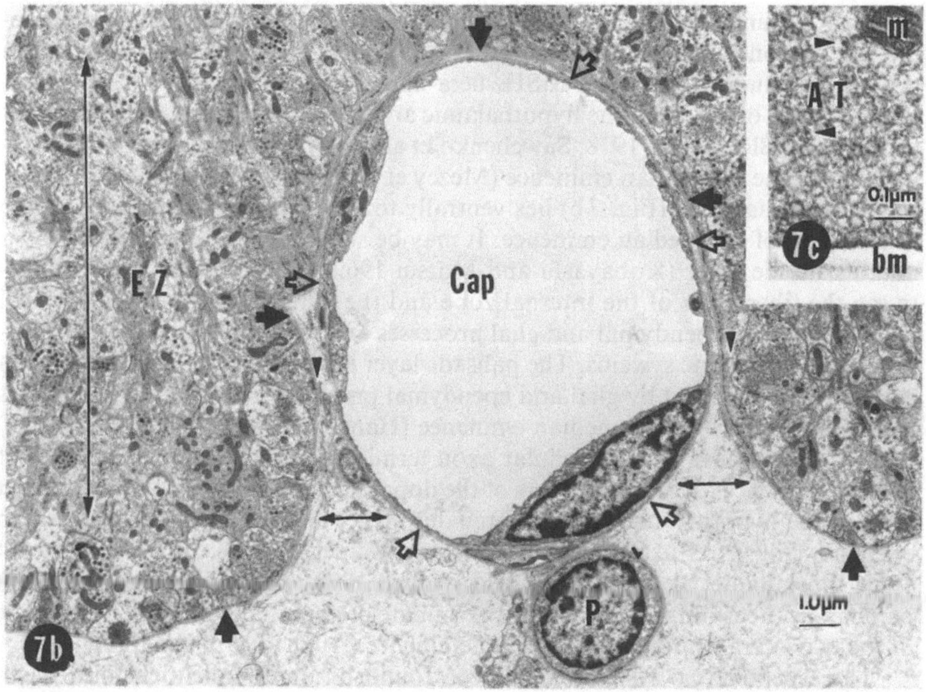

Fig. 7 a–c. TEM of rabbit median eminence (coronal section). **a** Ependymal layer and internal zone. Nonciliated ependymal cells (*E*) separate the median eminence from the infundibular recess (*IR*) of the third ventricle. In the hypependymal zone subependymal cells (*S*), neurites containing granular vesicles (*black arrowheads*), glial processes, and Herring bodies (*HB*) are present. The fibrous layer is made up of the supraopticohypophyseal tract (*SOHT*), whose axons contain large dense-core vesicles. The supraopticohypophyseal tract passes by a fenestrated capillary (*Cap*), which is surrounded by a cuff of axon terminals (*brackets*). These terminals contain osmophilic vesicles which are smaller than vesicles in Herring bodies and in axons of the SOHT. **b** External zone. Terminals containing lucent and granular vesicles make neurohemal contact with a fenestrated capillary (*Cap*) in the external zone (*EZ*). *Hollow arrows* designate the inner basement membrane applied to the capillary; *black arrows* designate the outer basement membrane applied to the capillary. *Double-headed arrows* demonstrate the divergence of the inner and outer layer of the basal laminae at the surface of the median eminence. The extracellular space ventral to the median eminence is continuous with the perivascular space about the capillary. *Arrowheads indicate condensations of the plasma membrane of axon terminals facing the perivascular space. P*, pericyte **c** axon terminal (*AT*) in the perivascular space of the external plexus of a rabbit pretreated with 5-OHDA. Many small (lucent) vesicles contain eccentrically placed dense cores (*arrowheads*). *m*, mitochondrion; *bm*, basement membrane. From Page and Dovey-Hartman (1984 a)

brain stem are found. The latter system is believed to be noradrenergic and is called the reticuloinfundibular tract (Björklund et al. 1973; Ajika 1980). The fiber layer is defined by axons of the supraopticohypophyseal tract (Wingstrand 1966; Hanstrom 1953). Although axon terminals of the supraopticohypophyseal tract are not found in the internal zone, other fiber systems terminate there (Page and Dovey-Hartman 1984 a). The aminergic reticuloinfundibular tract terminates in the internal zone of the rat median eminence, principally in its medial third. The mode of termination, whether neurohemal or synaptic, is not yet known. Peptidergic neurons which contain the processed products of pro-opiomelanocortin (POMC), including ACTH, α-MSH, beta-lipotropin (β-LPH), and beta-endorphin (β-End), originate in the hypothalamic arcuate (tuberal) nucleus (Bugnon et al. 1979a, b; Bloch et al. 1978; Sawchenko et al. 1982) and terminate in the internal zone of the rat median eminence (Mezey et al. 1985).

The external zone (Fig. 7b) lies ventrally to the internal zone and forms the oral surface of the median eminence. It may be subdivided into a reticular layer and a palisade layer (Kobayashi and Matsui 1969). The reticular layer lies between the fiber layer of the internal zone and the palisade layer of the external zone. It contains ependymal and glial processes and axons of aminergic and parvicellular peptidergic systems. The palisade layer lies at the surface of the median eminence; it is defined by glial and ependymal processes oriented at right angles to the oral surface of the median eminence (Hanstrom 1953; Wingstrand 1966) and by the presence of parvicellular axon terminals in the perivascular space of median eminence capillaries. Fibers of the dopaminergic tuberoinfundibular tract arise in the hypothalamic tuberal (arcuate) nuclei and terminate in the median eminence (Fig. 7c; Fuxe 1963, 1964; Szentagothai et al. 1968; Fuxe and Hökfelt 1969; Björklund et al. 1970; Ajika and Hökfelt 1973). In the laboratory rat, dopaminergic tuberoinfundibular fibers terminate preferentially in the external zone of the two lateral thirds of the median eminence (Ajika 1979; Selmanoff 1981).

Three peptidergic systems have been found by immunohistochemical techniques to terminate in the external zone of the median eminence. The first arises from the paraventricular nuclei and carries OT and AVP (in separate neurons) to the external zone of the median eminence (Antunes et al. 1977; Vandesande et al. 1977). The second system is comprised of TRH- (Lechan and Jackson 1982), corticolibrin- (CRF; Bugnon et al. 1982; Kawata et al. 1982, 1983; Merchenthaler et al. 1982), and somatostatin-containing neurons (Vigh et al. 1978; Ajika 1980). Palkovits (1982) proposes that in the rat these two systems originate in the periventricular region of the hypothalamus, particularly in the paraventricular nuclei, projecting first into the lateral retrochiasmatic area, and then inferomedially to enter the median eminence anteriorly and then to distribute to its medial third. The GnRH-containing fibers constitute a third system which originates principally in the preoptic region (Ajika 1979; Ibata et al. 1979) and projects directly posteriorly to terminate in the external zone of the two lateral thirds of the median eminence (Ajika 1980; Palkovits 1982). Thus within the median eminence of the rat there is a further segregation of functional systems. Systems concerned with the response to stress or with survival of the individual are segregated in the medial third of thc median eminence, whereas systems related to reproduction or to survival of the species are aggregated in the two lateral thirds of the median eminence.

The infundibular stem lies interposed between the infundibulum and the infundibular process. It is classically held to be made up of only fibers of the supraopticohypophyseal tract passing on to the infundibular process. However, some axons of the dopaminergic tuberohypophyseal tract terminate in this region (Björklund et al. 1970). In addition, peptidergic systems terminate in the infundibular stem (Hökfelt et al. 1978a; Lechan and Jackson 1982) of the rat.

The infundibular process, or neural lobe, is the caudal region of the neurohypophysis, and is made up of axons and terminals of the magnocellular supraopticohypophyseal tract (Zimmerman 1981), which contain OT and AVP, and of pituicytes. The axons and terminals of the supraopticohypophyseal tract arise principally from the supraoptic nuclei, although some magnocellular neurons in the paraventricular nuclei contribute to it (Sherlock et al. 1975). With hypophysectomy, 90% of the cells in the supraoptic nuclei degenerate, whereas only about 15% of the cells in the paraventricular nuclei degenerate (Rasmussen 1940). In the rat, the paraventricular nuclei contain only 16.5% of the total number of labelled cells in the hypothalamus after injection of the infundibular process with horseradish peroxidase. The supraoptic nuclei contain twice as many labelled cells (Fisher et al. 1979). The neurohypophyseal projections of the supraoptic nuclei of the rat are predominantly to the neural lobe, and constitute the majority of its axons and terminals. The neurohypophyseal projections of the paraventricular nuclei are to the median eminence, infundibular stem, and neural lobe. In the neural lobe, axons and terminals arising from the paraventricular nuclei are in the minority. Terminals of the dopaminergic tuberohypophyseal tract (Björklund et al. 1970) may also be found in this region, and some parvicellular peptidergic systems also extend into the neural lobe (Hökfelt et al. 1978a). The secretions of the neural lobe are carried to distant target organs to regulate their function.

It is of more than passing interest that the specific role of a neurohormone may be different in each of the sites in which it is released, but the different roles are functionally related. For example, AVP released from the neural lobe increases blood pressure and decreases urine output, while AVP released from the median eminence potentiates the release of ACTH from the pars distalis. The function of AVP released from nerve terminals in the brain stem remains to be clarified, but it appears to be involved in a response to stress (Swanson and Sawchenko 1980). This concept, that the function of a neurohormone may vary with its site of release, is analogous to the concept that the role of a hormone molecule may depend upon the niche in the phylogenetic tree occupied by the organism in which it is found. For example, OT-/AVP-like immunoreactivity has recently been found in the hydra (Grimmelikhuijzen et al. 1982).

The pattern of regional segregation of neurosecretory systems within the neurohypophysis is best worked out in the rat. This pattern is generally accepted as a prototype for the mammalian neurohypophysis. However, species variations may be marked. For instance, the distinction between the internal and external zone is blurred in some species, including man, by intrusion of the palisade layer into the internal zone (Hanstrom 1953). The sites of termination of neurosecretory systems also vary with the species studied. Anthony et al. (1984) report that GnRH fibers preferentially terminate in the internal zone (not the external zone) of the bat median eminence, whereas in humans, ferrets, and monkeys luteinizing hormone-releasing hormone (LHRH) terminals are abundant in both zones. In the latter species, LHRH fibers project well into the neural lobe. The prototype presented presumes that the finding of one peptide (or peptide family) in a neurosecretory neuron precludes the finding of another peptide or of a biogenic amine. This presumption may not be strictly true, and neurosecretory cells containing (or capable of synthesizing) more than one hormone have been reported (Watson et al. 1982; Vanderhaeghen et al. 1983; Whitnall et al. 1983; Kiss et al. 1984). However, reports of colocalization of AVP and CRF (Roth et al. 1982; Burlet et al. 1983) have recently been challenged (Taniguchi 1984). In addition, colocalization of one peptide hormone (substance P) and a neurotransmitter (serotonin) with neurons in the rat brain stem has been reported (Hökfelt et al. 1978 a, b, 1980).

Notwithstanding reservations, the basic organizational pattern within the vertebrate neurohypophysis stands out clearly from the background of species differences in detail. It is the relationship of individual axon terminals to capillaries and the regional segregation of systems of neurosecretory terminals within the neurohypophysis. Blood vessels, not other nerves, carry the information released from axon terminals in the median eminence and contribute to the organization of the neurohypophysis. In general, it may be said that fiber systems terminating caudally (in the neural lobe) will have their secretions delivered to the systemic circulation or to restricted regions of the adjacent adenohypophysis by capillaries and short portal vessels. Fiber systems which terminate rostrally (in the median eminence) have their secretions delivered to the pars distalis of the adenohypophysis by long portal vessels.

4 Organization of the Portal System

4.1 Arterial Supply and Venous Drainage of the Neurohypophyseal Capillary Bed

Although it is regionally specialized, the entire neurohypophysis is subserved by a single capillary bed (Fig. 8; Wislocki and King 1936). It is supplied rostrally by superior hypophyseal arteries and caudally by inferior hypophyseal arteries (Fig. 9; Page and Bergland 1977). A third source of arterial supply has been identified and variously termed the "anterior hypophyseal" artery, "peduncular" artery, "loral" artery, or the "trabecular" artery, depending upon the species studied (Harris 1947; Landsmeer 1951; McConnell 1953; Xuereb et al. 1954; Daniel and Prichard 1975). Because this artery arises between the inferior and the superior hypophyseal arteries and supplies the infundibular stem lying between the infundibulum and the infundibular process, the name "middle hypophyseal" artery seems appropriate. Each of the arteries supplying the neurohypophysis arises from the internal carotid artery. The inferior hypophyseal artery arises from the cavernous segment of the internal carotid artery. The middle and superior hypophyseal arteries arise from its supraclinoid segment. The superior and inferior hypophyseal arteries are innervated by sympathetic postganglionic fibers during their course through the subarachnoid space (Green 1966; Fuxe 1963, 1964). The innervation of the middle hypophyseal artery has not to the author's knowledge been reported. Sympathetic fibers also innervate arterioles supplying the neural lobe of the pig and the rat (Björklund 1968), but have not been found to innervate arterioles supplying the rat median eminence (Fuxe 1963, 1964).

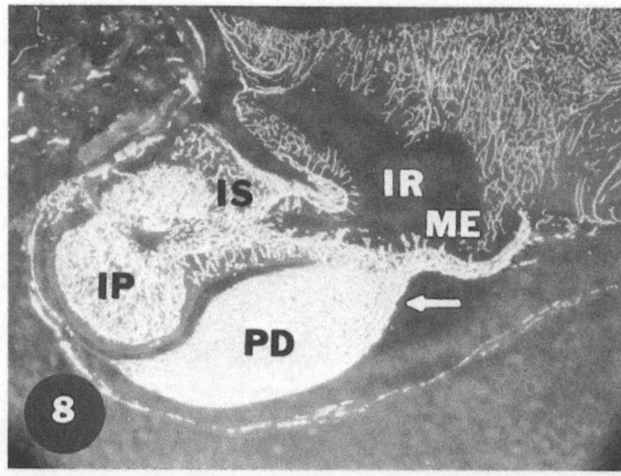

Fig. 8. LM of hypothalamus median eminence pituitary complex in the cat in parasagittal section. The rostral aspect is to the reader's *right*. The median eminence (*ME*), infundibular stem (*IS*), and infundibular process (*IP*) share a common capillary bed. In the floor of the infundibular recess (*IR*), an internal plexus of long capillary loops arises from the external plexus. Long portal vessels draining the median eminence are designated by an *arrow*. *PD*, pars distalis. (From Page et al. 1978) Intravascular injection with Microfil; Zeiss Tessovar reflected lighting (×9)

The venous drainage of the caudal region of the neurohypophysis (the neural lobe) mirrors the arterial supply. Inferior hypophyseal veins course from the neural lobe to the adjacent cavernous sinuses, which lie lateral to the midline (Bergland and Page 1978). A less prominent route of drainage is from the neural lobe into the adjacent pars distalis by capillary and "short portal" routes. Short portal vessels uniting the neural lobe and the pars distalis are abundant in some species (rat, mouse, rabbit), but scarce in others (dog, pig, monkey; Page et al. 1976, 1978). The presence of the hypophyseal cleft, which separates the pars intermedia, lower infundibular stem, and infundibular process from the pars distalis, limits the number and location of vascular communications between the pars distalis and the "neurointermediate" lobe. The neural lobe has ample venous drainage routes to the systemic circulation and limited drainage routes to the pars distalis (Bergland and Page 1978).

The venous drainage of the rostral region of the neurohypophysis (the median eminence) does not mirror the arterial supply (Wislocki and King 1936; Green 1951; Xuereb et al. 1954). Blood is collected from the primary capillary plexus in the median eminence into long fenestrated portal vessels which course to the pars distalis of the adenohypophysis, where the long portal vessels break up into a secondary capillary plexus (Duffy and Menefee 1965; Monroe 1967; Page et al. 1978). Drainage routes from the primary capillary plexus in the median eminence to the veins at the surface of the brain are less well-developed. However, Duvernoy et al. (1971) have demonstrated systemic venous drainage routes from the postinfundibular eminence in man, and Ambach et al. (1976) have demonstrated drainage routes from the median eminence to the chiasmatic and tuberal veins at

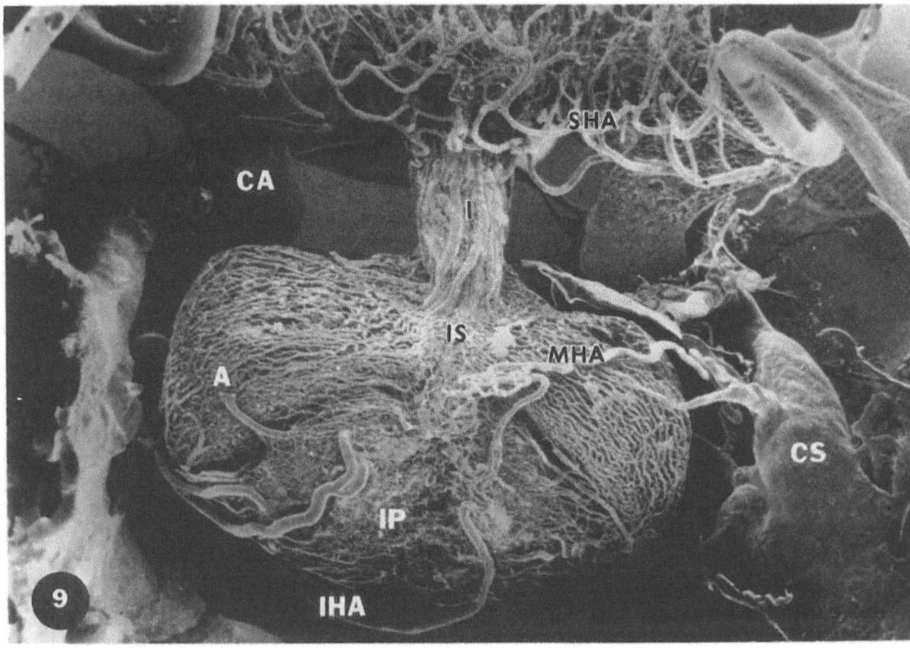

Fig. 9. SEM of a vascular cast of the pituitary of a rhesus monkey (posterior view). Numerous superior hypophyseal arteries (*SHA*) form an arterial arcade below the circle of Willis which extends to the infundibulum (*I*). Note the continuity of the neurohypophyseal capillary bed which extends through the infundibulum, the infundibular stem (*IS*), and the infundibular process (*IP*). Whereas the superior hypophyseal arteries supply the rostral region of the neurohypophyseal capillary bed, the inferior hypophyseal artery (*IHA*) supplies the caudal regions. Note that the middle hypophyseal artery (*MHA*), which others have called the trabecular or loral artery, joins the neurohypophyseal capillary bed at the junction of the infundibular stem and process. Note the paucity of vascular connections between the adenohypophysis (*A*) and the adjacent cavernous sinus (*CS*). Note also the long portal vessels coursing from the infundibulum to the adenohypophysis. *CA*, carotid artery. From Bergland and Page (1978)

the cerebral base in the rat. Venous drainage routes from the rostral region of the neurohypophysis, the median eminence, to the pars distalis are ample, whereas drainage routes to the systemic circulation at the base of the brain are scant.

4.2 Venous Supply and Venous Drainage of the Adenohypophyseal Capillary Bed

The adenohypophysis does not receive a direct arterial supply (Green 1951; Page and Bergland 1977). Routes for the passage of blood from the neurohypophysis to the adenohypophysis are provided by capillaries and portal vessels (Wislocki and King 1936; Green 1951; Xuereb et al. 1954; Page et al. 1976, 1978). The adenohypophysis is applied to the neurohypophysis, and the plexus of fenestrated capillaries that invades the neurohypophysis of the developing vertebrate also invades the adenohypophysis.

The pars tuberalis is applied to the median eminence (infundibulum) and the upper infundibular stem. Together, the pars tuberalis and the upper infundibular stem are called the pituitary stalk. In the rat, the pars tuberalis contains gonadotrophs and thyrotrophs (Baker 1977; Gross and Page 1979). Capillary connections between the median eminence and the pars tuberalis are abundant, and the capillary bed of the pars tuberalis is dense. Synaptoid contact between parvicellular axon terminals crossing from the median eminence to the pars tuberalis has been observed, but such contacts are not abundant (Cameron and Foster, 1972).

The pars intermedia is also applied to the neurohypophysis – to the lower infundibular stem and the neural lobe. It is comprised principally of melanotrophs (Baker 1974). Capillary connections between the pars intermedia and the adjacent lower infundibular stem and neural lobe are not plentiful, and the pars intermedia is relatively avascular in comparison to the pars tuberalis and pars distalis. However, synaptoid contact between parvicellular axon terminals and glandular pituitary cells (melanotrophs) is frequently found (Baumgarten et al. 1972; Naik 1973; Tilders and Smelik 1977). Although the evidence that melanotrophs are regulated by direct neural input through the dopaminergic tuberohypophyseal tract is compelling (Tilders and Smelik 1977), the story of neural control of pars intermedia function is likely to be more complicated. There is evidence that hypothalamic peptides released into neurohypophyseal capillaries may also regulate MSH secretion. The side chain of oxytocin (Pro-Leu-Gly-NH_2) has been proposed as a melanocyte hormone-inhibiting factor (MIF; Celis 1977). More recently, Vale et al. (1981) characterized a 41-amino acid hypothalamic peptide (CRF) which releases ACTH from corticotrophs and β-End and MSH from melanotrophs (Vale et al. 1981; Proulx-Ferland et al. 1982). The complex relationship between neural and humoral factors has been stressed by Antakly et al. (1985), who found that glucocorticoid receptors (normally not evident on melanotrophs) are expressed on melanotrophs after stalk section or in organ culture. Although the relative lack of vessels in the intermediate lobe suggests that synaptoid relationships are more important than humoral mechanisms, studies of the rate of exchange between blood and the intermediate lobe have not as yet been reported. Until they are, the relative roles played by each mechanism (neural and humoral) under physiologic conditions remain to be clarified.

The pars distalis makes up the bulk of the adenohypophysis, but is spacially separated from the neurohypophysis. It is displaced from the median eminence, and the degree of displacement, the length of the pituitary stalk, determines the prominence of the long portal vessels (Green 1951). It is separated from the neurointermediate lobe by the hypophyseal cleft. No nerves innervate its glandular cells (Green 1951, 1966), but its capillary bed is dense. The distribution of cell types within the pars distalis of the rat and of man has been revealed by immunocytochemistry. Lactotrophs (prolactin-secreting cells) and somatotrophs (growth hormone-secreting cells) lie in the lateral wings. Thyrotrophs and gonadotrophs lie in the medial third of the gland to form the "mucoid wedge." This term is employed because the secreted hormones – TSH, follicle-stimulating hormone (FSH), and LH – are glycoproteins. Because the mucoid wedge is continuous with the pars tuberalis and contains the same cell types, it is also referred to as the zona tuberalis. Corticotrophs (ACTH-secreting cells) lie anteriorly in the mucoid wedge and over the surface of the lateral wings (Baker and Drummond 1972; Baker 1977). The study of the distribution of cell types throughout the adenohypophysis by immunohistochemistry has not been extended to a large number of species. However, the concept of an "endocrinotopic" distribution of glandular secretory cells has been useful in clinical settings to predict the sites in which functional adenomas can be expected to arise.

The venous routes of egress from the pars distalis course to the adjacent paired cavernous sinuses, which form the lateral walls of the sella turcica. There are only scant venous drainage routes to the anterior and posterior intercavernous sinuses (Green 1957; Bergland and Page 1978); these venous channels lie within the dura mater on the anterior and posterior surfaces of the sella turcica. They connect the right and left cavernous sinuses, but their development is highly variable among, and even within, species. A few lateral hypophyseal veins drain the wings of the pars distalis to the adjacent left and right cavernous sinuses (Wislocki 1938 a, b). The largest drainage routes are the adenohypophyseal limbs of the Y-shaped pituitary (inferior hypophyseal) veins. One limb of the Y drains the pars distalis, while the other drains the adjacent neural lobe. These two limbs unite to form a common trunk, which passes to the adjacent cavernous sinus (Bergland and Page 1978).

5 Blood Flow in the Portal System

5.1 The View from Below

Observations of blood flow in the pituitary portal system of living animals have been few (Houssay et al. 1935; Green 1947; Green and Harris 1949; Barrnett and Greep 1951; Worthington 1955, 1960; Torok 1954, 1964; Page 1983). Houssay et al. (1935) reported that blood flowed from the median eminence to the pars distalis in the portal vessels of amphibians. Green (1947) confirmed this observation in amphibians and subsequently extended it to mammals (rats; Green and Harris 1949). The observations of Green and Harris (1949) became the cornerstone for their theory that the brain regulates pituitary function by means of a "humoral relay through portal vessels" (Green and Harris 1947). Their observations were subsequently confirmed by Barrnett and Greep (1951). These investigators all employed a parapharyngeal approach which carried them to the anterior surface of the pituitary stalk. Although the approach was satisfactory to answer the pressing question of the time, the question of the direction of flow in the portal vessels, it could not permit evaluation of the circulation of the pituitary gland as a whole.

5.2 The View from Behind

To circumvent the above-mentioned limitation, Torok (1954, 1964) employed a transtemporal approach to observe the patterns of blood flow in the canine and feline pituitary gland. He confirmed the observations of Green and Harris (1947) that blood flowed down the pituitary stalk from the median eminence to the pars distalis. He could not confirm the predictions of Wislocki (Wislocki and King 1936; Wislocki 1937a, b, 1938a, b) or of Green (1957) that the route of venous drainage of the pars distalis lay via lateral hypophyseal veins to the adjacent cavernous sinus. Blood appeared to course around the wings of the pars distalis to the dorsomedial zone, the junction of the pars distalis and the neural lobe, and then to drain to the cavernous sinus. The present author has employed a similar approach to observe and record with high-speed cinephotography the patterns of blood flow in the porcine pituitary gland following the intracarotid bolus injection of a vital dye, Lissamine Green (Page 1983). This approach permitted visualization of blood flow in the entire neurohypophysis and in the lateral wings of the pars distalis; only the ventral surface of the gland could not be seen (Fig. 10a).

Blood entered the two ends of the neurohypophysis almost simultaneously, but because the inferior hypophyseal arteries arise along the course of the carotid artery closer to the heart than to the superior hypophyseal arteries, blood entered the neural lobe before entering the median eminence (Fig. 10b). Blood flow within the neurohypophysis was not rigidly compartmentalized, and occurred between different neurohypophyseal regions: from the median eminence anterogradely into the upper infundibular stem, and from the neural lobe retrogradely into the lower infundibular stem. The site in the neurohypophysis where the two wave fronts met constituted a watershed zone, and its location varied. In the 14 normotensive animals studied, the watershed zone lay at the junction of the upper infundibular stem and lower infundibular stem in 6, in the middle of the lower infundibular stem in 4, and at the junction of the lower infundibular stem and infundibular process (neural lobe) in the remaining 4.

Blood drained from the neural lobe into the adjacent cavernous sinus via Y-shaped pituitary veins. The most caudal regions of the neurohypophysis were cleared first, followed by successively rostral regions. In this manner, blood was cleared from the neural lobe and lower infundibular stem in a caudorostral sequence into the Y-shaped pituitary veins, while at the same time blood was cleared from the median eminence by portal vessels. In 5 of the 14 normotensive animals, blood flowed into a small zone of adenohypophyseal territory from the adjacent neural lobe and lower infundibular stem.

Blood drained from the rostral end of the neurohypophysis (the median eminence) by portal and capillary routes to the pars distalis in 13 of the 14 normotensive animals (Fig. 10c). Drainage of the median eminence to veins at the cerebral base was not observed. Within the pars distalis, blood swept around the lateral wings to enter the dorsomedial zone. A small amount drained from the lateral wings to the adjacent cavernous sinuses via lateral hypophyseal veins to the adjacent cavernous sinus. Most coursed into the dorsomedial zone (which was on occasion also supplied from the neural lobe), and drained to the ipsilateral cav-

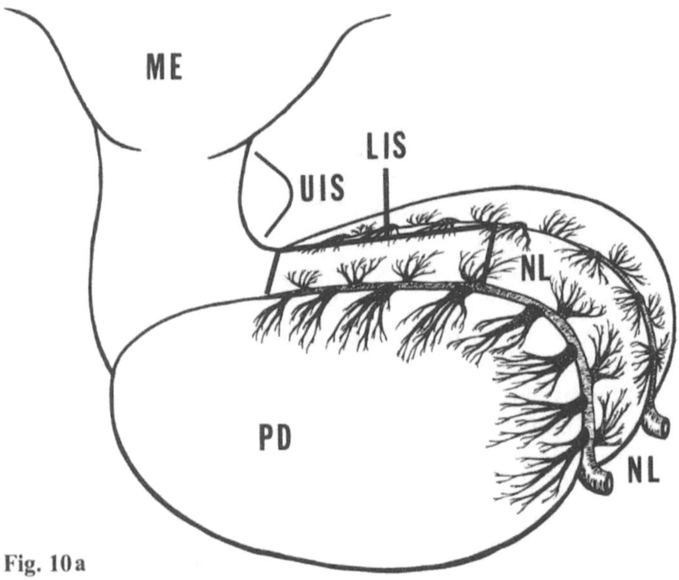

Fig. 10a

Fig. 10. a The neurohypophysis extends from the median eminence (*ME*) through the infundibular stem to the neural lobe (*NL*). The infundibular stem is subdivided into an upper (*UIS*) and a lower (*LIS*) infundibular stem. The ME is the most rostral region of the neurohypophysis. The UIS unites the ME with the LIS. The UIS comprises the posterior portion of the pituitary stalk and lies in the subarachnoid space. The LIS lies on the dorsum of the pituitary gland, within the sella turcica. The border between the UIS and LIS is demarcated by a plane extending between the rostral limit of the Y-shaped pituitary veins lying on each side of the LIS and NL. The border between the LIS and NL is demarcated by a visible thickening of neurohypophyseal tissue, which can be visualized through the operating microscope. *PD*, pars distalis. **b** The pattern of neurohypophyseal filling from inferior and superior hypophyseal arteries is demonstrated in 14 normotensive animals; blood flowed from inferior hypophyseal arteries into the neural lobe in all 14 animals. The most rostral extent of the Y-shaped veins demarcated the most rostral extent of retrograde flow. In each instance, blood flowed through superior hypophyseal arteries into the ME. Anterograde flow from the ME proceeded caudally through the UIS to meet the retrograde flow of blood proceeding rostrally. The *horizontal plane* designates the portion of the gland that can be visualized. **c** The pattern of adenohypophyseal filling from the neurohypophyseal capillary bed is illustrated in 14 normotensive animals. In 13 animals, the PD was filled from the ME. In 5 animals, a small region of the adenohypophysis (indicated by *dots*) was filled from the LIS and NL. From Page (1983)

ernous sinus through Y-shaped veins. With stalk section, the pattern changed. Within the neurohypophysis blood entered the median eminence and (presumably) drained from it to the cerebral base. Blood entered the neural lobe and passed retrogradely into the lower infundibular stem to the site of the section. The pars distalis could not be supplied by flow through the several long portal vessels. Although the territory supplied from the adjacent neural lobe and lower infundibular stem expanded if the blood pressure was sustained, on no occasion did it expand enough to support the entire pars distalis. With a fall in mean arterial blood pressure, no blood flowed from the neural lobe to the adjacent adenohypophysis.

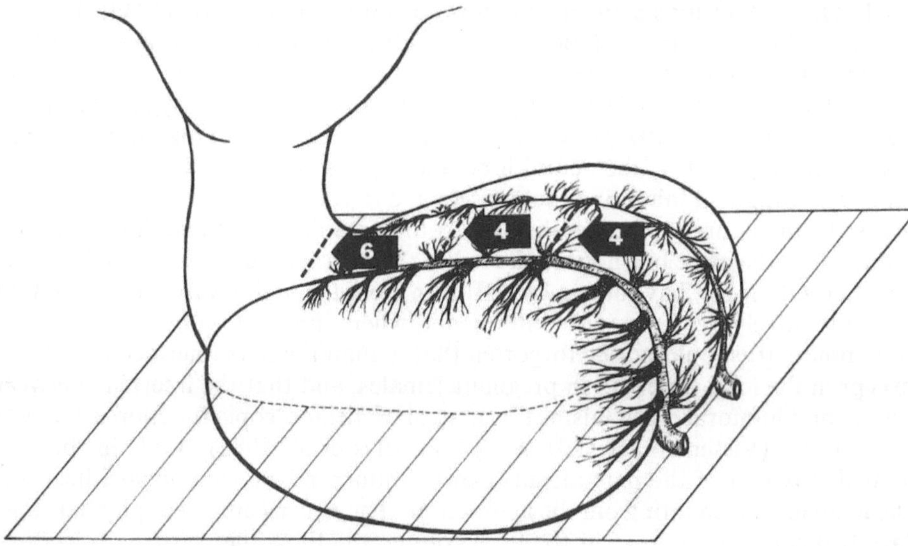

Fig. 10b

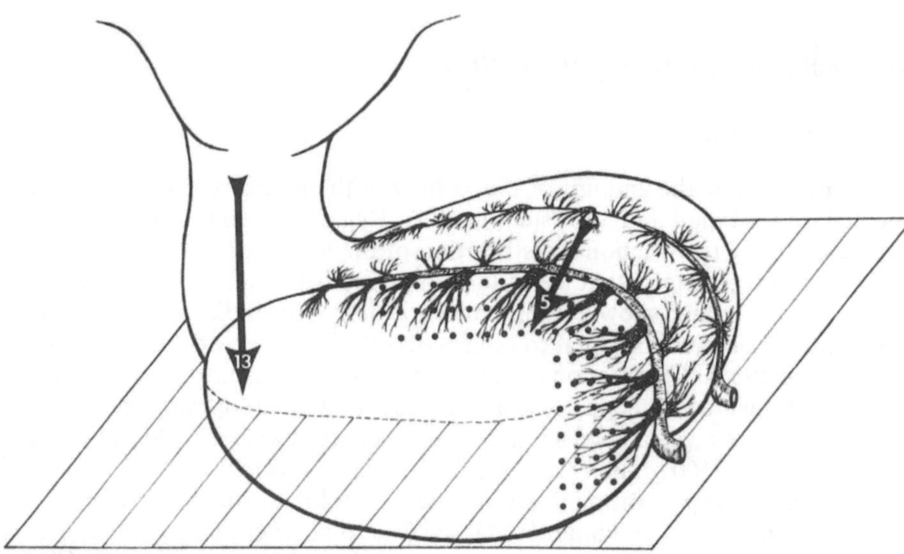

Fig. 10c

These observations supplement those of Green and Harris (1949), because they record the pattern of blood flow within the neurohypophysis as well as that of blood flow between the neural lobe and the adenohypophysis. They demonstrate that the pattern of survival of pars distalis tissue after hypophysectomy (Daniel and Prichard 1975) does not accurately predict which regions of the pars distalis are supplied by the neural lobe under physiologic conditions. Hypothalamic releasing and inhibiting factors secreted in the median eminence thus have access to the entire pars tuberalis and to the pars distalis, where innervation is scant. Hypothalamic hormones released in the neural lobe (such as AVP, OT, or Pro-Leu-Gly-NH_2) have access by diffusion from the highly vascular neural lobe to the relatively avascular pars intermedia, where innervation of glandular cells is common. [It should not be forgotten that in man the pars intermedia is absent except in the fetal state and in pregnant females, and that the interplay between neural and humoral mechanisms for control of melanotropic function is likely to be complex (Visser and Swaab 1977; Lamberts et al. 1982)]. Cells in the dorsomedial region of the pars distalis can be influenced by secretions either from the median eminence or from the neural lobe. The entire neurohypophyseal capillary bed can serve as part of the final common path to the glandular pituitary, with the median eminence supplying the entire pars tuberalis and the vast majority of the pars distalis, whereas the neural lobe supplied the relatively avascular pars intermedia and a small region of the adjacent pars distalis.

6 Regulation of Pituitary Blood Flow

6.1 Measurement

Measurements of the amount of blood flow in the pituitary gland were first reported by Goldman and Sapirstein in 1958. Regional blood flow was measured by determining the fractional distribution of the indicator ^{86}Rb. The method is based upon the premise that the uptake of an indicator injected into the vasculature by organs which show a stable concentration of the indicator during the time of recirculation must be equal to their blood flow fraction of the cardiac output (Sapirstein 1958). The fraction of cardiac output to the neural lobe and pars distalis was measured in the rat. Assuming a rate of 225 ml/kg/min for cardiac output based upon measurements made in another set of animals, blood flow values of 360 ± 100 ml/100 g/min (mean $\pm$ SD) for the neural lobe and of 59 ± 100 ml/100 g/min for the pars distalis were calculated (Goldman and Sapirstein 1958). In a subsequent report (Goldman 1961), the fractional distribution of the indicator Rb^{86} and the cardiac output were calculated in the same animals. Rates of 369 ± 149 and 41 ± 11 ml/100 g/min were reported for the neural lobe and the adenohypophysis, respectively. As the technique employed is not applicable to the study of cerebral blood flow (Sapirstein 1958), comparisons with brain blood flow could not be made. Subsequently, thermal and hydrogen ion washout techniques have yielded adenohypophyseal blood flow rates of about 70 ml/100 g/min in the rat and the dog (Porter et al. 1967; Kopaniky and Gann 1975), a rate not greatly different from that calculated by Goldman and Sapirstein (1958).

The study of neurohypophyseal blood flow is of special interest because the neurohypophysis is the gateway to the adenohypophysis, and all blood destined for the adenohypophysis must first pass through it. The initial entry of blood into the pituitary gland is regulated by resistance vessels in the neurohypophysis. Worthington (1960) observed blood flow through the long portal vessels on the anterior surface of the pituitary stalk of anesthetized albino mice. He reported that the calibre of these vessels failed to change with the depth of pentobarbital anesthesia or with the application of various drugs such as morphine, epinephrine, norepinephrine or acetyl-beta-methylcholine. The application of these drugs to arterioles supplying the median eminence caused their dilatation in the case of morphine and acetyl-beta-methylcholine and their constriction in the case of epinephrine and norepinephrine. He concluded that blood flow into the pars distalis was regulated solely by arterioles supplying the neurohypophysis, and that portal vessels served merely as passive conduits (Worthington 1960).

Within the neurohypophysis, hormones are released from nerve terminals to be carried by vascular routes to neighboring and distant sites to regulate their function. Hence, blood flow into the median eminence coupled with the rate of secretion of hypophysiotropic hormones will determine the concentration of these neurosecretory products reaching the pars distalis in a given time. Viewed in this light, control of blood flow into the median eminence becomes an important factor in neural control of pituitary function. As stressed in this review, the neurohypophysis is a diverticulum of brain supplied at each end by hypophyseal arteries and drained at each end by veins (portal venuoles rostrally and Y-shaped hypophyseal veins caudally). Neurohypophyseal blood flow could thus reasonably be expected to be regulated in a fashion similar to that in the rest of the brain.

The measurement of regional neurohypophyseal and of regional cerebral blood flow in large animals, such as sheep, can be carried out with the radiolabelled microsphere technique. This procedure is based upon the repeatedly affirmed premise that a bolus of microspheres injected into the left cardiac atrium or ventricle is distributed to the various organs in proportion to the fraction of cardiac output delivered to them (provided that the microspheres are trapped within the organ and do not escape into the venous system). If a measured sample of blood is withdrawn from the aorta at a known rate during radiolabelled microsphere injection, blood flow to each sampled organ can be calculated according to the relationship

$$\text{Blood flow} = \frac{\text{cpm tissue/g tissue}}{\text{cpm reference blood sample}} \times \text{rate of withdrawal of reference sample}$$

(Hoffbrand and Forsyth 1969; Marcus et al. 1976; Heistad et al. 1977; Heymann et al. 1977). Large animals are required, as the neurohypophyseal sample must be of sufficient size to contain at least 400 microspheres in order to assure accurate and reproducible measurements (Buckberg et al. 1971). This method allows determination of regional cerebral and regional neurohypophyseal blood flow, but not of adenohypophyseal blood flow. The absence of a direct arterial supply and the presence of a vascular supply through portal vessels prohibits the application of the technique to the determination of adenohypophyseal blood flow.

6.2 Mechanisms of Blood Flow Regulation

The radiolabelled microsphere technique has been employed to investigate neurohypophyseal blood flow in barbiturate-anesthethized sheep (Page et al. 1981). We were able to confirm the reports of Goldman (Goldman 1961; Goldman and Sapirstein 1958) that neural lobe blood flow is high. We found neural lobe blood flow to be 436 ± 39 ml/100 g/min (mean $\pm$ SE), and median eminence blood flow to be 461 ± 41 ml/100 g/min. These values were not significantly different. Hence blood flow throughout the neurohypophysis was uniform under the conditions employed and was about eight times the blood flow in the cerebral gray matter of these animals. In subsequent studies, hypothalamic blood flow was found to average 40 ml/100 g/min, in contrast to neurohypophyseal blood flow values of about 400 ml/100 g/min. Blood flow in the neurohypophysis decreased with hypocarbia, as did regional cerebral blood flow. In contrast to regional cerebral blood flow, it did not increase with hypercarbia. Blood flow in the neural lobe and median eminence autoregulated with changes in mean arterial blood pressure (MABP). At the two ends of the ovine neurohypophysis, blood flow response to changes in $PaCO_2$ and MABP was similar; as under basal conditions, blood flow in the neural lobe and median eminence did not significantly differ. The neurohypophysis regulates its blood flow in a manner similar to the brain in response to hypocarbia and changes in MABP. It differs from the cerebral cortex and the hypothalamus in the amount of its blood flow and in its lack of a blood–brain barrier, its organization for neurosecretion, and its unique pattern of arterial supply and venous drainage.

In the brain, regional blood flow, function, and metabolism are coupled (Reivich and Sokoloff 1976; Sokoloff 1977). Studies attempting to relate regional blood flow to regional functional activity in the neurohypophysis are few and are limited to the neural lobe. Lichardus et al. (1977) reported an increase in rat neural lobe blood flow following 24 h of water deprivation, and concluded that blood flow was related to function. However, these authors employed the $[_{125}I]$ antipyrine technique (Sakurada et al. 1978), a diffusible indicator technique which seriously underestimates blood flow in regions of high flow (Eklof et al. 1974) and is of limited usefulness in studies of the neurohypophysis. Further, AVP levels were not measured and neural lobe function could not be accurately assessed. Kapitola et al. (1977) observed increases in neural lobe blood flow during dehydration of laboratory rats with and without hereditary hypothalamic diabetes insipidus. They concluded that the observed increase in blood flow with dehydration could not be related to increased neural lobe function, as these animals cannot synthesize AVP. This conclusion may be erroneous, as plasma levels of OT, as well as of AVP, increase in rats (and in dogs) with increased plasma osmolality (Dyball 1971; Weitzman et al. 1978).

We recently tested the hypothesis that regional neurohypophyseal blood flow and function are coupled in the neural lobe (Ziedonis et al. 1986) employing the protocol of Verney (1947) to exteriorize the right carotid artery in each of five sheep. After 4–6 months, vascular catheters were implanted for regional blood flow measurements using radiolabelled microspheres. The animals were acclimated to stanchions. We allowed at least 4 days to pass before making blood flow measurements in these awake and apparently unstressed animals prior to and following the intracarotid infusion of hypertonic (0.517-*M*) sodium chloride at a rate of 4 ml/min for 30 min. Plasma AVP measurements were made prior to and 15 and 30 min following the start of the hypertonic sodium chloride infusion. We observed a significant increase in plasma AVP levels from 2.3 to 6.6 pg/ml and a 26% increase and in neural lobe blood flow following intracarotid hypertonic sodium chloride infusion ($P<0.05$, Student's paired *t* test). Interestingly, the animals became more alert but not disturbed, salivated, and looked about for water as the infusion was in progress. If water was proffered to them, they would seek to drink it. Concurrent with this behavioral change was an increase in regional cortical (but not white matter, cerebellar vermis, or choroid plexus) blood flow. We conclude that the complex behavioral response elicited by intracarotid hypertonic sodium chloride infusion is accompanied by an (appropriate) increase in regional cerebral blood flow, and that in the neurohypophysis, as in the rest of the brain, increases in regional blood flow accompany increases in regional functional activity.

At least in the caudal region of the neurohypophysis, the neural lobe, the exchange of materials between neural terminals and the blood stream with increased neurosecretion may be facilitated by a more than regional increase in blood flow. In the neural lobe, the relationship between magnocellular axon terminals and pituicytes is not a static one (Wittkowski and Brinkmann 1974). Under control conditions, Tweedle and Hatton (1980a) found that axon terminals were often enclosed by pituicyte cytoplasm, not opposed to the outer basal lamina of the double basement membrane surrounding fenestrated capillaries (Tweedle and Hatton 1980b). With dehydration, the number of neurosecretory terminals enclosed by pituicytes decreased (Tweedle and Hatton 1980a). This change in axon glial relationships could be replicated if the neural lobe was incubated in vitro in hyperosmotic solutions (Perlmutter et al. 1984). The number of exposed terminals was proportional to the osmolality of the incubation medium. Similarly, the number of enclosed nerve terminals decreased after parturition and with lactation (Tweedle and Hatton 1982). Tweedle and Hatton propose that with appropriate stimulation the dynamic interaction between axon terminals and glial cells facilitates the movement of oxytocin and vasopressin from nerve terminals into capillaries of the neural lobe. Conversely, under resting conditions, the movement of these hormones between axon terminal and capillary lumen would be impeded. Support for this concept comes from a recent report by Gross et al. (1985), who demonstrated a marked increase in permeability of the rat neural lobe to the neutral amino acid α-aminoisobutyric acid (AIB) when rats were deprived of water for 5 days. The authors calculate that the permeability-surface area product (PS), as well as the blood flow, must have increased markedly to permit the observed increase in permeability. In the caudal region of the neurohypophysis, blood flow and permeability increase with increasing functional demand.

6.3 Blood Flow and Metabolism

In the brain, regional blood flow and function are coupled to glucose utilization (Reivich and Sokoloff 1976; Sokoloff 1977). Because glucose is the primary source of fuel for brain metabolism, the estimates of glucose utilization are accurate indicators of energy expenditure (Reivich and Sokoloff 1976). The neurohypophysis is a diverticulum of the brain. Schwartz et al. (1979) presented qualitative evidence that 2-deoxyglucose (2-DG) incorporation into the rat neural lobe is increased after 5 days of dehydration. In subsequent reports, uptake of 2-DG by the neural lobe has been related not to neurosecretion but to sodium pump activity (Mata et al. 1980; Kadekaro et al. 1983). Acceptance of this conclusion depends upon the assumption that glucose is the primary fuel employed by the neurohypophysis as it is the primary fuel employed by the rest of the brain. As the neurohypophysis is organized differently from the rest of the brain and lacks a blood–brain barrier, this assumption may not be warranted. Indeed, Vannucci and Hawkins (1983) have found that in the rat, medium-chain fatty acids and not glucose provide the primary fuel for energy metabolism in the pituitary and pineal glands (Vannucci and Hawkins 1983). Further, in the porcine pituitary glucose is oxidized almost entirely to lactate and not to carbon dioxide and water (Vina et al. 1984). Hence, measurement of glucose metabolism will not provide an accurate measure of total energy expenditure. The neurohypophysis is made up of axon terminals whose associated cell bodies lie behind the blood–brain barrier in the hypothalamus and (presumably) consume glucose as their major fuel. One is thus faced with the situation that mitochondria in the cell body behind the blood–brain barrier metabolize glucose to CO_2 and H_2O to obtain ATP and hence energy to accomplish protein hormone synthesis, whereas the mitochondria in the terminal of the same cell (lying outside the blood–brain barrier) do not perform this task. They metabolize medium chain fats and fatty acids to CO_2 and H_2O to obtain ATP and hence energy to accomplish the task of protein hormone secretion.

7 Microcirculation in the Median Eminence

7.1 Morphology

At its rostral pole, the organization of the capillary bed and of the neuropil of the mammalian neurohypophysis is regionally specialized. The primary capillary plexus of the median eminence is subdivided into an external and an internal plexus (Fig. 11 and Page et al. 1976, 1978). The external plexus [mantle plexus of Romeis (1940) or superficial network of Duvernoy (1972)] lies interposed between the median eminence and the pars tuberalis. Its capillaries lie on the surface of the medium eminence, it is continuous with the capillaries of the infundibular stem, and its organization varies little among the several species studied by the author (Page et al. 1976, 1978). It is the external plexus which receives the arterial supply to the median eminence (Page and Bergland 1977).

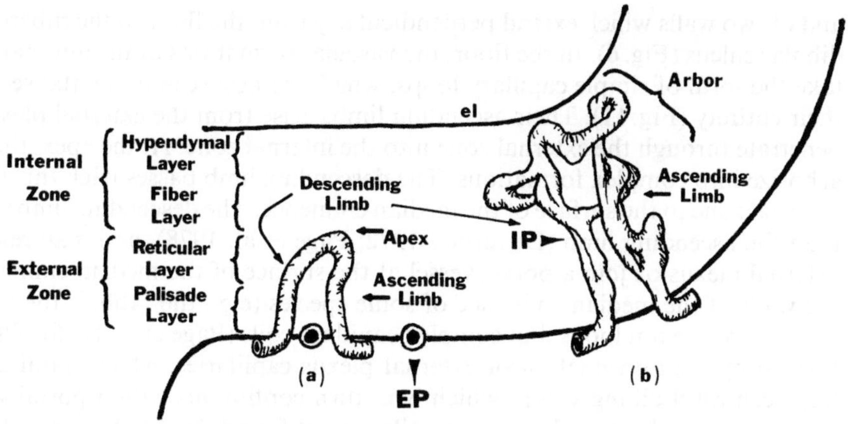

Fig. 11. The rabbit median eminence in sagittal section. The internal plexus (*IP*) is comprised of (*a*) short and (*b*) long capillary loops, which arise from the external plexus (*EP*). Each loop is made up of an ascending limb, an apex which may arborize in the internal zone, and a descending limb. *el*, ependymal layer

A direct arterial supply to vascular formations in the internal plexus has been claimed by some investigators and denied by others. Green (1948) concluded on the basis of histologic examination of the human pituitary stalk that the largest vessel of a vascular formation was its afferent vessel. Xuereb et al. (1954) examined human pituitary glands by light microscopy (LM) after intravascular injection with neoprene latex, and, making the same assumption, stated that the individual vascular formations of the internal plexus received an arterial supply. Similar conclusions were reached after study of the ovine pituitary by the same group of investigators (reviewed in Daniel and Prichard 1975). Murakami (1976) employed scanning detection microscopy of vascular casts of the rat median eminence and arrived at the same conclusion. However, Duvernoy (1972) demonstrated that the larger limb of each vascular formation was efferent, not afferent. Employing LM examination of intravascularly injected specimens, Landsmeer (1951), Akmayev (1971), and Duvernoy and Koritke (1968) were unable to demonstrate a separate arterial supply to internal plexus vascular formations of the median eminence in the rat, cat, dog, rabbit, badger, fox, and hedgehog. Page et al. (1976, 1978) employed scanning electron microscopic examination of vascular casts as well as LM examination of serial sections and whole mounts, and could not demonstrate a separate arterial supply to the internal plexus in the rabbit, rat, mouse, cat, dog, sheep, and monkey. The author and his colleagues demonstrated that the internal plexus of capillaries arose from the external capillary plexus of the median eminence in these species.

The internal plexus [deep network of Duvernoy (1972)] is made up of capillary loops and coils which penetrate the substance of the median eminence and may reach almost to the ventricular surface. Its angioarchitecture differs in detail among species, but some helpful generalizations can be made. When viewed in coronal section, the infundibular recess can be seen to be composed of a floor whose surface is oriented parallel to the floor of the skull in the horizontal plane, and of two walls which extend perpendicularly from the floor to the tuberoinfundibular sulcus (Fig. 6). In the floor, the vascular formations of the internal plexus take the form of simple capillary loops, which are best seen in sagittal section in their entirety (Fig. 11). Their ascending limbs arise from the external plexus and penetrate through the external zone into the internal zone. At the apex, they may arborize into complex formations. The descending limb passes back through the external zone to the surface of the median eminence. The descending limb is larger than the ascending limb (Duvernoy 1972; Page et al. 1978), and may rejoin the external plexus or join a portal vessel at the surface of the median eminence. In the walls of the median eminence of some species (e.g., the rabbit), the vascular formations are not loops, but complex capillary coils (Page et al. 1976). These are formed by the invagination of external plexus capillaries, which spiral about a large central draining vessel, which is in turn continuous with a portal vein. In the dog, cat, and rat, such complex coils are not found. In the monkey, where the palisade layer has migrated into the internal zone and where the neurohypophysis is oriented in a vertical, not horizontal, plane, the angioarchitecture of the internal plexus is much more complex (Page et al. 1978). However, the essential building blocks remain capillary loops with or without extensive arborization at their apices and capillary coils consisting of a small (afferent) capillary entwined about a larger (efferent) one.

In some species there is also a subependymal network of vessels that courses parallel with the floor of the infundibular recess (Duvernoy and Koritke 1968) and unites the apices of capillary loops with hypothalamic capillaries. Such capillaries have been found by the author in the rat, dog, and sheep, but not in the rabbit, cat, or monkey (Page et al. 1976, 1978). Ambach et al. (1976) argue that this plexus receives a direct arterial supply, an observation not confirmed by the author (Page et al. 1978). Both propose that the direction of blood flow is from external plexus to internal plexus to hypothalamus, as did Torok (1954) on the basis of observation of blood flow in the canine median eminence.

Blood delivered to the external plexus of the median eminence can be drained into the pars distalis by portal routes, can be diverted into the internal plexus before returning to the external plexus and hence to the pars distalis, or can, at least in some species, be carried from the external plexus to the internal plexus to the hypothalamus. The capillary bed of the median eminence is not completely isolated from the capillary bed as proposed by Wislocki and King (1936). In several species (for example, the rat, sheep, and dog), subependymal capillaries coursing parallel to the infundibulum unite the apices of internal plexus capillary loops and join hypothalamic capillaries (Torok 1964; Ambach et al. 1976; Page et al. 1978). In all species examined by the author, capillaries were found to course between the median eminence and the hypothalamus in the periventricular region at the edge of the median eminence. Such routes are not visualized on LM examination of midsagittal sections of injected specimens (Fig. 8), and so were not appreciated by earlier observers. However, when SEM of vascular casts is employed they can easily be seen (Fig. 12).

Although blood flow in these periventricular capillaries cannot be viewed in living animals, blood flow in the subependymal capillaries can. Torok (1954, 1964) observed flow in these vessels to be from the median eminence to the hypo-

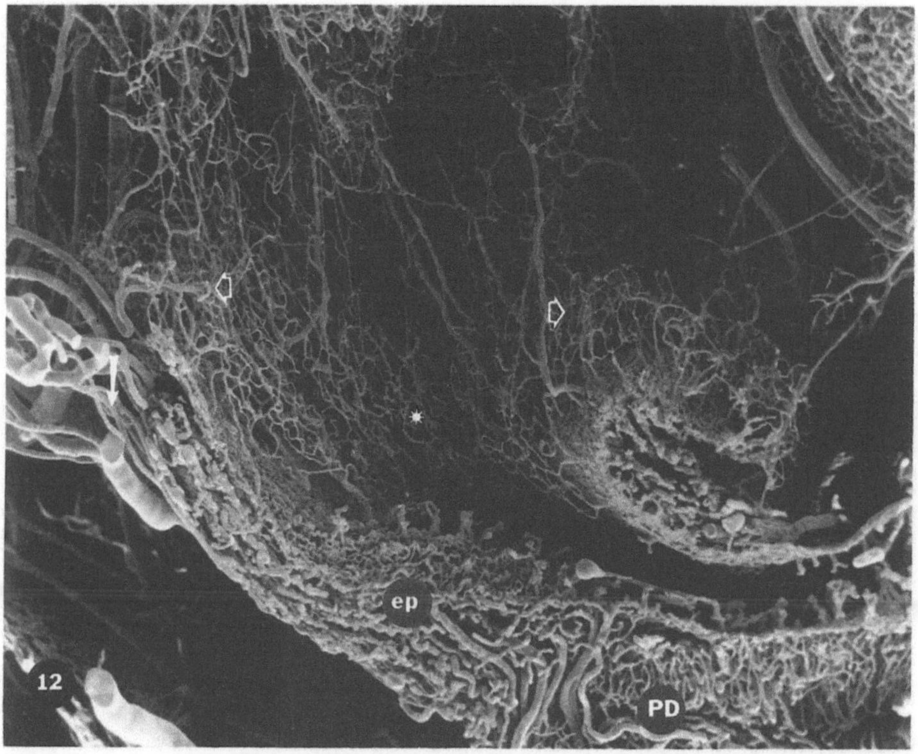

Fig. 12. SEM of a vascular cast of the cat hypothalamus and median eminence in sagittal section. The median eminence external plexus (*ep*) lies interposed between pars distalis (*PD*) capillaries below and internal plexus capillary loops above. The *vertical arrow* designates superior hypophyseal arteries which vascularize the external plexus. External plexus capillaries unite to form portal vessels coursing to the pars distalis. Capillary connections between the median eminence and the hypothalamus are present anteriorly and posteriorly (*hollow arrowheads*) and in the periventricular zone (*asterisk*). Note that periventricular capillary connections uniting the lateral wall of the median eminence with the hypothalamus cannot be appreciated on Microfil section (compare with Fig. 8). From Page et al. (1978)

thalamus. These same vessels increase in size and number following hypophysectomy in the rat (Moll 1958), and infarction of the median eminence does not occur (Raisman 1973). The division of portal routes of egress which occurs with hypophysectomy stimulates the hypertrophy of the internal and subependymal plexus in the rat (Fig. 13). The conclusion that this system provides an alternative route of egress from the median eminence seems inescapable, but the significance of these vascular links to the hypothalamus remains unknown. Given the presence of a blood–brain barrier in the hypothalamus, how could peptide hormones in the blood influence neuronal function?

Whereas the evidence for vascular links between the neurohypophysis (the median eminence) and the hypothalamus seems sound, the evidence for vascular routes from the adenohypophysis to the neurohypophysis and thence to the hy-

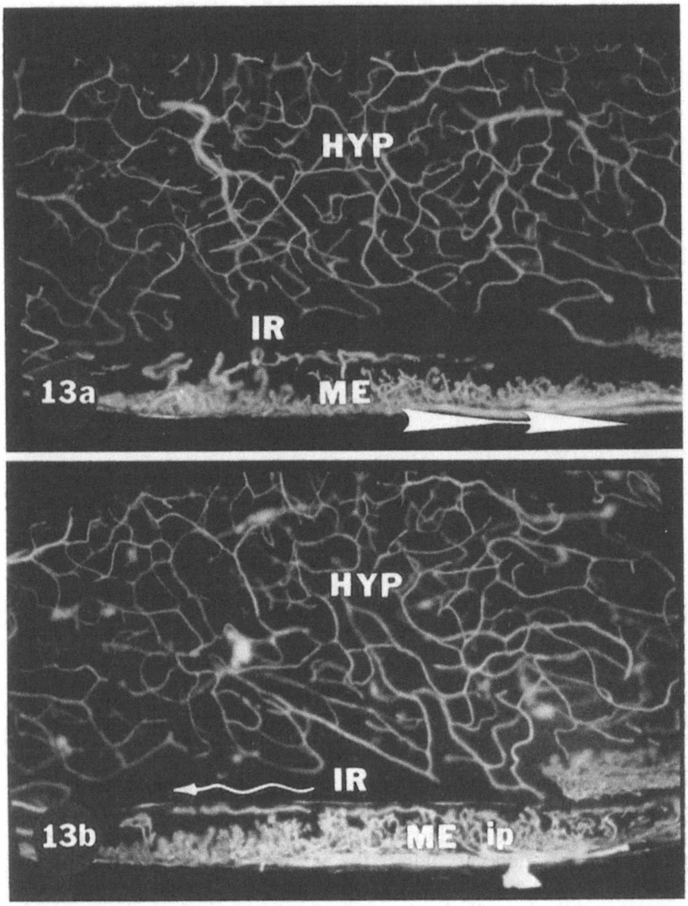

Fig. 13 a, b. LM of rat median eminence after vascular injection with Microfil (sagittal section). Rostral aspect to reader's *left*. **a** Control animal. External, internal, and subependymal plexus capillaries are filled with Microfil. *HYP*, hypothalamus; *IR*, infundibular recess; *ME*, median eminence. *White arrow* parallel to portal vessels. **b** Hypoxed animal 1 month after hypophysectomy, age-matched with animal depicted in **a**. Note marked hypertrophy of internal plexus capillary loops (*ip*) and of subependymal vessels. *Arrow* parallel to subependymal plexus

pothalamus is more difficult to evaluate. In their original description of the portal vascular system, Popa and Fielding (1930 a, b) proposed that blood flowed from the pars distalis to the hypothalamus. Torok (1954, 1964) observed blood flow in the canine pituitary. While acknowledging that blood flowed from the median eminence to the pars distalis through portal vessels on the surface of the pituitary stalk, Torok reported that it flowed from the adenohypophysis to the median eminence through vessels deep in the pituitary stalk, near the hypophyseal cleft. He further reported that blood flowed from the pars distalis through these vessels to the capillary loops of the internal plexus and to the hypothalamus through the subependymal plexus (see Szentagothai et al. 1968 for review). His experiments

(Torok 1954, 1964) have not been repeated, and his findings have generally been dismissed. Bergland and Page (1978) described several other routes by which pituitary secretions might reach the brain. A proposal that blood might flow from the pars distalis to the neural lobe and then retrogradely to the median eminence could not be substantiated by the visualization and recording of pituitary blood flow (Page 1983).

Still, the suggestion that hormones can be carried by vascular routes from the pars distalis to the brain cannot be summarily dismissed. Labelled analogues of pituitary peptides, $[^3H]$ACTH(4–9) and $[^3H]\beta$-LPH, injected into the pituitary can be detected in the hypothalamus (Mezey et al. 1978; Mezey and Palkovits 1982), but if the pituitary stalk is severed prior to the injection, label is not found in the hypothalamus. If the wound is permitted to heal for 8 days, sufficient time for vascular but not for neural regeneration, label is again detectable in the brain after injection into the pituitary. Injection of neurotensin into the pituitary of rats produced a significant decrease in body temperature, an effect produced only by the action of neurotensin upon hypothalamic structures. The effect was abolished if stalk transection was carried out prior to neurotensin injection, but was regained if the vessels were permitted to regenerate (Dorsa et al. 1979). While it is possible that the act of injection could have reversed blood flow in the (passive) adenohypophyseal vasculature and its connections with the neurohypophysis, the possibility of "two-way transport in the hypothalamohypophyseal system" remains open.

7.2 Modelling

To build a conceptual model of the angioarchitecture of the median eminence, it is necessary to begin with the external plexus. This is comprised of a reticular network of fenestrated capillaries which lie partially embedded within the oral (ventral) surface of the median eminence (Page et al. 1976; Page 1982; Page and Dovey-Hartman 1984a). These capillaries are arranged in the form of hexagons, and when viewed from above resemble the surface of a honeycomb; this capillary mosaic, made up of multiple hexagonal units, lies parallel to the surface of the median eminence. On TEM examination, the capillaries are found to be partially embedded in the external zone so that their inferior surface is flush with that of the external zone. The central region of each hexagonal capillary array is filled with "posts" of median eminence tissue (Fig. 14). The mosaic pattern of capillaries in the external plexus of the median eminence resembles the pattern of those in the gastric muscosa and in the lung – two other organs where a rapid transport of substances into the blood occurs (Baez 1977; Sobin and Tremer 1977). Sobin et al. (1970) portray the pulmonary vascular space as resembling "an underground garage consisting of a floor, ceiling and supportive pillars all covered with endothelium," and argue that blood flows as a sheet through an essentially continuous space lined by endothelium. A similar flow pattern may be present in the external plexus, because its angioarchitecture resembles that in the lung.

The organization of the external zone and of the external plexus leads to several conclusions. Each hexagonal capillary unit (microvascular module) encloses

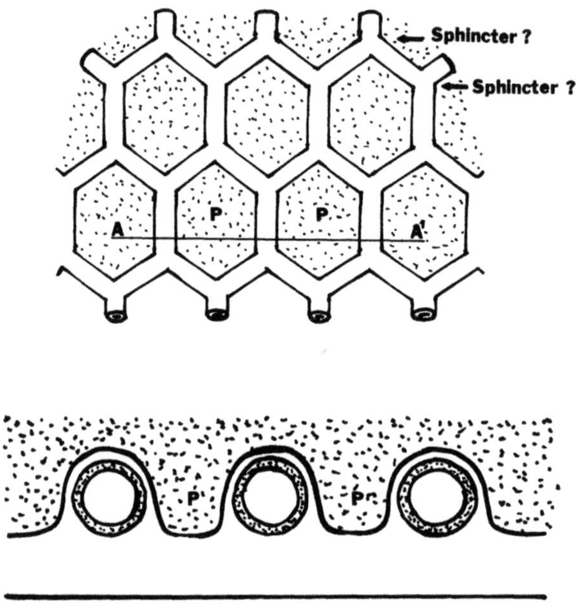

Fig. 14. External plexus from below. Hexagonal arrays of capillaries are separated by posts (*P*) of median eminence tissue. This architectural arrangement could permit "sheet flow" through the external plexus. Alternatively, if sphincters are present at the entry to each hexagonal unit, fine regional control of external plexus blood flow could be obtained on the median eminence surface. *Lower panel:* schematic cross-sectional view of the median eminence external plexus along plane *A-A'*, in *upper panel:* capillaries lie embedded in the neuropil, separated by posts of median eminence tissue. From Page (1982)

a post, or microdomain, of neurosecretory axon terminals and supportive tissue. With increased functional demand glial processes retract, making more capillary surface area available for the transport of neurosecretory material into the portal system (Wittkowski and Scheuer 1974). Within each microdomain, amines or neurosecretions released from one terminal are free to act upon the terminals of neighboring neurons if appropriate receptors are present; a paracrine action of secretions released from terminals in the median eminence external zone upon neighboring nerve terminals is possible, as is an endocrine action upon glandular cells of the pars distalis. Should sheet flow occur throughout the external plexus, the secretions of terminals in one microdomain would be free to interact with terminals in other microdomains. Such a pattern could permit a wide range of interactions between neurosecretory products and axon terminals at the surface of the median eminence. Alternatively, should further TEM studies reveal smooth-muscle sphincters arrayed at the entrance to each hexagonal unit, a different model would have to be constructed. In such a case, each hexagonal unit could be seen as a microvascular module, and very regional adjustments of blood flow at the surface of the median eminence would be possible. The opportunities for neurons terminating in one microdomain to influence neurons terminating in another could be limited or regulated.

The external plexus receives the arterial supply to the median eminence. It distributes that blood into portal vessels destined for the pars distalis, or into the internal plexus, which invaginates into the neuropil of the median eminence.

The capillary loops of the internal plexus also form a series of microvascular modules (Fig. 8). When the median eminence is sectioned in the sagittal plane after intravascular injection with India ink (Wislocki and King 1936; Green 1951; Duvernoy 1969, 1972; Duvernoy and Koritke 1964, 1968) or Microfil (Page et al. 1978), the capillary loops which form the internal plexus in the floor of the median eminence can be seen in their entirety.

A better appreciation of their angioarchitecture is gained from SEM studies of median eminence vascular casts (Page et al. 1976, 1978). Such studies reveal that each loop arises from and returns to the external plexus (Figs. 8 and 12). The perivascular space between the ascending and descending limbs of each loop is continuous with the perivascular space at the (oral) surface of the median eminence. It lies between the widely split basement membrane characterized by an external basal lamina surrounding each loop in its entirety and the inner basal lamina closely apposed to the endothelial tubes that make up each formation.

Recently, neurohemal contact has been documented on the capillary loops in the floor of the rabbit median eminence (Page and Dovey-Hartman 1984 a). Columns of capillary loops surrounded by parvicellular axon terminals stand upon a base made up of hexagonal arrays of capillaries, each of which encloses a microdomain of neurosecretory terminals. As each loop invaginated from the surface into the depths of the median eminence, it appears to have carried with it axon terminals from the surface. Hence, each vertical column could be expected to contain terminals functionally similar to those at its base. The area available for neurohemal contact would be increased, but the endocrinotopic organization of the median eminence (Ajika 1980) would be preserved by such an arrangement. Whether smooth-muscle cells are present at the origin of each loop has yet to be determined; thus it is not known whether flow into the capillary loops is regulated by sphincters. The recruitment of capillary loops (or coils) in functionally specific regions of the median eminence – in addition to an increase in blood flow, an increase in permeability, a retraction of glial processes (Wittkowski and Scheuer 1974), and an increase in neurosecretion – would markedly enhance the response to a functional demand. Such mechanisms are the subject of ongoing and future investigations (Fig. 15).

The unique organization of the capillary bed in the median eminence raises additional questions as to the means by which blood flow is regulated. Arterioles, supplying the external plexus, lie in close proximity to it and to axon terminals on the surface of the median eminence (Page and Dovey-Hartman 1984 b). Hence they will be exposed to high levels of dopamine and neurohormones released at the ventral surface of the median eminence into external plexus capillaries (Ben-Jonathan et al. 1977; Gibbs and Neill 1978). In addition, the descending limbs of capillary loops carry blood to the surface of the median eminence from deep in the internal zone. Neurohemal contact by aminergic and peptidergic terminals occurs about these capillary loops in the internal zone (Page and Dovey-Hartman 1984 a). The noradrenergic reticuloinfundibular tract terminates in this zone, and blood returning to the surface of the median eminence may contain norepineph-

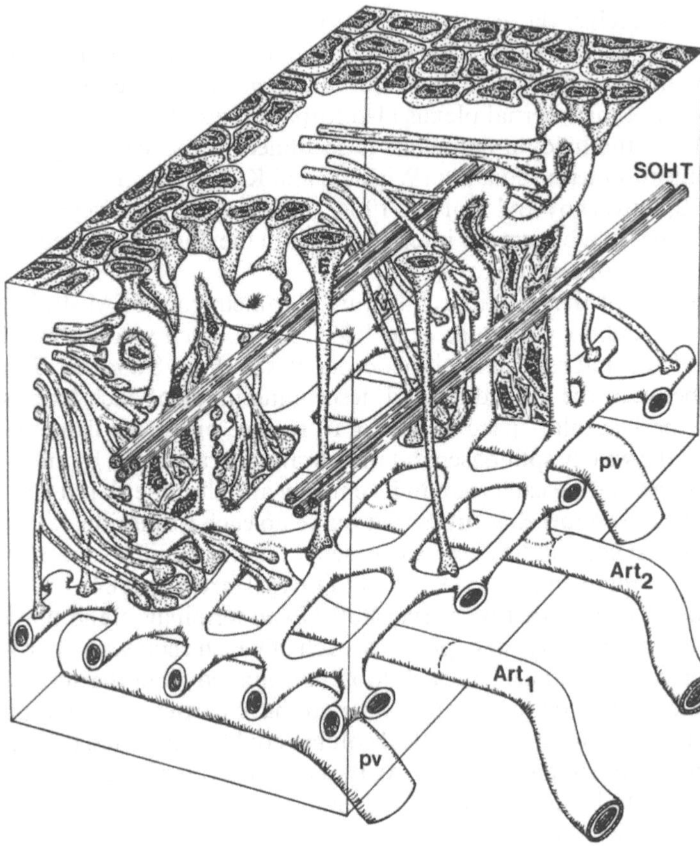

Fig. 15. Neurovascular organization in the medial third of the rabbit median eminence. The external plexus forms a hexagonal array oriented parallel to the median eminence external surface, and receives arterial blood from arteriolar branches of the superior hypophyseal arteries (Art_1, Art_2). Internal plexus capillary loops arise from the external plexus and arborize in the subependymal zone. Connective tissue fills the ascending and descending limbs of the internal plexus loops, and the spaces between the surface of the external zone and the pars tuberalis at the ventral aspect of the median eminence. Nonciliated ependymal cells line the infundibular recess (*E*). The basal processes of some ependymal cells end in the perivascular space about the apex of capillary loops and form an ependymal cuff about them in the hypependymal layer. Some ependymal processes extend to the external plexus. Parvicellular axons terminate on external plexus capillaries and about internal plexus capillary loops to form an axonal cuff about them deep in the fiber layer and in the palisade zone. These axons are peptidergic and aminergic. Magnocellular systems (*SOHT*) pass through the fiber layer and do not send collaterals to the capillary loops. Arterioles which originate from superior hypophyseal arteries lie on the median eminence surface and supply the external plexus either by tapering down from terminal arterioles to precapillary arterioles (Art_1) and/or by sending small (precapillary arteriolar) branches at right angles from the terminal arterioles (Art_2). Arterioles lie in close proximity to the external plexus and to posts (microdomains) of neurosecretory terminals. Portal vessels (*pv*) are fenestrated and do not contain a continuous layer of smooth muscle in their walls. Modified from Page and Dovey-Hartman (1984a)

rine and epinephrine (Gibbs 1985) as well as dopamine. The resistance vessels at the surface of the median eminence regulate blood flow into its external plexus, but are in turn exposed to vasoactive amines and peptides secreted by the median eminence (Page and Dovey-Hartman 1984b). The secretions of neural systems terminating in the median eminence may regulate blood flow into it.

8 Summary and Conclusions

The pituitary portal system develops from mesenchyme entrapped between the advancing evaginations of the stomodeum and the floor of the diencephalon. It is comprised of feeding arteries (the inferior, middle, and superior hypophyseal arteries), the neurohypophyseal capillary bed, portal vessels which pass from the neurohypophysis and adenohypophysis, the adenohypophyseal capillary bed, and draining veins – the inferior hypophyseal veins. A single capillary bed extends throughout the neurohypophysis: the median eminence, the infundibular stem, and the neural lobe. Blood flows between different regions within the neurohypophysis (from the infundibular process retrogradely into the lower infundibular stem and from the median eminence anterogradely into the upper infundibular stem), and also from the neurohypophysis to the adenohypophysis. The major drainage route of the rostral region of the neurohypophysis, the median eminence, is to the pars distalis via fenestrated portal vessels and capillaries. The major drainage route of the caudal region of the neurohypophysis, the neural lobe, is via hypophyseal veins to the cavernous sinus, but some drainage occurs across short portal vessels and capillaries to the adjacent adenohypophysis.

Blood flow into the pituitary gland is regulated by arterioles supplying the neurohypophysis, and blood flow in the neurohypophysis is higher than that in cerebral gray and white matter. The arterioles supplying the median eminence and neural lobe respond to hypocarbia and to changes in MABP and in functional demand similarly to arterioles in the brain. However, unlike the brain, the pituitary gland oxidizes glucose only as far as lactate and relies upon the oxidation of medium-chain fats for its primary fuel.

The neurohypophysis is a diverticulum of brain organized for neurosecretion. In it, nerves terminate on fenestrated blood vessels. Rostrally, in the median eminence, the organization of the neuropil and of the capillary bed is specialized. Pillars of vertically oriented capillary loops surrounded by nerve terminals stand on a horizontal mosaic of hexagonally arrayed capillaries. Investigators now seek to understand the role of the microcirculation in the trafficking of neurohormones from the neuro- to the adenohypophysis. The mechanisms to increase the efficiency of neurohormone delivery in response to increased functional demand are being studied. In addition to an increase in the rate of neurosecretion, mechanisms such as regional increases in blood flow, increases in transport across the capillary wall, increases in the number of exposed (as opposed to enclosed) nerve endings, and recruitment of repeating microvascular modules are being investigated.

References

Ajika K (1979) Simultaneous localization of LHRH and catecholamines in rat hypothalamus. J Anat 128:331–347

Ajika K (1980) Relationship between catecholaminergic neurons and hypothalamic hormone-containing neurons in the hypothalamus. In: Martini L, Ganong WF (eds) Frontiers of neuroendocrinology, vol 6. Raven, New York, pp 1–32

Ajika K, Hökfelt T (1973) Ultrastructural identification of catecholamine neurones in the hypothalamic periventricular–arcuate nucleus–median eminence complex with special references to quantitative aspects. Brain Res 57:97–117

Akmayev IG (1969) Morphological aspects of the hypothalamic-hypophyseal system. I. Fibers terminating in the neurohypophysis of mammals. Z Zellforsch 96:609–624

Akmayev IG (1971) Morphological aspects of the hypothalamic hypophyseal system. II. Functional morphology of pituitary microcirculation. Z Zellforsch 116:178–194

Ambach G, Palkovits M, Szentagothai J (1976) Blood supply of the rat hypothalamus. IV. Retrochiasmatic area, median eminence, arcuate nucleus. Acta Morphol Hung 24:93–119

Antakly T, Sasaki A, Liotta AS, Palkovits M, Kreiger DT (1985) Induced expression of the glucocorticoid receptor in the rat intermediate pituitary lobe. Science 229:277–279

Anthony ELP, King JC, Stopa EG (1984) Immunocytochemical localization of LHRH in the median eminence, infundibular stalk and neurohypophysis. Evidence for multiple sites of releasing hormone secretion in humans and other mammals. Cell Tissue Res 236:5–14

Antunes JL, Zimmerman EA (1978) The hypothalamic magnocellular system of the rhesus monkey: an immunocytochemical study. J Comp Neurol 181:539–565

Antunes JL, Carmel PW, Zimmerman EA (1977) Projections from the paraventricular nucleus to the zona externa of the median eminence of the rhesus monkey: an immunohistochemical study. Brain Res 137:1–10

Atwell WJ (1926) The development of the hypophysis cerebri in man, with special reference to the pars tuberalis. Am J Anat 37:159–193

Baez S (1977) Skeletal muscle and gastrointestinal microvascular morphology. In: Kaley G, Altura BM (eds) Microcirculation, vol 1. Univ Park Press, Baltimore, pp 69–94

Baker BL (1974) Functional cytology of the hypophyseal pars distalis and pars intermedia. In: Knobil E, Sawyer WH (eds) The pituitary gland and its neuroendocrine control, part 1. Williams and Wilkins, Baltimore, pp 45–80 (Handbook of Physiology, vol 4)

Baker BL (1977) Cellular composition of the human pituitary pars tuberalis as revealed by immunocytochemistry. Cell Tissue Res 182:151–163

Baker BL, Drummond ST (1972) The cellular origins of corticotropin and melanotropin as revealed by immunochemical staining. Am J Anat 134:395–409

Bargmann W (1966) Neurosecretion. Int Rev Cytol 19:183–201

Bargmann W, Scharrer E (1951) The site of origin of the hormones of the posterior pituitary. Am Sci 39:255–259

Barrnett RJ, Greep RO (1951) The direction of flow in the blood vessels of the infundibular stalk. Science 113:185

Baumgarten HG, Björklund A, Holstein AF, Nobin A (1972) Organization and ultrastructural identification of the catecholamine nerve terminals in the neural lobe and pars intermedia of the rat pituitary. Z Zellforsch 126:483–517

Begeot M, Dupouy JP, Dubois MP, Dubois PM (1981) Immunocytological determination of gonadotropic and thyrotropic cells in fetal rat anterior pituitary during normal development and under experimental conditions. Neuroendocrinology 32:285–294

Ben-Jonathan N, Oliver C, Weiner HJ, Mical RS, Porter JC (1977) Dopamine in hypophyseal portal plasma of the rat during the estrous cycle and throughout pregnancy. Endocrinology 100:452–458

Bergland RM, Page RB (1978) Can the pituitary secrete directly to the brain? (Affirmative anatomical evidence). Endocrinology 102:1325–1338

Bergland RM, Torack RM (1969) An electron microscopic study of the human infundibulum. Z Zellforsch 99:1–12
Bern HA, Knowles FGW (1966) Neurosecretion. In: Martini L, Ganong WF (eds) Neuroendocrinology. Academic, New York, pp 139–186
Björklund A (1968) Monoamine-containing fibers in the pituitary neuro-intermediate lobe of the pig and rat. Z Zellforsch 89:573–589
Björklund A, Falck B, Hromek F, Owman C, West K (1970) Identification and terminal distribution of the tubero-hypophyseal monoamine fibre systems in the rat by means of stereotaxic and microspectrofluorimetric techniques. Brain Res 17:1–23
Björklund A, Moore RY, Nobin A, Stenevi U (1973) The organization of the tubero-hypophyseal and reticulo-infundibular catecholamine neuron systems in the rat brain. Brain Res 51:171–191
Bloch B, Bugnon C, Fellman D, Lenys D (1978) Immunocytochemical evidence that the same neurons in the human infundibular nucleus are stained with anti-endorphins and antisera of other related peptides. Neurosci Lett 10:147–152
Bloom FE (1970) The fine structural localization of biogenic amines in nervous tissue. Int Rev Neurobiol 13:27–66
Boler J, Enzmann F, Folkers K, Bowers CY, Schally AV (1969) The identity of chemical and hormonal properties of the thyrotropin-releasing hormone and pyroglutamyl-histidyl-proline amide. Biochem Biophys Res Commun 37:705–710
Brightman MW, Reese TS (1969) Junctions between intimately apposed cell membranes in the vertebrate brain. J Cell Biol 40:648–677
Broadwell RD, Cataldo AM, Balin BJ (1984) Further studies of the secretory process in hypothalamo-neurohypophyseal neurons: an analysis using immunocytochemistry, wheat germ agglutinin-peroxidase and native peroxidase. J Comp Neurol 228:155–167
Buckberg GD, Luck JC, Payne DB, Hoffman JIE, Archie JP, Fisler DE (1971) Some sources of error in measuring regional blood flow with radioactive microspheres. J Appl Physiol 31:598–604
Bugnon C, Bloch B, Lenys D, Fellmann D (1979 a) Infundibular neurons of the human hypothalamus simultaneously reactive with antisera against endorphins, ACTH, MSH and β-LPH. Cell Tissue Res 199:177–196
Bugnon C, Bloch B, Lenys D, Gouget A, Fellmann D (1979 b) Comparative study of the neuronal populations containing β-endorphin, corticotropin and dopamine in the arcuate nucleus of the rat hypothalamus. Neursci Lett 14:43–48
Bugnon C, Fellmann D, Gouget A, Cardot J (1982) Corticoliberin in rat brain: immunocytochemical identification and localization of a novel neuroglandular system. Neurosci Lett 30:25–30
Burgus R, Dunn TF, Desiderio D, Ward DN, Vale W, Guillemin R (1970) Characterization of ovine hypothalamic hypophysiotropic TSH-releasing factor. Nature (Lond) 226:321–325
Burlet A, Tonon M-C, Tankosic P, Coy D, Vaudry H (1983) Comparative immunocytochemical localization of corticotropin releasing factor (CRF-41) and neurohypophyseal peptides in the brain of Brattleboro and Long-Evans rats. Neuroendocrinology 37:64–72
Cameron E, Foster CL (1972) Some light- and electron-microscopical observations on the pars tuberalis of the pituitary gland of the rabbit. J Endocrinol 54:505–511
Campbell HJ (1966) The development of the primary portal plexus in the median eminence of the rabbit. J Anat 100:381–387
Carson KA, Nemcroff CB, Rone MS, Youngblood WW, Prange AJ, Hanber JS, Kizer JS (1977) Biochemical and histochemical evidence for the existence of a tubero-infundibular cholinergic pathway in the rat. Brain Res 129:169–173
Celis ME (1977) Hypothalamic peptides involved in the control of MSH secretion: identity, biosynthesis and regulation of their release. Front Horm Res 4:69–79
Chatelain A, Dubois MP, Dupouy JP (1976) Hypothalamus and cytodifferentiation of the foetal pituitary gland. Cell Tissue Res 169:335–344

Daniel PM, Prichard MML (1975) Studies of the hypothalamus and the pituitary gland. With special reference to the effects of transection of the pituitary stalk. Acta Endocrinol (Copenh) 80 [Suppl 201]:1–216
Dempsey EW, Wislocki GB (1955) An electron microscopic study of the blood brain barrier in the rat employing silver nitrate as a vital stain. J Biophys Biochem Cytol 1:245–256
Dorsa DM, deKloet ER, Mezey E, deWied D (1979) Pituitary–brain transport of neurotensin: functional significance of retrograde transport. Endocrinology 104:1663–1666
Duffy PE, Menefee M (1965) Electron microscopic observations of neurosecretory granules, nerves and glial fibers and blood vessels of the rabbit. Am J Anat 117:251–286
Duvernoy H (1969) Considérations sur la vascularisation de l'hypophyse. Acta Neurol Belg 69:469–481
Duvernoy H (1972) The vascular architecture of the median eminence. In: Knigge KM, Scott DE, Weindl A (eds) Brain-endocrine interaction. Median eminence: structure and function. Karger, Basel, pp 79–108
Duvernoy H, Koritke JG (1964) Contribution a l'étude de l'angioarchitectonie des organes circumventriculaires. Arch Biol 75:849–904
Duvernoy H, Koritke JG (1968) Les vaisseaux sous-épendymaires du recessus hypophysaire. J Hirnforsch 10:227–245
Duvernoy H, Koritke JG, Monnier G (1971) Sur la vascularisation du tuber postérieur chez l'homme et sur les relations vasculaires tubérohypophysaires. J Neuro Visc Relat 32:112–142
Dyball REJ (1971) Oxytocin and ADH secretion in relation to electrical activity in antidromically identified supraoptic and paraventricular units. J Physiol (Lond) 214:245–256
Eklof B, Lassen NA, Nilsson L, Norberg K, Siesjo BK, Tarlof P (1974) Regional cerebral blood flow in the rat measured by the tissue sampling technique: a critical evaluation using four indicators C^{14} antipyrine, C^{14}-ethanol, H^3-water and xenon^{133}. Acta Physiol Scand 91:1–10
Enemar A (1961) The structure and development of the hypophyseal portal system in the laboratory mouse, with particular regard on the primary plexus. Arch Zool 13:203–252
Eurenius L, Jarskar R (1971) Electron microscope studies on the development of the external zone of the mouse median eminence. Z Zellforsch 122:488–502
Fisher AWF, Price PG, Burford GD, Lederis K (1979) A 3-dimensional reconstruction of the hypothalamo-neurohypophysial system of the rat. Cell Tissue Res 204:343–354
Flament-Durand J, Couck A-M, Dustin P (1975) Studies on the transport of secretory granules in the magnocellular hypothalamic neurons of the rat. Cell Tissue Res 164:1–9
Fuxe K (1963) Cellular localization of monoamines in the median eminence and in the infundibular stem of some mammals. Acta Physiol Scand 58:383–384
Fuxe K (1964) Cellular localization of monoamines in the median eminence and infundibular stem of some mammals. Z Zellforsch 61:710–724
Fuxe K, Hökfelt T (1969) Catecholamines in the hypothalamus and the pituitary gland. In: Martini L, Ganong WF (eds) Frontiers of neuroendocrinology. New York Univ Press, New York London Toronto, pp 47–96
Gainer H, Sarne Y, Brownstein MJ (1977) Biosynthesis and axonal transport of rat neurohypophysial proteins and peptides. J Cell Biol 73:366–381
Gash D, Ahmad N, Schechter J (1977) Cytodifferentiation of pituitary primordia transplanted to the kidney capsule of adult hosts. Anat Rec 189:149–160
Gibbs DM (1985) Hypothalamic epinephrine is released into hypophyseal portal blood during stress. Brain Res 335:360–364
Gibbs DM, Neill JD (1978) Dopamine levels in hypophyseal stalk blood in the rat are sufficient to inhibit prolactin secretion in vivo. Endocrinology 102:1895–1900
Glydon RJ (1957) The development of the blood supply of the pituitary in the albino rat, with special reference to portal vessels. J Anat 91:237–244
Goldman H (1961) Endocrine gland blood flow in the unanesthetized, unrestrained rat. J Appl Physiol 16:762–764

Goldman H, Saperstein LA (1958) Determination of blood flow to the rat pituitary gland. Am J Physiol 194:433–435
Green HT (1957) The venous drainage of the human hypophysis cerebri. Am J Anat 100:435–469
Green JD (1947) Vessels and nerves of the amphibian hypophyses. A study of the living circulation and of the histology of the hypophyseal vessels and nerves. Anat Res 99:21–54
Green JD (1948) The histology of the hypophyseal stalk and median eminence in man with special reference to blood vessels, nerve fibers and a peculiar neurovascular zone in this region. Anat Rec 100:273–295
Green JD (1951) The comparative anatomy of the hypophysis with special reference to its blood supply and innervation. Am J Anat 88:225–311
Green JD (1966) The comparative anatomy of the portal vascular system and of the innervation of the hypophysis. In: Harris GW, Donovan BT (eds) The pituitary gland. Butterworths, London, pp 126–146
Green JD, Harris GW (1947) The neurovascular link between the neurohypophysis and the adenohypophysis. J Endocrinol 5:136–146
Green JD, Harris GW (1949) Observation of the hypophysio-portal vessels of the living rat. J Physiol (Lond) 108:359–361
Grimmelikhuijzen CJP, Dierickx K, Boer GJ (1982) Oxytocin vasopressin-like immunoreactivity is present in the nervous system of hydra. Neuroscience 7:3191–3199
Gross DS, Baker BL (1977) Immunohistochemical localization of gonadotropin-releasing hormone (GnRH) in fetal and early postnatal mouse brain. Am J Anat 148:195–215
Gross DS, Baker BL (1979) Development correlation between hypothalamic gonadotropin-releasing hormone and hypophyseal luteinizing hormone. Am J Anat 154:1–10
Gross DS, Page RB (1979) Luteinizing hormone and follicle stimulating hormone production in the pars tuberalis of the hypophysectomized rats. Am J Anat 156:285–291
Gross PM, Blasberg RG, Fenstermacher JD, Patlak CS (1985) Rapid amino acid uptake in rat pituitary neural lobe during functional stimulation by chronic dehydration. J Cereb Blood Flow Metab 5:151–155
Halasz B, Kosaras B, Lengvari I (1972) Ontogenesis of the neurovascular link between the hypothalamus and anterior pituitary in the rat. In: Knigge KM, Scott DE, Weindl A (eds) Brain-endocrine interaction, median eminence structure and function. Karger, Basel, pp 27–34
Hanstrom B (1953) The neurohypophysis in a series of mammals. Z Zellforsch 39:241–259
Harris GW (1947) The blood vessels of the rabbit's pituitary gland and the significance of the pars and zona tuberalis. J Anat (Lond) 81:343–351
Harris GW (1955) Neural control of the pituitary gland. Arnold, London
Hatton GI, Hutton UE, Hoblitzell ER, Armstrong WE (1976) Morphological evidence for two populations of magnocellular elements in the rat paraventricular nucleus. Brain Res 108:187–193
Heistad DD, Marcus ML, Mueller S (1977) Measurement of cerebral blood flow with microspheres. Arch Neurol 34:657–659
Hemming FJ, Begeot M, Dubois MP, Dubois PM (1984) Fetal rat somatotropes in vitro: effects of insulin, cortisol and growth hormone-releasing factor on their differentiation: a light and electron microscopic study. Endocrinology 114:2107–2113
Heymann MA, Payne BD, Hoffman JIE, Rudolph AM (1977) Blood flow measurements with radionuclide-labelled particles. Prog Cardiovasc Dis 20:55–79
Hoffbrand BI, Forsyth RP (1969) Validity studies of the radioactive microsphere method for the study of the distribution of cardiac output, organ blood flow and resistance in the conscious rhesus monkey. Cardiovasc Res 3:426–432
Hökfelt T, Elde R, Fuxe K, Johansson O, Ljungdahl A, Goldstein M, Luft R, Efendic S, Nilsson G, Terenius L, Ganten D, Jeffcoate SL, Rehfeld J, Said S, Perez de la Mora, Possani L, Tapia R, Teran L, Palacios R (1978 a) Aminergic and peptidergic pathways in the nervous system with special reference to the hypothalamus. In: Reichlin S, Baldessarini RJ, Martin JB (eds) The hypothalamus. Raven, New York pp 69–135

Hökfelt T, Ljungdahl A, Steinbusch H, Verhofstad A, Nilsson G, Brodin E, Pernow B, Goldstein M (1978 b) Immunohistochemical evidence of substance P-like immunoreactivity in some 5-hydroxytryptamine-containing neurons in the rat central nervous system. Neuroscience 3:517–538

Hökfelt T, Lundberg JM, Schultzberg M, Johansson O, Ljungdahl A, Rehfeld J (1980) Coexistence of peptides and putative transmitter in neurons. Adv Biochem Psychopharmacol 22:1–23

Houssay BA, Biasotti A, Sammartino R (1935) Modifications fonctionnelles de l'hypophyse après les lésions infundibulo-tubériennes chez le crapaud. C R Soc Biol 120:725–727

Ibata Y, Watanabe K, Kinoshita H, Kubo S, Sano Y, Sin S, Hashimura E, Imagawa K (1979) The localization of LH-RH neurons in the rat hypothalamus and their pathways to the median eminence. Cell Tissue Res 198:381–395

Ishikawa H, Shiino M, Arimura A, Rennels EG (1977) Functional clones of pituitary cells derived from Rathke's pouch epithelium of fetal rats. Endocrinology 100:1227–1230

Johansson O, Hökfelt T, Jeffcoate SL, White N, Sternberger LA (1980) Ultrastructural localization of TRH-like immunoreactivity. Exp Brain Res 38:1–10

Kadekaro M, Gross PM, Sokoloff L, Holcomb HH, Saavedra JM (1983) Elevated glucose utilization in subfornical organ and pituitary neural lobe of the Brattleboro rat. Brain Res 275:189–193

Kapitola J, Dlouha H, Krecek J, Zicha J (1977) The effect of dehydration on the neurohypophyseal blood flow in rats with hereditary diabetes insipidus. Experientia 33:1615–1616

Kawata M, Hashimoto K, Takahara J, Sano Y (1982) Immunohistochemical demonstration of corticotropin-releasing factor containing nerve fibers in the median eminence of the rat and monkey. Histochemistry 76:15–19

Kawata M, Hashimoto K, Takahara J, Sano Y (1983) Immunohistochemical identification of neurons containing corticotropin-releasing factor in the rat hypothalamus. Cell Tissue Res 230:239–246

Kiss JZ, Mezey E, Skirboll L (1984) Corticotropin-releasing factor – immunoreactive neurons of the paraventricular nucleus become vasopressin positive after adrenalectomy. Proc Natl Acad Sci USA 81:1854–1858

Knigge KM, Scott DE (1970) Structure and function of the median eminence. Am J Anat 129:223–244

Knigge KM, Scott DE, Kobayashi H, Ishii S (eds) (1974) Brain endocrine interaction II. The ventricular system in neuroendocrine mechanisms. Karger, Basel

Kobayashi H, Matsui T (1969) Fine structure of the median eminence and its functional significance. In: Martini L, Ganong WF (eds) Frontiers in neuroendocrinology. Oxford Univ Press, New York London Toronto, pp 3–46

Kobayashi H, Oota Y, Uemura H, Hirano T (1966) Electron microscopic and pharmacologic studies on the rat median eminence. Z Zellforsch 71:387–404

Kopaniky DR, Gann DS (1975) Anterior pituitary vasodilatation after hemorrhage in the dog. Endocrinology 97:630–635

Lamberts SW, deLange SA, Stefanko SZ (1982) Adrenocorticotropin-secreting pituitary adenomas originate from the anterior or the intermediate lobe in Cushing's disease: differences in the regulation of hormone secretion. J Clin Endocrinol Metab 54:286–291

Landsmeer JMF (1951) Vessels of the rat's hypophysis. Acta Anat 12:82–109

Lechan RM, Jackson I (1982) Immunohistochemical localization of thyrotropin-releasing hormone in the rat hypothalamus and pituitary. Endocrinology 111:55–65

Leclerc R, Pelletier G (1974) Electron microscope immunohistochemical localization of vasopressin in the hypothalamus and neurohypophysis of the normal and Brattleboro rat. Am J Anat 140:583–588

Lederis K (1965) An electron microscopical study of the human neurohypophysis. Z Zellforsch 65:847–868

Lichardus B, Albrecht I, Ponec J, Linhart L (1977) Water deprivation for 24 hours increases selectively blood flow in the posterior pituitary of conscious rats. Endocrinol Exp 11:99–104

Marcus ML, Heistad DD, Ehrhardt JC, Abboud FM (1976) Total and regional cerebral blood flow measurement with 7-, 10-, 15-, 25-, and 50-μm microspheres. J Appl Physiol 40:501–507
Mata M, Fink DJ, Gainer H, Smith CB, Davidsen L, Savaki H, Schwartz WJ, Sokoloff L (1980) Activity-dependent energy metabolism in rat posterior pituitary primarily reflects sodium pump activity. J Neurochem 34:213–215
McConnell EM (1953) The arterial supply of the human hypophysis cerebri. Anat Rec 115:175–204
Merchenthaler I, Vigh S, Petrusz P, Schally AV (1982) Immunocytochemical localization of corticotropin-releasing factor (CRF) in the rat brain. Am J Anat 165:385–396
Mezey E, Palkovits M (1982) Two way transport in the hypothalamo-hypophyseal system. In: Ganong WF, Martini L (eds) Frontiers in neuroendocrinology, vol 7. Raven, New York pp 1–29
Mezey E, Palkovits M, deKloet ER, Verhoef J, deWied D (1978) Evidence for pituitary-brain transport of a behaviorally potent ACTH analog. Life Sci 22:831–838
Mezey E, Kiss JZ, Mueller GP, Eskay R, O'Donohue TL, Palkovits M (1985) Distribution of the pro-opiomelanocortin derived peptides, adrenocorticotropic hormone, α-melanocyte-stimulating hormone and β-endorphin (ACTH, -MSH, β-End) in the rat hypothalamus. Brain Res 328:341–347
Moll J (1958) The effect of hypophysectomy on the pituitary vascular system of the rat. J Morphol 102:1–21
Monroe BG (1967) A comparative study of the ultrastructure of the median eminence, infundibular stem and neural lobe of the hypophysis of the rat. Z Zellforsch 76:405–432
Murakami T (1976) Pliable methacrylate cast of blood vessels: use of a scanning electron microscope study of the microcirculation in the rat hypophysis. Arch Histol Jpn 38:151–168
Naik DV (1973) Immunohistochemical localization of adrenocorticotropin and melanocyte-stimulating hormone in the pars intermedia of the rat hypophysis. Z Zellforsch 142:289–304
Nakai Y, Shioda S, Ochiai H, Kudo J, Hashimoto A (1983) Ultrastructural relationship between monoamine and TRH-containing axons in the rat median eminence as revealed by combined autoradiography and immunocytochemistry in the same section. Cell Tissue Res 230:1–14
Norstrom A, Hansson H-A, Sjostrand J (1971) Effects of colchicine on axonal transport and ultrastructure of the hypothalamo-neurohypophyseal system of the rat. Z Zellforsch 113:271–293
Page RB (1982) Pituitary blood flow. Am J Physiol 243:E427–442
Page RB (1983) Directional pituitary blood flow: a microcinephotographic study. Endocrinology 112:157–165
Page RB, Bergland RM (1977) The neurohypophyseal capillary bed. Part I. Anatomy and arterial supply. Am J Anat 148:345–358
Page RB, Dovey-Hartman BJ (1984a) Neurohemal contact in the internal zone of the rabbit median eminence. J Comp Neurol 226:274–288
Page RB, Dovey-Hartman BJ (1984b) Resistance vessels supplying the median eminence of the rabbit, rat and cat. Anat Rec 210:647–655
Page RB, Munger BL, Bergland RM (1976) Scanning microscopy of pituitary vascular casts: the rabbit pituitary portal system revisited. Am J Anat 146:273–301
Page RB, Leure-duPree AE, Bergland RM (1978) The neurohypophyseal capillary bed. II. Specializations within median eminence. Am J Anat 153:33–66
Page RB, Funsch DJ, Brennan RW, Hernandez MJ (1981) Regional neurohypophyseal blood flow and its control in adult sheep. Am J Physiol 241:R36–43
Palay SL (1945) Neurosecretion VII: the preoptico-hypophysial pathway in fishes. J Comp Neurol 82:129–143
Palay SL (1955) An electron microscopic study of the neurohypophysis in normal, hydrated and dehydrated rats. Anat Rec 121:348
Palay SL (1960) The fine structure of secretory neurons in the preoptic nucleus of the goldfish (*Crassius auratus*). Anat Rec 138:417–444

Palkovits M (1982) Neuropeptides in the median eminence: their sources and destinations. Peptides 3:299–303

Pearse AGE, Takor TT (1976) Neuroendocrine embryology and the APUD concept. Clin Endocrinol [Suppl] 5:229S–244S

Pelletier G, Dube D (1977) Electron microscopic immunohistochemical localization of α-MSH in the rat brain. Am J Anat 150:201–204

Pelletier G, Labrie F, Arimura A, Schalley AV (1974) Electron microscopic immunohistochemical localization of growth hormone release inhibiting hormone (somatostatin) in the rat median eminence. Am J Anat 140:445–450

Perlmutter LS, Hatton GI, Tweedle CD (1984) Plasticity in the in vitro neurohypophysis: effects of osmotic change on pituicytes. Neuroscience 12:503–511

Popa G, Fielding U (1930a) The vascular link between the pituitary and the hypothalmus. Lancet 2:238–240

Popa G, Fielding U (1930b) A portal circulation from the pituitary to the hypothalamic region. J Anat 65:88–91

Porter JC, Hines MFM, Smith KR, Repass RL, Smith AJK (1967) Quantitative evaluation of local blood flow of the adenohypophysis in rats. Endocrinology 80:583–598

Proulx-Ferland L, Labrie F, Dumont D, Cote J, Coy DH, Sveiraf J (1982) Corticotropin-releasing factor stimulates secretion of melanocyte-stimulating hormone from the rat pituitaries. Science 217:62–63

Raisman G (1973) Electron microscopic studies of the development of new neurohaemal contacts in the median eminence of the rat after hypophysectomy. Brain Res 55:245–261

Rasmussen AT (1940) Effects of hypophysectomy and hypophyseal stalk resection on the hypothalamic nuclei of animals and man. Res Publ Assoc Res Nerv Ment Dis 20:245–269

Reivich M, Sokoloff L (1976) Application of the 2-deoxy-D-glucose method to the coupling of cerebral metabolism and blood flow. Neurosci Res Prog Bull 14:474–475

Richards JG, Tranzer JP (1970) The ultrastructural localisation of amine storage sites in the central nervous system with the aid of a specific marker, 5-hydroxydopamine. Brain Res 17:463–469

Rinne UK (1966) Ultrastructure of the median eminence of the rat. Z Zellforsch 74:98–122

Rioch DM, Wislocki GB, O'Leary JL (1940) A precis of preoptic, hypothalamic and hypophyseal terminology with atlas. Res Publ Assoc Res Nerv Ment Dis 20:3–30

Rodriguez EM (1969a) Ependymal specializations. I. Fine structure of the neural (internal) region of the toad median eminence, with particular reference to the connections between the ependymal cells and subependymal capillary loops. Z Zellforsch 92:153–171

Rodriguez EM (1969b) Ultrastructure of the neurohemal region of the toad median eminence. Z Zellforsch 93:182–212

Romeis B (1940) Innersekretorische Drüsen. II. Hypophyse. In: van Mollendorf W (Hrsg) Handbuch der mikroskopischen Anatomie des Menschen. Springer, Berlin

Roth KA, Weber E, Barchas JD (1982) Immunoreactive corticotropin releasing factor (CRF) and vasopressin are colocalized in a subpopulation of the immunoreactive vasopressin cells in the paraventricular nucleus of the hypothalamus. Life Sci 31:1857–1860

Sakurada O, Kennedy C, Jehle J, Brown JD, Carbin GL, Sokoloff L (1978) Measurement of local cerebral blood flow with iodo[^{14}C]-antipyrine. Am J Physiol 234:3, H59–66

Saperstein LA (1958) Regional blood flow by functional distribution of indicators. Am J Physiol 193:161–168

Sawchenko PE, Swanson LW, Joseph SA (1982) The distribution and cells of origin of ACTH (1–39) stained varicosities in the paraventricular and supraoptic nuclei. Brain Res 232:365–374

Scharrer B, Scharrer E (1944) Neurosecretion. VI. A comparison between the intercerebralis-cardiacum-allatum system of the insects and the hypothalamo-hypophyseal system of the vertebrates. Biol Bull 87:242–251

Scharrer E (1965) The final common path in neuroendocrine integration. Arch Anat Microscop Morphol Exp 54:359–370
Scharrer E, Scharrer B (1940) Secretory cells within the hypothalamus. Res Publ Assoc Res Nerv Ment Dis 20:170–194
Schwab ME, Thoenen H (1983) Mechanism of uptake and retrograde axonal transport of noradrenaline in sympathetic neurons in culture: reserpine-resistant large dense-core vesicles as transport vehicles. J Cell Biol 96:1538–1547
Schwartz WJ, Smith CB, Davidsen L, Savaki H, Sokoloff L (1979) Metabolic mapping of functional activity in the hypothalamo-neurohypophyseal system of the rat. Science 205:723–725
Selmanoff M (1981) The lateral and medial median eminence: distribution of dopamine, norepinephrine, and luteinizing hormone-releasing hormone and the effect of prolactin on catecholamine turnover. Endocrinology 108:1716–1722
Sherlock DA, Field PM, Raisman G (1975) Retrograde transport of horseradish peroxidase in the magnocellular neurosecretory system of the rat. Brain Res 88:403–414
Shiino M, Ishikawa H, Rennels EG (1978) Accumulation of secretory granules in pituitary clonal cells derived from the epithelium of Rathke's pouch. Cell Tissue Res 186:53–61
Shioda S, Nakai Y (1983) Immunocytochemical localization of TRH and autoradiographic determination of 3H-TRH binding sites in the arcuate nucleus–median eminence of the rat. Cell Tissue Res 228:475–487
Silverman AJ, Zimmerman EA (1975) Ultrastructural immunocytochemical localization of neurophysin and vasopressin in the median eminence and posterior pituitary of the guinea pig. Cell Tissue Res 159:291–301
Sloper JC, Arnott DJ, King BC (1960) Sulphur metabolism in the pituitary and hypothalamus of the rat: a study of radioisotope-uptake after the injection of ^{35}S dl-cysteine, methionine and sodium sulphate. J Endocrinol 20:9–23
Sobin SS, Tremer HM (1977) Three-dimensional organization of microvascular beds as related to function. In: Kaley G, Altura B (eds) Microcirculation, vol 1. Univ Park Press, Baltimore, pp 43–67
Sobin SS, Tremer HM, Fung YC (1970) The morphometric basis of the sheet-flow concept of the pulmonary alveolar microcirculation in the cat. Circ Res 26:397–414
Sokoloff L (1977) Relation between physiological function and energy metabolism in the central nervous system. J Neurochem 29:13–26
Swanson LW, Sawchenko PE (1980) Paraventricular nucleus: a site for the integration of neuroendocrine and autonomic mechanisms. Neuroendocrinol 31:410–417
Szentagothai J, Flerko B, Mess B, Halasz B (1968) Hypothalamic control of the anterior pituitary. Akademiai, Budapest
Takor TT, Pearce AGE (1975) Neuroectodermal origin of avian hypothalamo-hypophyseal complex: the role of the ventral neural ridge. J Embryol Exp Morphol 34:311–325
Taniguchi Y (1984) Immunohistochemical evidence against the coexistence of a corticotropin-releasing factor and oxytocin or vasopressin in the rat paraventricular nucleus. Arch Histol Jpn 47:475–483
Thureson-Klein A (1983) Exocytosis from large and small dense-core vesicles in noradrenergic nerve terminals. Neuroscience 10:245–252
Tilders FJH, Smelik PG (1977) Direct neural control of MSH secretion in mammals: the involvement of dopaminergic tubero-hypophyseal neurones. Front Horm Res 4:80–93
Torok B (1954) Lebendbeobachtung des Hypophysenkreislaufes an Hunden. Acta Morphol Hung 4:83–89
Torok B (1964) Structure of the vascular connections of the hypothalamo-hypophysial region. Acta Anat 59:84–99
Tranzer JP, Richards JG (1976) Ultrastructural cytochemistry of biogenic amines in nervous tissue: methodologic improvements. J Histochem Cytochem 24:1178–1193
Tranzer JP, Theonen H (1967) Electronmicroscopic localization of 5-hydroxydopamine (3,4,5-trihydroxyphenyl-ethylamine) a new "false" sympathetic transmitter. Experentia 23:743–745

Tweedle CD, Hatton GI (1980 a) Evidence for dynamic interactions between pituicytes and neurosecretory axons in the rat. Neuroscience 5:661–667
Tweedle CD, Hatton GI (1980 b) Glial cell enclosure of neurosecretory endings in neurohypophysis of the rat. Brain Res 192:555–559
Tweedle CD, Hatton GI (1982) Magnocellular neuropeptidergic terminals in the neurohypophysis: rapid glial release of enclosed axons during parturition. Brain Res Bull 8:205–209
Vale W, Spiess J, Rivier C, Rivier J (1981) Characterization of a 41-residue ovine hypothalamic peptide that stimulates secretion of corticotropin and β-endorphin. Science 213:1394–1397
Vanderhaeghen JJ, Lotstra F, Liston DR, Rossier J (1983) Proenkephalin, (Met)enkephalin and oxytocin immunoreactivities are colocalized in bovine hypothalamic magnocellular neurons. Proc Natl Acad Sci USA 80:5139–5143
Vandesande F, Dierickx K (1975) Identification of the vasopressin- and of the oxytocin-producing neurons in the hypothalamic magnocellular neurosecretory system of the rat. Cell Tissue Res 164:153–162
Vandesande F, Dierickx K, DeMey J (1977) The origin of vasopressinergic and oxytocinergic fibers of the external region of the median eminence of the rat hypophysis. Cell Tissue Res 180:443–452
VanDyke HB, Adamsons K, Engel SL (1955) Aspects of the biochemistry and physiology of the neurohypophyseal hormones. Recent Prog Horm Res 11:1–41
Vannucci S, Hawkins R (1983) Substrates of energy metabolism of the pituitary and pineal glands. J Neurochem 41:1718–1725
Verney EB (1947) The antidiuretic hormone and the factors which determine its release. Proc R Soc Lond [Biol] 135:25–106
Vigh S, Horvath J, Schally AV, Arimura A, Setalo G (1978) Simultaneous localization of luteinizing hormone-releasing hormone (LH-RH) and somatostatin-containing nerve fibres in the brain of the rat. Acta Biol Acad Sci Hung 29:291–294
Vigneaud duV, Lawler HC, Popenoe EA (1953) Enzymatic cleavage of glycinamide from vasopressin and a proposed structure for this pressor-antidiuretic hormone of the posterior pituitary. J Am Chem Soc 75:4880–4881
Vina JR, Page RB, Davis DW, Hawkins RA (1984) Aerobic glycolysis by the pituitary gland in vivo. J Neurochem 42:1479–1482
Visser M, Swaab DF (1977) Alpha-MSH in the human pituitary. Front Horm Res 4:42–45
Watanabe YG, Daikoku S (1976) Immunohistochemical study of adenohypophyseal primordia in organ culture. Cell Tissue Res 166:407–412
Watson SJ, Akil H, Fischli W, Goldstein A, Zimmerman E, Nilaver G, Greidanus TBV (1982) Dynorphin and vasopressin. Common localization in magnocellular neurons. Science 216:85–87
Weitzman RE, Glatz TH, Fisher DA (1978) The effect of hemorrhage and hypertonic saline upon plasma oxytocin and arginine vasopressin in conscious dogs. Endocrinology 103:2154–2160
Whitnall MH, Gainer H, Cox BM, Molineaux CJ (1983) Dynorphin-A-(1–8) is contained within vasopressin neurosecretory vesicles in rat pituitary. Science 222:1137–1139
Wingstrand KG (1966) Comparative antomy and evolution of the hypophysis. In: Harris GW, Donovan BT (eds) The pituitary gland, vol 1. Butterworths, London, pp 58–126
Wislocki GB (1937 a) The vascular supply of the hypophysis cerebri of the cat. Anat Rec 69:361–387
Wislocki GB (1937 b) The meningeal relations of the hypophysis cerebri. II. An embryological study of the meninges and blood vessels of the human hypophysis. Am J Anat 61:95–129
Wislocki GB (1938 a) The vascular supply of the hypophysis cerebri of the rhesus monkey and man. Res Publ Assoc Res Nerv Ment Dis 17:48–68
Wislocki GB (1938 b) Further observations on the blood supply of the hypophysis cerebri of the rhesus monkey. Anat Rec 72:137–150

Wislocki GB, King LS (1936) The permeability of the hypophysis and hypothalamus to vital dyes, with a study of the hypophyseal vascular supply. Am J Anat 58:421–472
Wittkowski W, Brinkmann H (1974) Changes of extent of neuro-vascular contacts and number of neuro-glial synaptoid contacts in the pituitary posterior lobe of dehydrated rats. Anat Embryol (Berlin) 146:157–165
Wittkowski W, Scheuer A (1974) Functional changes of the neuronal and glial elements at the surface of the external layer of the median eminence. Z Anat Entwickl Gesch 143:255–262
Worthington WC (1955) Some observations on the hypophyseal portal system in the living mouse. Bull Johns Hopkins Hosp 97:343–357
Worthington WC (1960) Vascular responses in the pituitary stalk. Endocrinology 66:19–31
Xuereb GP, Prichard M, Daniel PM (1954) The arterial supply and venous drainage of the human hypophysis cerebri. Q J Exp Physiol 39:199–217
Ziedonis DM, Severs WB, Brennan RW, Page RB (1986) Blood flow and functional responses correlate in the ovine neural lobe. Brain Res 373:27–34
Zimmerman EA (1981) The organization of oxytocin and vasopressin pathways. In: Martin JB, Reichlin S, Bick KL (eds) Neurosecretion and brain peptides. Raven, New York, pp 63–74
Zimmerman EA, Antunes JL (1976) Organization of the hypothalamic-pituitary system: current concepts from immunohistochemical studies. J Histochem Cytochem 24:807–815
Zimmerman EA, Robinson AG, Husain MK, Acosta M, Frantz A, Sawyer WH (1974) Neurohypophyseal peptides in the bovine hypothalamus: the relationship of neurophysin I to oxytocin and neurophysin II to vasopressin in the supraoptic and paraventricular regions. Endocrinology 95:931–936

Functional Ultrastructure of Gonadotropes: a Review *

G.V. Childs

Contents

1 Identification of Gonadotrope Subtypes

1.1 Identification of the Gonadotrope by Morphological Criteria

The earliest electron microscopic studies of gonadotropes were reported by Farquhar and Rinehart in 1954. The design of their study followed that of pioneering light microscopic studies of the previous 3 decades, which showed distinct changes in each pituitary cell type after removal of its respective target organ. These studies had allowed the identification of most of the pituitary cells by linking their affinity for unique tinctorial stains with striking cytologic changes observed upon alteration of target organ hormonal feedback (Purves and Griesbach 1954). In the case of the gonadotrope, it was already known that the removal of

* The research described in this review was supported by National Foundation March of Dimes Basil O'Connor and Basic Research Grants 1-551, and by NIH grants RO1 HD00842 (University of Nebraska), RO1 HD10930 (Northwestern University) and RO1 HD15472 (University of Texas Medical Branch). Dr. Childs was supported by RCDA 1 KO4 HD AM-00306 (Northwestern) and HD00395 (UTMB) between 1979 and 1984

Department of Anatomy, The University of Texas, Medical Branch, Galveston, TX 77550, USA

Current Topics in Neuroendocrinology, Vol. 7

the gonads caused the cells to enlarge dramatically and become filled with vacuoles or one large vacuole, giving them the appearance of "swiss cheese," or "signet rings."

Farquhar and Rinehart identified two types of cell that exhibited cytologic changes following ovariectomy. Six days after surgery, one of these cell types was larger than all the others, round, and adjacent to a vessel. The cytoplasm was filled with granules and dilated vesicles of rough endoplasmic reticulum. The Golgi complex was prominent and surrounded by mitochondria and small vesicles. The cells continued to enlarge with time after gonadectomy, and by 20 days after surgery the Golgi complex was more prominent. There were fewer granules and more dilated profiles of rough endoplasmic reticulum (rer). The rer appeared to coalesce so that by 35 days after surgery some cells contained only one large vacuole, giving them the signet ring appearance seen in the light microscopic studies. However, morphological heterogeneity was also evident in the population, since some of the cells resembled those of earlier periods following gonadectomy.

Farquhar and Rinehart (1954) also identified a second cell type that responded later than the first. These cells were characterized by a dense nucleus and cytoplasm filled with vesicles of rer that had a filigreed appearance. They also had a higher concentration of secretion granules than the other cell type.

In discussing the distinction between luteinizing hormone (LH) and follicle-stimulating hormone (FSH) cells, Farquhar and Rinehart referred to earlier studies by Purves and Greisbach in which the two types of gonadotrope were differentiated on the basis of their response to testosterone treatment and their prominence prior to the onset of puberty. The first type was more centrally located than the second. It degranulated prior to first ovulation and involuted following testosterone treatment, which linked it to the production of LH. The more peripherally located second type did not involute completely following testosterone treatment, and showed increased granulation following estrogen treatment. This cell type was linked to the production of FSH.

Purves and Greisbach had also shown that pituitary LH activity was highest at later periods after castration, when the presumptive LH cells showed the most intense tinctorial stain for glycoproteins (discussed in Purves and Greisbach 1954). Farquhar and Rinehart (1954) correlated this finding with the late appearance of the filigreed castration cell and suggested that this was the LH cell, whereas the earlier appearing cell was the FSH cell.

During the subsequent decade, workers continued to link each of the cell types with the production of one of the gonadotropins. Barnes (1962) reported cytologic changes in mice following gonadectomy that suggested that the ovoid gonadotropes with two types of granule predominated when there were high FSH levels (beta cells), while a more angular gonadotrope predominated when LH levels were higher (gamma cells).

Kurosumi and Oota (1968) later applied Barnes' classification criteria to the rat and reported similar findings. They described the FSH cell as round, large, and situated along blood capillaries. It was well-granulated, with granules of two sizes: a population of large granules (700–1000 nm in diameter) and a population of granules 200 nm in diameter that were more electron-dense. The second cell type also lay along capillaries, was more angular or polygonal, and contained secretion granules averaging 250 nm in diameter. Thereafter, the ovoid gonadotrope was named the "FSH cell" and the more angular gonadotrope was named the "LH cell."

However, a number of cytophysiological studies during the same period showed that there were gonadotropes that exhibited morphological features of both types: studies by Rennels et al. 1971; Shiino et al. 1972; Mendoza et al. 1973; and Debeljuk et al. 1973. These workers suggested that functional differentiation of LH and FSH cells based solely on morphology may not always be possible, and that each of the subtypes may be in one stage of a synthesis or secretory cycle belonging to a common LH-FSH cell. While this may have presented difficulties to those who espoused the theory that each pituitary cell type produces only one hormone, it did prepare the field for the confirmatory and surprising findings presented by the immunocytochemists.

1.2 Identification of Gonadotropes by Immunocytochemistry: Early Studies

The earliest immunocytochemical studies of gonadotropes presented the field with several new findings. First, the clever application of double stains for LH and FSH by Nakane (1970) showed that a high percentage of the gonadotropes contained both hormones. However, during the same period biochemists found that the gonadotropin molecules each had a subunit with identical amino acids, the alpha chain (reviewed in Pierce and Parsons 1981). Thus, the initial reaction to the early findings was cautious, primarily because antisera to the unique beta chain subunits were not available to distinguish the gonadotropin storage sites. Critics reasoned that the joint staining could be due to antibodies to sequences in the identical alpha subunits (Phifer et al. 1973; Moriarty 1973).

In man, light microscopic studies of serial sections through cells stained for LH or FSH also showed that the two hormones could be found in the same cell. The studies employed careful immunoabsorption experiments to prove specific staining with each antiserum (Phifer et al. 1973).

Nakane (1971) also presented the first ultrastructural view of immunocytochemically stained gonadotropes, and was able to confirm the findings of Kurosumi and Oota (1968) as to the morphology of one of the cell types. The ovoid cell that had been called the FSH cell was stained for either LH or FSH (Nakane's type A cell). The angular gonadotropes were not evident in his stained cell population. Another cell type that stained for FSH alone was stellate and contained peripheral granules (Nakane's type B cell); it could not be linked to any of the known cell types at the time.

Tougard and Tixier-Vidal and their colleagues confirmed Nakane's findings with the use of antisera to the alpha and beta subunits of LH (Tougard et al. 1971, 1973, 1980; Tixier-Vidal 1972; Tixier-Vidal et al. 1975). They reported staining for LH and its subunits in both of the classical types of gonadotrope described by Kurosumi and Oota (1968), again indicating that one could not distinguish the two morphological subtypes by their hormone content. In addition, they reported that a third angular cell type stained for anti-LH-beta, and cell types with morphological features shared by all three were also described.

Tougard et al. were also the first to apply the stains to sections of pituitaries prior to embedding (Tougard et al. 1971, 1973). This new method further validated the findings reported by Nakane (1970), who had applied the stains to paraffin- or plastic-embedded sections. In a later report, the two methods were found to be comparable for studies of the two major gonadotropic cell types (Tougard et al. 1980). When the stain was applied to embedded sections, the analysis showed that LH and FSH were present in the same cells and sometimes in the same granules.

Tougard et al. (1973) and Tixier-Vidal et al. (1975) also observed changes in the two gonadotrope subtypes following ovariectomy. Their studies agreed in part with the pioneering work of Farquhar and Rinehart (1954). The type A gonadotrope (FSH cell) changed following surgery as described by Farquhar and Rinehart. However, their type B gonadotrope, which resembled the LH cell, did not change. In agreement with Farquhar and Rinehart (1954), the immunocytochemists demonstrated that the filigreed ovariectomy cells appeared later after surgery and contained LH. The serial sections through ovariectomy cells showed that they contained both LH and FSH.

Tixier-Vidal's group was also the first to stain pituitary monolayers for LH and FSH (Tixier-Vidal 1972), correlating the stain with the LH and FSH content of the medium. Tixier-Vidal (1972) noted that the gonadotropin content of the medium fell rapidly during the 1st week and then continued to decline slowly until it was undetectable. The percentage of cells reactive with antisera to whole LH or FSH was 2%–5%, whereas the percentage of cells immunoreactive with antisera against LH-beta subunits was four times higher. This may reflect the types of antigen exposed to the antibody, since there is little free beta subunit stored in gonadotropes (Grotjan et al. 1984).

As late as 5 days after dispersion, Tixier-Vidal reported that the gonadotropes still resembled those in the intact gland, apart from the fact that they contained more large lysosomes as determined by acid phosphatase cytochemistry. After 7 days in culture, the type A gonadotrope (with two types of secretory granule) disappeared, and a single cell type, which had one class of small secretory granule (125–150 nm in diameter), sparse rer cisternae, and a small Golgi complex, remained. This finding indicated that gonadotropin-releasing hormone (GnRH) from the hypothalamus and the steroid environment were necessary to maintain normal percentages of gonadotropes and their ultrastructural morphology. Intracellular LH immunoreactivity persisted longer than that of FSH, although FSH could still be measured in the media long after the stainable FSH had disappeared. Tixier-Vidal suggested that the gonadotrope was secreting FSH too rapidly for the buildup of stores.

In 1975, studies in this laboratory confirmed findings fom both of the above groups of workers when we applied antisera to LH-beta and the peroxidase–antiperoxidase complex (PAP) technique of Sternberger et al. (1970) to plastic sections and localized LH in cells that resembled both types of gonadotropes (Moriarty 1975, 1976a, b). A sex difference in the proportion of each of the two types was noted.

In the male rats, the ovoid gonadotrope with two sizes of granules predominated (type I gonadotrope). It was stained intensely with high dilutions of anti-

LH-beta; however, the absorption of this antiserum with high concentrations of related antigens [FSH and thyroid-stimulating hormone (TSH)] showed that the granule population was heterogeneous. The largest granules were unreactive if LH was added to the anti-LH-beta, and remained strongly stained following the addition of FSH or TSH (Moriarty 1975, 1976a). The smallest granules, however, were unreactive following the addition of 10 ng LH, or 1 μg TSH per milliliter. FSH had no effect on their staining intensity. This evidence suggested that the smaller granules might contain antigens shared by both LH and TSH, and that the polyclonal antisera to LH-beta were probably cross-reactive. For this reason, in subsequent studies the antisera to LH were preabsorbed with TSH and FSH prior to staining. Similar reactions were seen following TSH absorption of anti-FSH sera (Moriarty 1976a).

In female rats, the angular gonadotrope subtype (type II gonadotrope) predominated. The granules in this type of gonadotrope were homogeneous in their reactivity with the antisera, and only LH absorption abolished the stain for LH. In addition, different shapes of gonadotropes were noted in the female, and we therefore attempted to correlate the morphological heterogeneity with the stage of the estrous cycle (Moriarty 1975, 1976a, b). We noted that cells from proestrous and estrous rats (taken at 1 p.m.) were more variably shaped and granules were either sparse or at the periphery, in cellular processes (Figs. 1 and 2). The

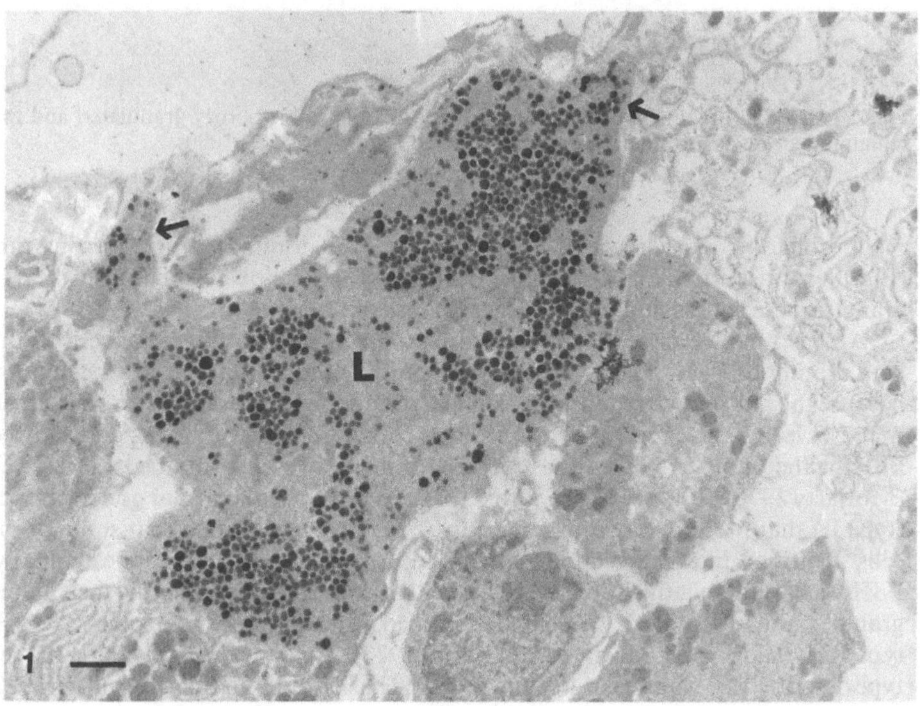

Fig. 1. LH cell (*L*) from proestrous rat (1000) stained with a 1:10000 dilution of anti-bLH-beta and the PAP technique (Moriarty and Halmi 1972) after fixation in 1% glutaraldehyde and embedding in Araldite 6005. Early in proestrus, the cells are well-granulated and send processes (*arrows*) to nearby vessels. *Bar*, 1 μm. No counterstains applied, × 9600

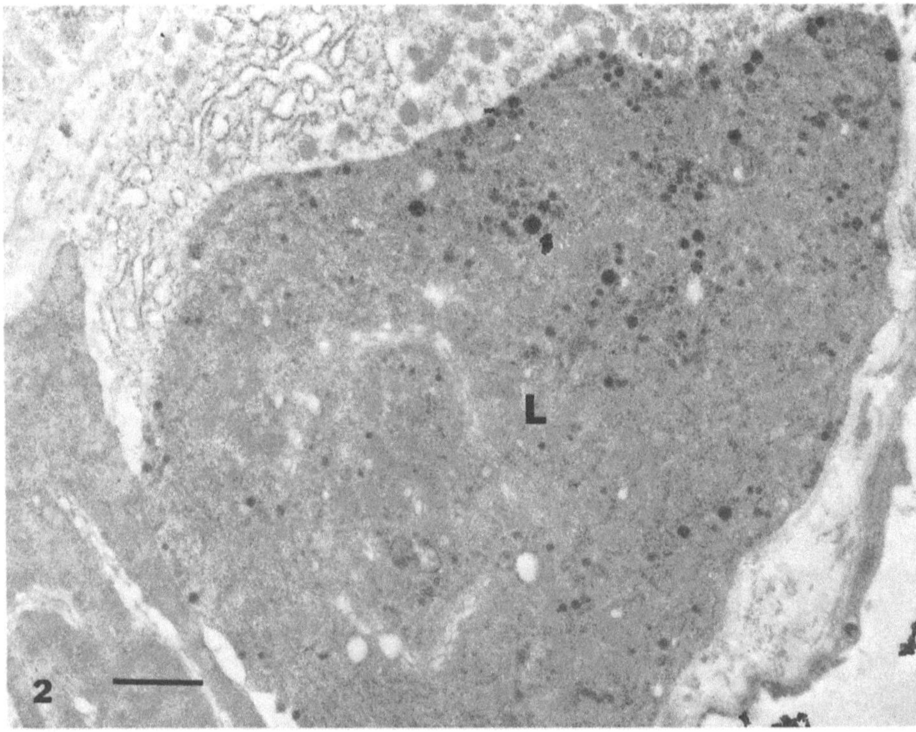

Fig. 2. LH cell (*L*) from rat at 1000 during estrous. The cell is poorly granulated and the remaining granules are small. *Bar*, 1 µm. Stained as in Fig. 1, × 14000

ovoid gonadotrope with two types of granule predominated in diestrous rats (Fig. 3). It was difficult to distinguish from type II cells, however, because its large granules were sparse and not as large as those in the male rat.

At this time we were also involved in studies with the use of antisera to the beta subunit of FSH. The FSH stains were problematic from the very beginning, because the antisera were more cross-reactive and the FSH antigens more difficult to preserve in tissue. In 1976, we confirmed Nakane's earlier findings that showed FSH in the ovoid gonadotropes (type I), some of which contained LH as well (Moriarty 1976 a, b). We also reported that FSH was in the angular gonadotropes (type II) that predominated in female rats. However, we found that some of the cells identified in serial sections contained only one of the hormones. Finally, FSH was found in a third stellate cell type that contained peripherally distributed granules and resembled the cell that stained for adrenocorticotropin (ACTH; type III gonadotrope; Moriarty and Halmi 1972). This was probably the same cell type described by Nakane in 1971 (type B FSH cell) and by Tougard et al. in 1980. Further studies of the identity of this cell type are described in Sect. 2.5.

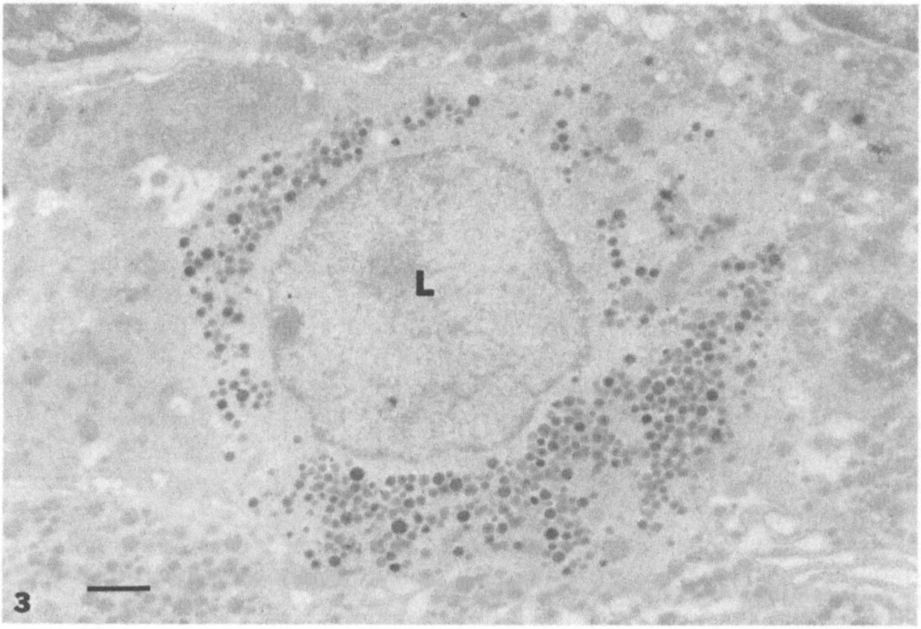

Fig. 3. LH cell (*L*) from diestrous rat taken at 1000. The cell contains large and small granules and is round; both features are characteristic of a type I gonadotrope (Moriarty 1975, 1976 a, b; Childs et al. 1980). The granule size distinction is not as great as that in the male, however. Note that the cell is not stained as intensely as that in proestrous rats. *Bar*, 1 μm. Stained as in Fig. 1, × 12300

1.3 Comparative Studies of Gonadotropes in a Variety of Mammalian Species

During the late 1970s, other laboratories were studying gonadotrope populations in many species with the use of ultrastructural immunocytochemical techniques. Dacheux and Dubois reported studies of bovine (1976), ovine (1978), and porcine (Dacheux 1978, 1980) gonadotropes with PAP complex immunocytochemistry. A complete series of fixation tests showed that LH and FSH were preserved well after fixation in either *p*-formaldehyde or glutaraldehyde and embedding in Epon. In addition, they reported the first stains for gonadotropins in pituitaries fixed in glutaraldehyde and postfixed in 1% osmium tetroxide.

In bovine pituitaries, LH was found in cells in the basophilic zone that were medium-sized, round, or oval, and that contained two populations of secretion granules. The small granules were 250–400 nm in diameter and the large granules were characterized by their irregular shape and lower electron density. Both types of granules stained for LH and its subunits (Dacheux and Dubois 1976). In ovine pituitaries, the gonadotropes were also ovoid and distributed along blood vessels. The Golgi complex was prominent and enlarged during the breeding season. As in the case of bovine pituitaries, the granule population included small, round granules (250 nm average diameter) and large round bodies 800 nm in diameter (Dacheux and Dubois 1978).

Gonadotropes from porcine pituitaries were identified with antisera to both LH-beta and FSH-beta (Dacheux 1978, 1980). The gonadotropes were cytologically similar to those in ovine and bovine pituitaries. In addition, many of them stained for both LH and FSH (multihormonal), while the remainder stained weakly or not at all for one of the hormones (monohormonal). The multihormonal and monohormonal gonadotropes could not be distinguished morphologically. Furthermore, the investigators reported that many of the granules in multihormonal gonadotropes contained both LH and FSH.

During the same period, Batten and Hopkins (1978) localized gonadotropins on serial sections of dissociated cells from porcine pituitaries. Their results essentially agreed with those of Dacheux (1978), except that they did not report the presence of monohormonal gonadotropes. Furthermore, they were unable to stain sections fixed in glutaraldehyde with or without osmium tetroxide postfixation. [This may have been due to the Spurr embedding resin employed in their studies that required more intense etching agents (sodium methoxide) and prolonged incubations in concentrated antisera (1:500 dilution for 1 week for successful staining).]

Another mammalian species that has been studied recently is the goat. Shirasawa et al. (1985) reported that goat gonadotropes were similar to those in bovine pituitaries, except that were there no large granules. In fact, the entire granule population was, on average, smaller than that of most other species (190–210 nm in diameter). The cells were more irregularly shaped, and the Golgi complex was not as prominent as that in gonadotropes in other species.

Ultrastructural immunocytochemical studies of primate pituitaries began with pioneering work by Pelletier et al. (1978), who focused on the localization of gonadotropes in man. The gonadotropes were large and ovoid, with sparsely distributed secretion granules scattered among dense vesicles. Only the granules were labeled, and serial sections showed that most contained both LH and FSH. In an earlier light microscopic study, Pelletier et al. (1976) had reported that 90% of the gonadotropes stained for both hormones; the remaining ones contained only FSH. However, in the electron microscopic study the monohormonal FSH cells were not found.

In the rhesus monkey, light microscopic studies by Herbert (1976, 1978) showed LH in ovoid cells that were randomly scattered throughout the gland. The only area of concentration was the pars tuberalis, in the most inferior regions. In 1980, Girod et al. focused on the pars tuberalis of the *Macaca irus* monkey with immunofluorescence techniques. They reported that whereas FSH cells were found elsewhere throughout the gland, only LH storage was seen in gonadotropes in the pars tuberalis. This finding contrasted with that reported earlier by Baker et al. (1977), who noted that pars tuberalis gonadotropes contained both LH and FSH in the same species.

In the *Papio cynocephalus,* baboon, the gonadotropes were distributed homogeneously in the gland, and their morphology resembled that in the other monkeys (Herbert and Silverman, 1983). Another genus of baboon, *Papio hamadryas,* and vervet monkeys have a higher percentage of gonadotropes in the median portion of the lateral wings (Girod et al. 1981). Therefore, it appears that the distribution of gonadotropes and their subtypes may vary from species to species in the lower primates.

1.4 Studies of Gonadotropes in the Pars Tuberalis

The pars tuberalis has been a primary area for the investigation of gonadotropes. Light microscopic studies by a number of investigators have demonstrated that the pars tuberalis contains both gonadotropes and thyrotropes (Baker and Yu 1975; Baker 1977; Baker et al. 1977; Herbert 1978; Gross 1978; Gross and Page 1979; Osamura and Watanabe 1978; Ordronneau and Petrusz 1980; Pearson and Licht 1982). The significance of this finding is intriguing in the light of the fact that the region also contains numerous GnRH-containing axonal endings that could stimulate the neighboring gonadotropes without an intervening blood supply (Gross 1984).

One of the most complete comparative studies of the pars tuberalis is found in a recent report by Gross (1984), who identified both gonadotropes and GnRH endings in close proximity in humans and the rhesus monkey. Species that had a pars tuberalis with relatively high percentages of gonadotropes included man, the baboon (75%), and sheep (85%–90%). The sheep pars tuberalis is unique in that it contains no thyrotropes (Gross et al. 1984).

Hormone storage in pars tuberalis gonadotropes varied from species to species. Some (10%–15%) of the ovine gonadotropes contained only LH. Most of the gonadotropes in the pars tuberalis from other species contained both hormones; however, a few monohormonal LH cells were also found in the pars tuberalis of primate pituitaries. As stated earlier, Girod et al. found only monohormonal LH gonadotropes in one species of monkey (1981).

The first ultrastructural immunocytochemical study of the human pars tuberalis was published in 1983, by Asa et al., who reported that the region contained mainly gonadotropes and occasional corticotropes and thyrotropes. Most of the gonadotropes were stained for both LH and FSH. The gonadotropes were characterized by the presence of scattered small secretion granules and several large dense bodies.

Pars tuberalis gonadotropes have been tested for their biological activity and significance by a number of workers (Legait 1969; Gross 1978; Gross and Page 1979). The cells contain biologically active LH (Legait 1969) and respond to fluctuations in gonadal steroids (Gross 1978; Gross and Page 1979). Recently, Aguado et al. (1982) reported changes in the LH content of the rat pars tuberalis. The cells showed the lowest LH content during the proestrous LH surge and higher values during estrus. Electron microscopic studies confirmed the morphological identity of the gonadotropes.

1.5 Studies of Human Pituitary Gonadotropin Adenomas

The ultrastructural studies of human pituitary gonadotropes have recently been extended to gonadotropin-secreting adenomas. Such tumors are rare (3%–4%) and difficult to diagnose because the cells may show little evidence of gonadotropin storage by their tinctorial staining properties. Serum or tumor levels of gonadotropins may or may not provide clues. The morphology of these tumor cells is also not typical of gonadotropin-producing cells, except in rare cases (Friend et al. 1976; Kovacs and Horvath 1975; Kovacs et al. 1978, 1980; Trouillas et al. 1981; Horvath and Kovacs 1984). The cells appear to contain very small secretory granules that are scattered sparsely in the cytoplasm. Only in a few cases do the tumor cells resemble the hypertrophied castration cell in the rat (Kovacs et al. 1978; Horvath and Kovacs 1984).

An interesting finding related to the concept of differential gonadotropin storage is the relatively frequent occurrence of monohormonal FSH or LH tumors. For example, Kovaks et al. 1978 and Trouillas et al. (1981) reported cases in which only FSH was secreted. Horvath and Kovacs (1984) reported FSH adenomas in 4/14 tumors in men and 2/13 tumors in women. Only one of the tumors displayed strong immunoreactivity for LH, with weak staining for FSH. The remaining stained for both FSH and LH or for neither gonadotropin. Horvath and Kovaks (1984) also noted an ultrastructural diagnostic feature present in gonadotropin adenomas in women in 14/15 cases: the cells exhibited vesicular dilation of the Golgi complex, giving the region a "honeycomb" appearance. It was not found in similar tumors in men.

1.6 Significance of the Immunocytochemical Studies of the 1970s

By 1978, immunocytochemists had agreed on the basic ultrastructural identity of the gonadotrope subtypes. However, the data that were presented to the field had shaken the "one-cell-one-hormone" hypothesis. This would not have been a problem had two sets of hypothalamic releasing hormones been defined for these cells. However, during this period only one releasing hormone had been discovered. The decapeptide luteinizing hormone releasing hormone (LHRH), or GnRH, was more potent for LH release; however, it clearly stimulated the release of FSH as well (Schally et al. 1971; Arimura et al. 1972, 1973). Thus the question was posed: If gonadotropes produce and secrete both hormones, how is nonparallel release of gonadotropins brought about by only one releasing hormone?

At the same time, Denef et al. (1978) were studying the responses of gonadotropes separated by size with unit gravity sedimentation techniques. They reported differences in the secretory responses to GnRH and steroids by the different cell fractions. They also showed that each fraction responded with different levels of LH and FSH secretion. For example, GnRH stimulation released more FSH than LH from the large cell fractions and more LH from the small cell fractions. This important study helped to introduce the concept that perhaps the structural heterogeneity exhibited by the gonadotropes could be correlated with functional heterogeneity.

Thus, by the late 1970s immunocytochemists were challenged to seek cytologic evidence for mechanisms underlying nonparallel release of gonadotropins. One postulate was that nonparallel release was brought about by storage in separate cells. However, not everyone agreed that monohormonal gonadotropes existed. As stated in Sects. 1.1 and 1.2, a number of workers had shown that most if not all gonadotropes contained both hormones. In contrast, earlier studies by Nakane (1970) showed some monohormonal cells in the central regions of the gland. Our early work confirmed that of Nakane (Moriarty 1976a, b); however, we had no percentages to support the findings, and the size of our sample was small. Thus, many subsequent studies were designed to resolve this problem by applying cytochemical stains to gonadotropes in a number of different physiological states. These will be discussed in the subsequent sections.

2 Cytophysiological Studies of Gonadotropes

2.1 Fetal and Neonatal Development

In 1972 and 1976, Sétáló and Nakane reported that cells that stained for LH-beta appeared on about the 17th day of gestation. The first electron microscopic study of these fetal gonadotropes was done by Tougard et al. in 1977. Stain for LH alpha and beta subunits appeared in rer and small, sparse secretion granules scattered throughout the cytoplasm. The fetal gonadotropes were round or angular, and even at birth they did not resemble any of the adult subtypes. A more recent study by Yoshimura et al. (1984) correlated granule size with fetal and postnatal development, reporting that granules were seldom larger than 200 nm in diameter before birth. However, there was a size increase to 200–400 nm after birth.

The immaturity of the gonadotrope at birth suggested that further development may coincide with the full development of the hypothalamohypophyseal axis, which occurs during the first 5 days of neonatal life (Dussault et al. 1977). This maturation includes an increase in hypothalamic GnRH content and in the number of pituitary GnRH receptors, changes which are coupled with increases in the LH and FSH content of the pituitary (reviewed in Childs et al. 1981). Therefore, our studies of gonadotrope development in the rat began at birth (Childs et al. 1981). They were designed to determine both the rate of gonadotrope development and the time of onset of joint storage of gonadotropins.

In agreement with Tougard et al. (1977), we found that the gonadotrope in the newborn rat is poorly granulated (Fig. 4). LH cells were more frequent than FSH cells (6% vs 3% of the cell population). This correlated well with the studies by Dussault et al. (1977), who measured a higher concentration of LH in neonatal pituitaries. Finally, the analysis of serially sectioned cells showed that 66% of the gonadotropes contained both hormones. As expected, there were more monohormonal LH cells than FSH cells (20% vs 14% of the gonadotropes) in neonatal rats (Fig. 5). Finally, some gonadotropes were multihormonal, in that they contained immunoreactive ACTH (Fig. 6).

During the 1st week of development, there were striking morphological changes in gonadotropes. Figures 7 and 8 show gonadotropes from 5- and 7-day-old male rats. The cells were better granulated, and began to resemble their counterparts in the adult. An even more striking change was seen in the cell percentages. By 1 week of age, the percentages of LH cells had increased from 6% to 14%; similarly, FSH cell percentages had increased from 3% to 14.2%. The serial sections showed that significantly more of the gonadotropes stored both hormones (79% of the gonadotropes; Fig. 5). By 11–15 days of age, the gonadotropes were large and morphologically similar to those in the adult (Figs. 9 and 10).

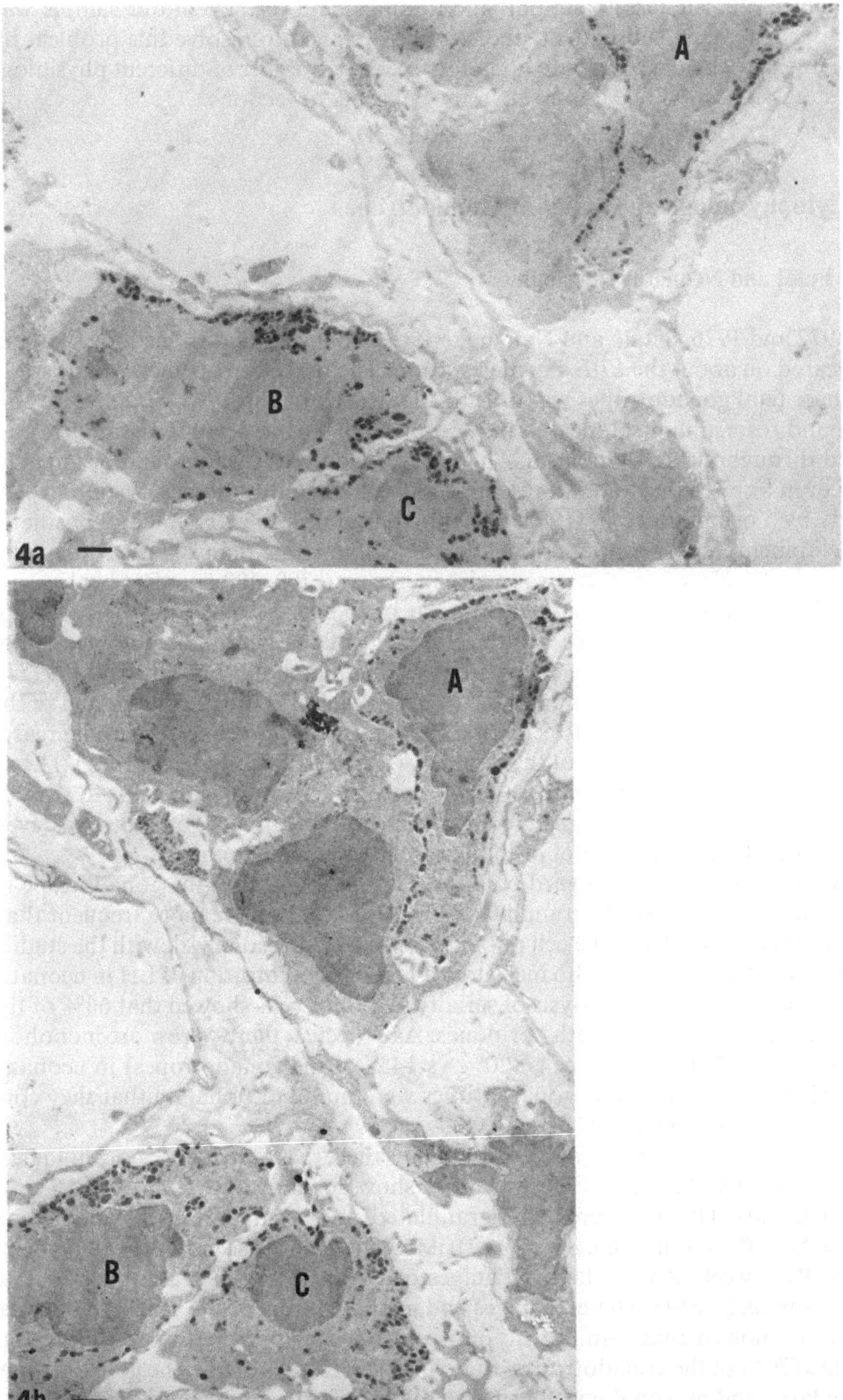
A
B
C
4a
A
B
C
4b

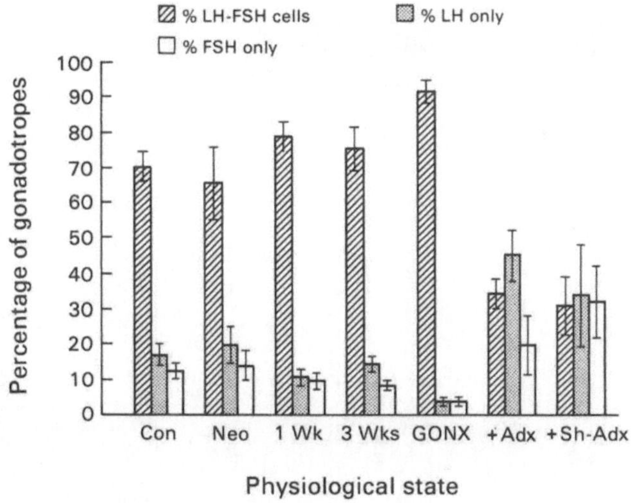

Fig. 5. Gonadotropin storage in male rats: analysis of over 3000 serially sectioned gonadotropes from different experimental groups of male rats. *Con*, intact, adult; *Neo*, neonatal; 1 and 3 wks, juvenile and prepubertal; *GONX*, castrated 12–24 h postsurgery; + *Adx*, castration and adrenalectomy 24 h previously; + *Sh-Adx*, castration and sham adrenalectomy 24 h previously. The sections were stained either for LH or FSH and fields matched to discover whether gonadotropes were monohormonal or multihormonal. The percentages of multihormonal cells increased significantly during the 1st week of development and 24 h after castration. There was a significant reduction in the percentage of multihormonal cells after castration and adrenalectomy or sham adrenalectomy. The percentage of monohormonal FSH cells decreased significantly at 3 weeks of age and 24 after castration. The percentage of monohormonal LH cells decreased at 3 weeks of age and after gonadectomy, and increased significantly after castration and adrenalectomy. These data show that there are higher percentages of multihormonal gonadotropes following physiological states that involve GnRH stimulation

The morphometric data have been compared to those collected in parallel studies of adult male pituitaries (Fig. 5; Childs et al. 1980; Childs 1984, 1985). The LH and FSH cells in the 1- to 2-week-old rat were more frequent than those in the adult rat. Our recent counts of individual LH and FSH cell populations in adults showed percentages of 9%–11% of the total cell population. The analysis of double stains showed that total gonadotropes constitute 16% of the population. (Childs et al. 1985; Childs 1984). Finally, serial sections of gonadotropes have indicated that the percentage of multihormonal gonadotropes in the adult male rat is 70%, which is significantly less than those in the 1- to 3-week-old male rats (79%; Fig. 5).

Thus, these studies showed that maturation of the hypothalamohypophyseal axis stimulated increases in relative percentages of LH or FSH gonadotropes along with increases in the percentages of multihormonal cells. This was early cy-

Fig. 4 a, b. Serial sections from 2-day-old male rat, showing three immature gonadotropes (*A, B, C*) staining for **a** LH and **b** FSH as in Fig. 1. The cells are sparsely granulated except in patches. Granules are pleomorphic and the stain is weaker for FSH. *Bar*, 1 μm × 6990

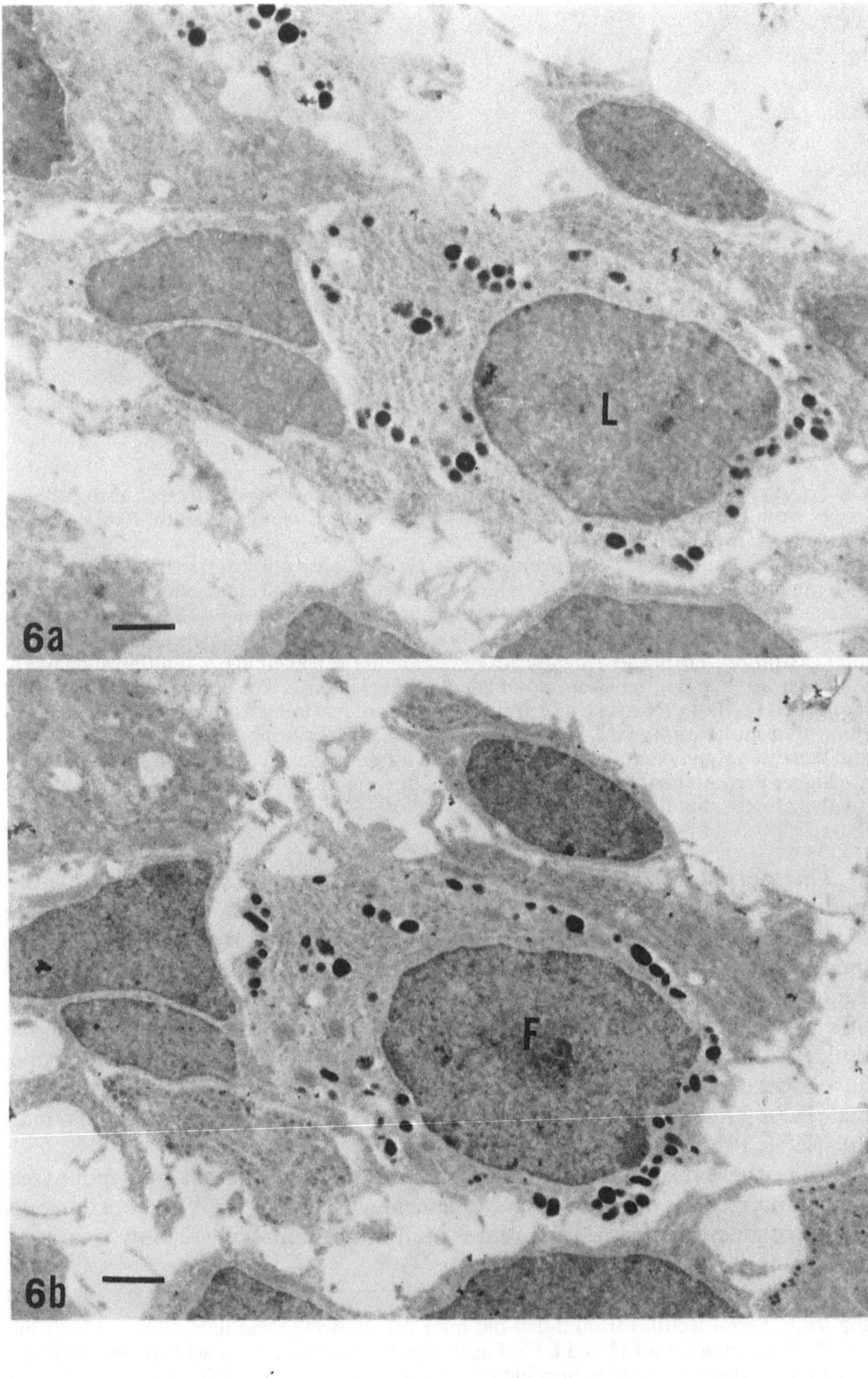
L
6a
F
6b

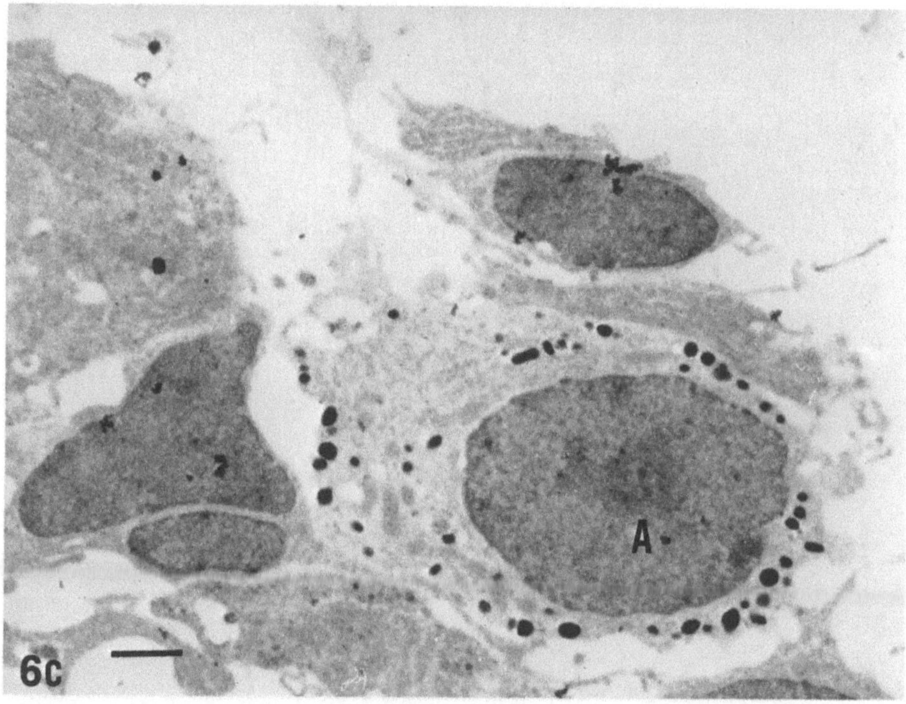

Fig. 6 a–c. Multihormonal gonadotrope from a 2-day-old rat, showing stain for **a** LH (*L*), **b** FSH (*F*) and **c** ACTH (25–39) (*A*) in the same cell. The cell is small and sparsely granulated and may be a precursor to gonadotropes and corticotropes (see Sect. 2.5 describing the gonadotrope as a multifunctional cell). *Bar*, 1 μm. Stain as in Fig. 1, × 10

tologic evidence for fluidity in the gonadotrope cell population. It also suggested that gonadotropes could shift storage patterns from cells that are 66% multihormonal to cells that are mostly multihormonal (80%), following stimulation by endogenous GnRH.

In 1981, Currie et al. published an electron microscopic study of fetal development of gonadotropes in man. They found both FSH and LH staining (with antisera to beta subunits) at as early as 9.5–10 weeks of gestation. The cells contained small granules that were lightly stained. In older fetuses (11–13 weeks) there were more granules per cell, especially in female rats. The granule staining intensity also increased progressively with age.

More recently, Dacheux (1984) reported an immunocytochemical study of fetal development in the porcine pituitary. There were changes in morphology similar to those seen in the rat as the hypothalamo pituitary axis matured. In the pig, however, this occurred during gestational days 70–113. It is important to note that alpha chain activity appeared before the beta subunit activity. This confirmed earlier studies of human pituitaries by Dubois et al. (1975, 1978) and Bugnon et al. (1976, 1977). Recent studies by Bégeot et al. (1984) have shown that the early appearance of the alpha subunit might be related to its role in the stimulation of prolactin cell development.

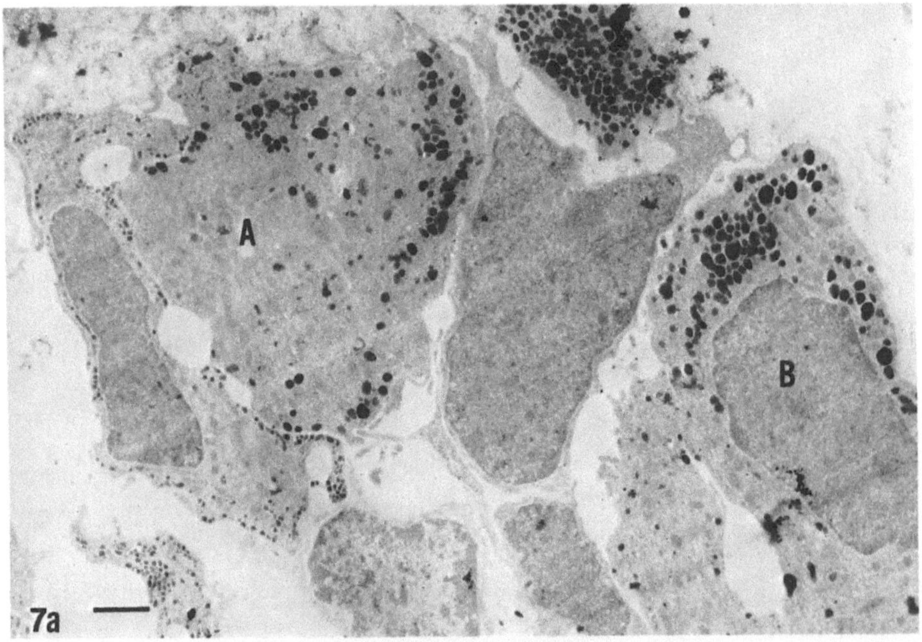

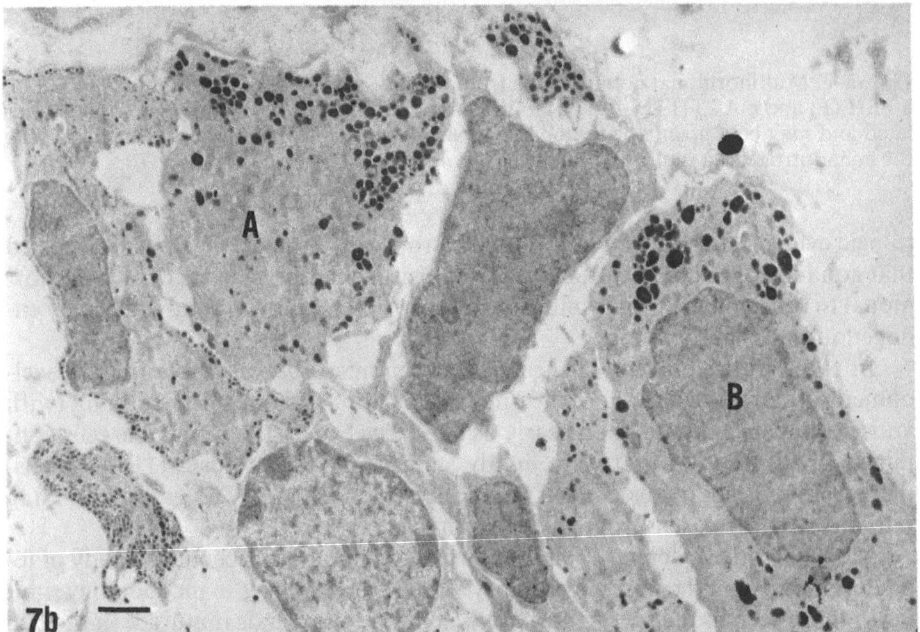

Fig. 7 a, b. By 5 days of age the gonadotropes (*A*, *B*) are better granulated; however, granules and patches of stain are still pleomorphic and not diagnostic of the adult subtypes. Cells in this cluster stain for both LH (**a**) and FSH (**b**). *Bar*, 1 μm. Stain as in Fig. 1, × 10 500

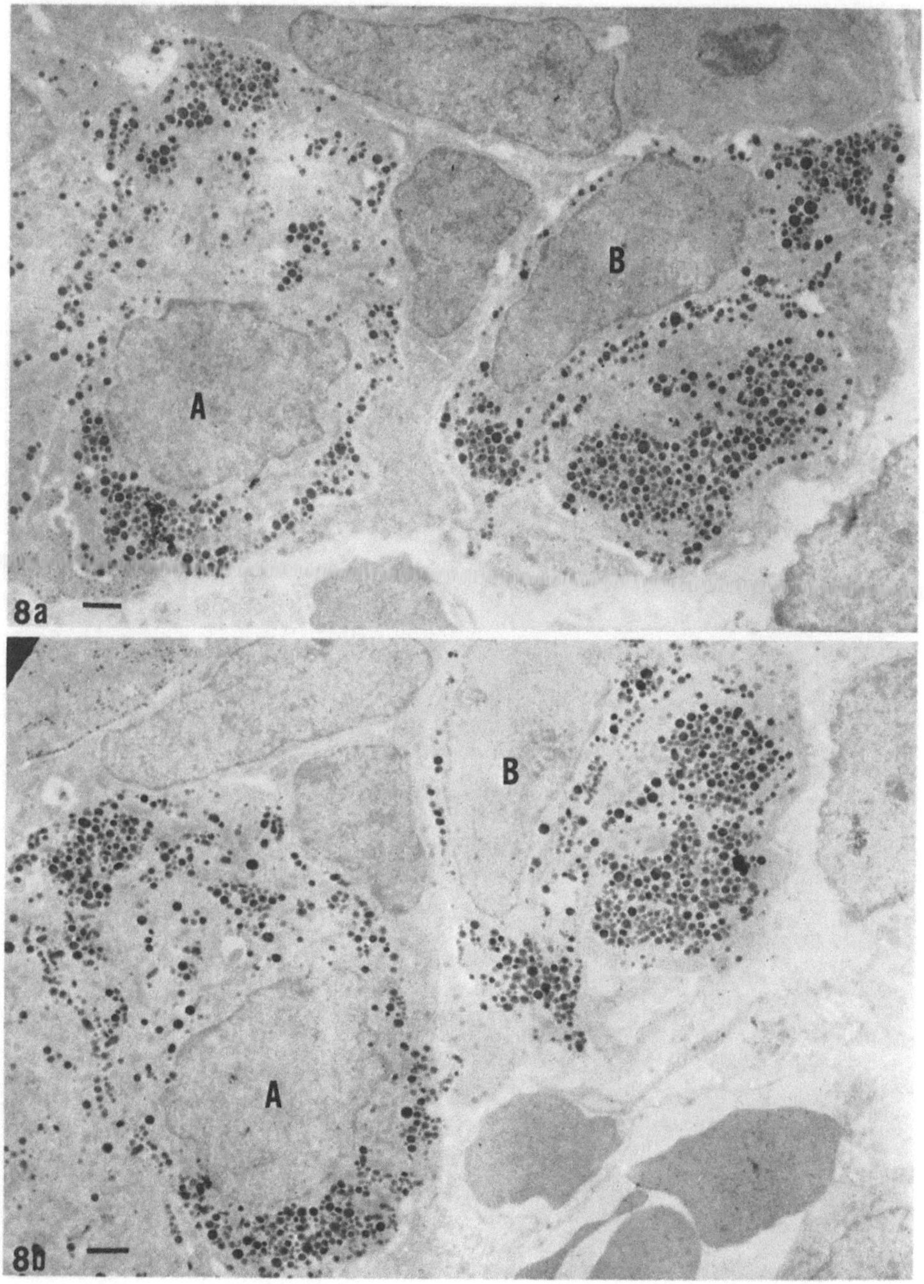

Fig. 8 a, b. The cells (*A, B*) begin to look like those in the adult by 7 days of age, and granules stain well for both hormones: LH (**a**) and FSH (**b**) as in Fig. 1. These cells look like typical type II (Moriarty 1975, 1976 a, b; Childs et al. 1980) gonadotropes. *Bar*, 1 μm × 7500

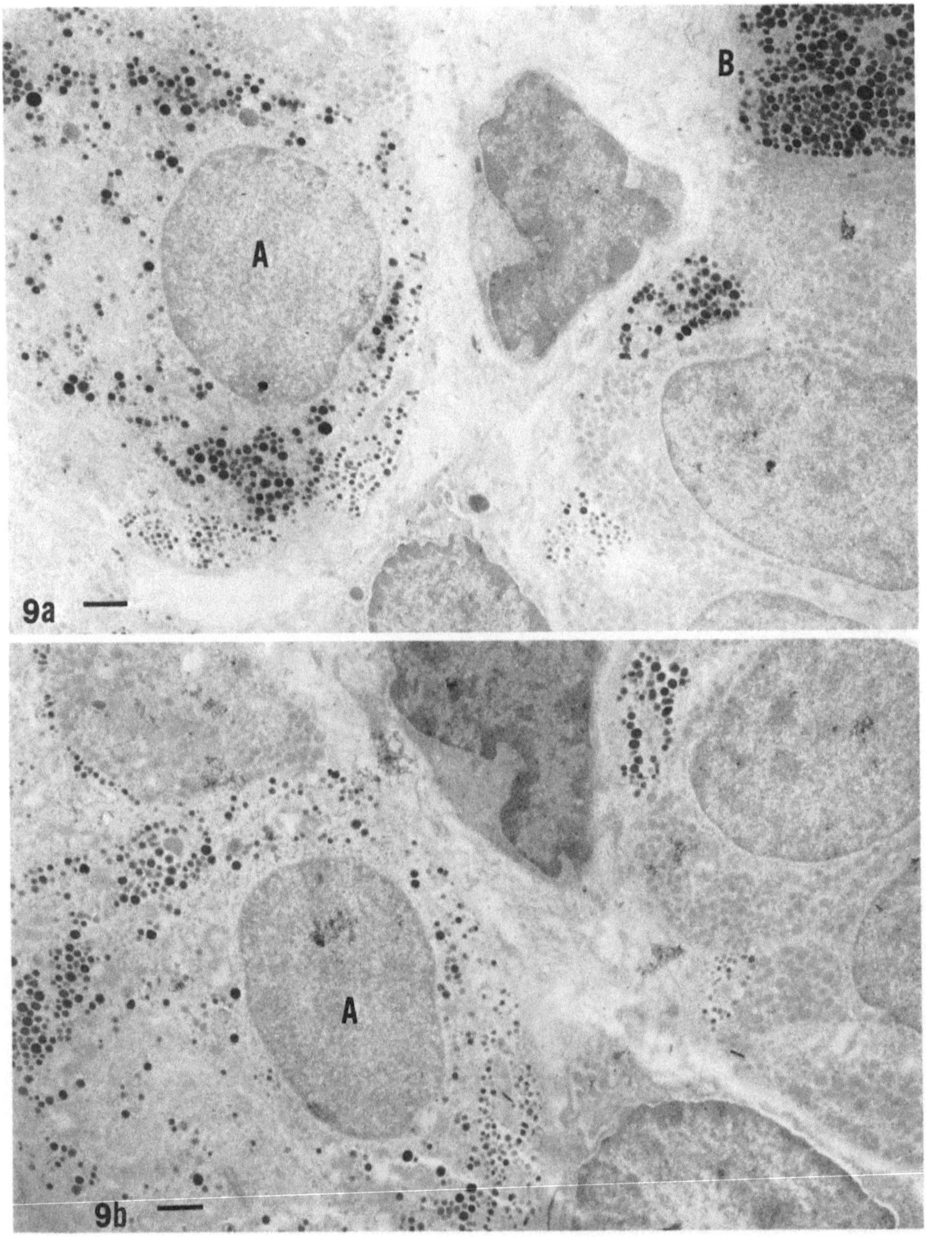
A
B
9a
A
9b

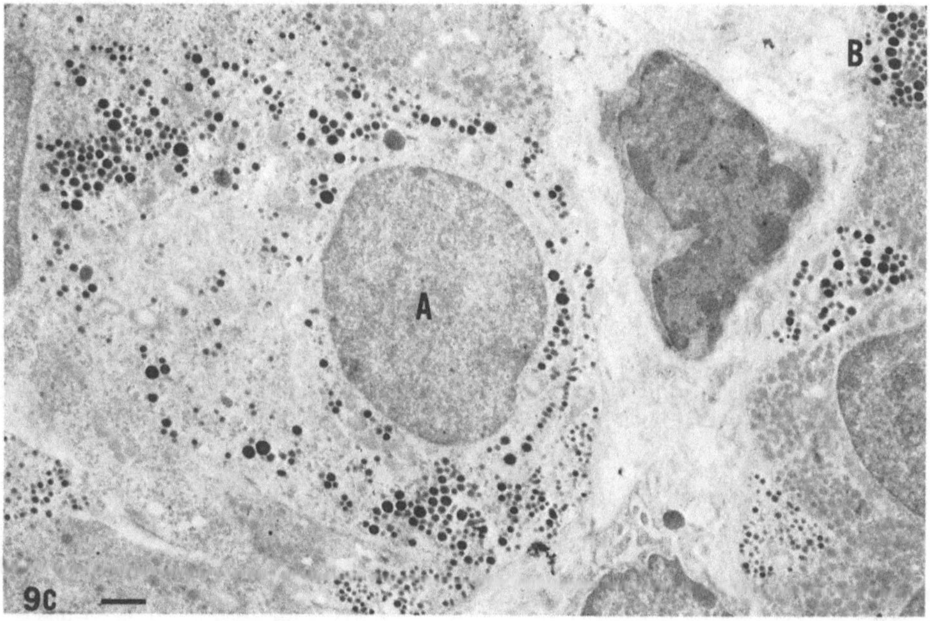

Fig. 9 a–c. In the 11-day-old rat, the LH cells *A* and *B* are very intensely stained. This field also shows that cell *A* is multifunctional in that it stores *a* LH, **b** FSH, and **c** ACTH. These ovoid ACTH-LH-FSH cells resemble those in the adult rat that represent 1%–3% of the cell population. *Bar,* 1 μm. Stain as in Fig. 1, × 8100

Thus, in many of the above studies the early appearance of the alpha subunit is followed by strong staining for the LH-beta subunit. Staining for FSH-beta appears later in development. This indicates that the gonadotropes have different schedules or requirements for timing the onset of production of each subunit.

2.2 Changes in Gonadotropes During the Estrous Cycle

Another important dynamic state that has been studied is the estrous cycle. In our early studies we reported that as rats approached the proestrous surge of LH, the cells were more irregularly shaped (Fig. 1) and showed accumulation of granules in cellular processes (Moriarty 1975, 1976a, b). After proestrus, some of the LH cells were degranulated (Fig. 2). Table 1 lists a more recent analysis of the frequency of stained, freshly dispersed gonadotropes taken at 1000 during estrus, diestrus II, and proestrus. Note that the percentage of LH cells is highest on the morning of proestrus in preparation for the LH surge later that day. The percentages of both LH and FSH cells are lower on the morning of estrus. Since FSH is secreted in a second rise during the night before, the immunocytochemical studies correlate with a partial exhaustion of hormone stores after the proestrous and estrous secretory activity.

In 1980; Blake reported a study of cycling rats in which he correlated changes in LH cell shape and morphology with the stage and time of the estrous cycle.

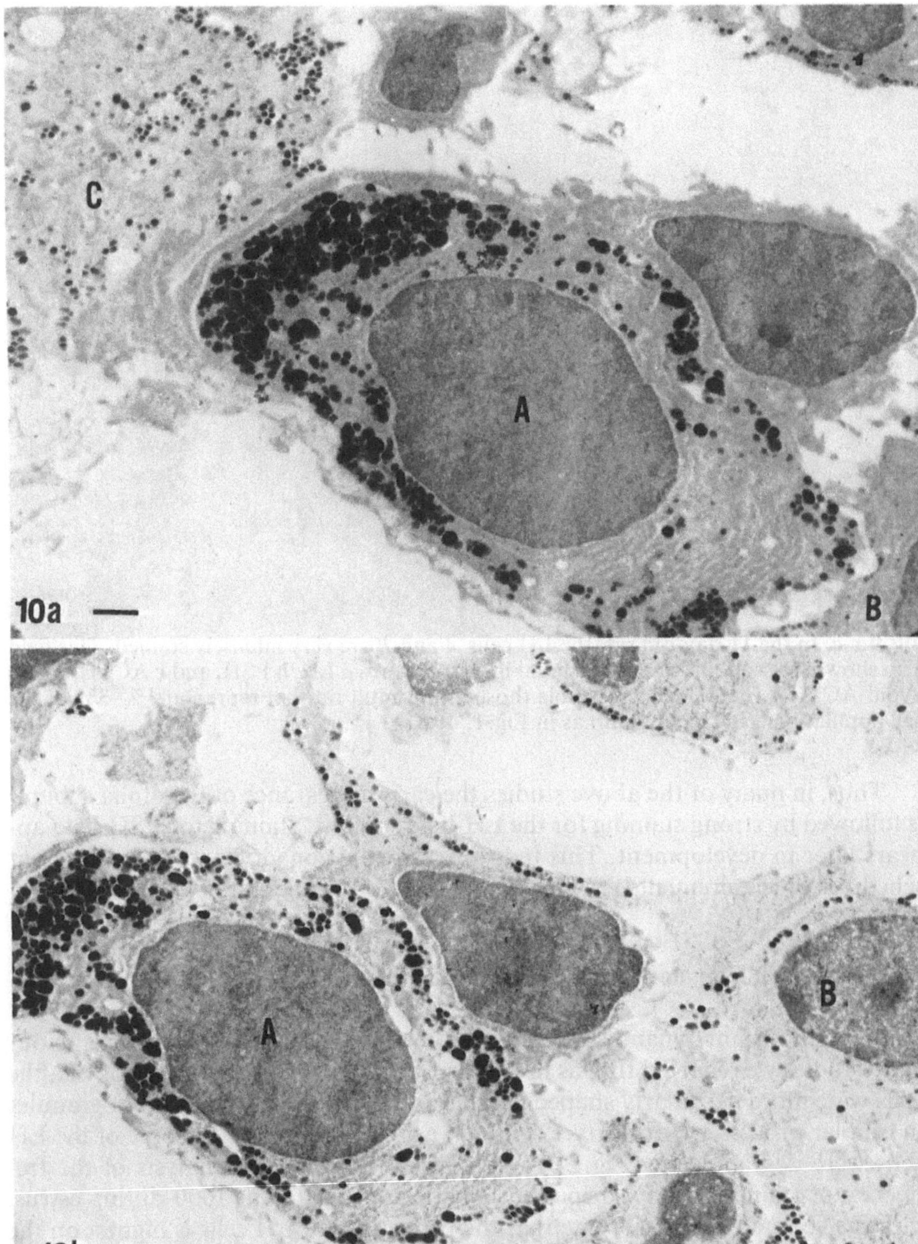
C
A
B
10a
A
B
10b

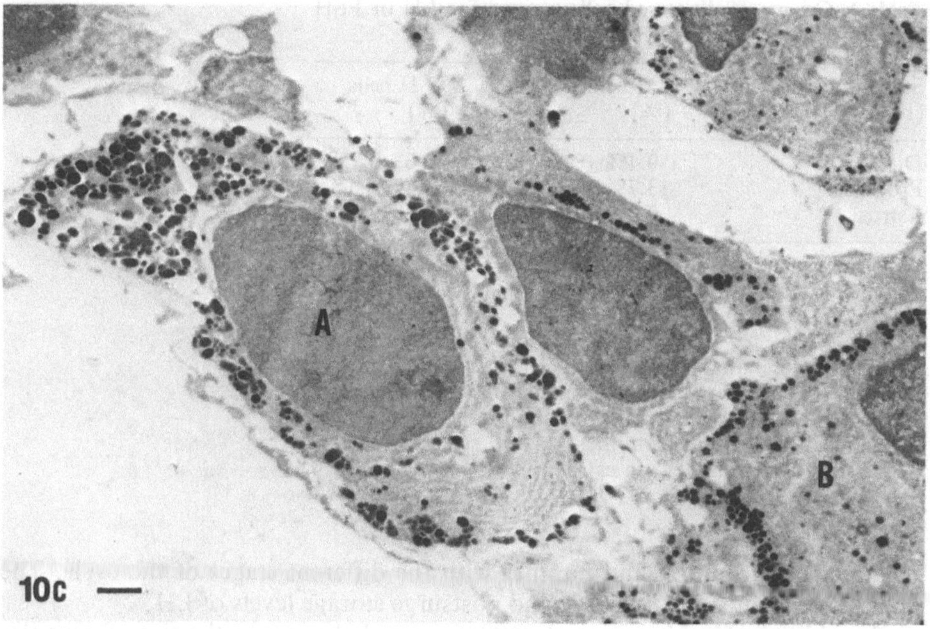

Fig. 10 a–c. Multifunctional cells (*A–C*) from 15-day-old rat, showing intense stain for **a** LH, **b** FSH, and **c** ACTH. Cell *A* is well granulated, with pleomorphic granules. Cell *B* stains well for ACTH and FSH; most of it is under a grid bar in the LH field. Note the striking differences in morphology between cells *A* and *B* and cell *C*. The latter are stellate with peripheral granules, features typical of type III cells (Moriarty 1976 a, b; Childs et al. 1980). *Bar*, 1 μm. Stain as in Fig. 1, × 8100

These data were also supported by serum and pituitary LH measurements. At 1400 proestrus, the concentration of anterior pituitary LH was highest and the cells were more polygonal. During the LH surge (1800–2000 on the day of proestrus), a fraction of the cells were more irregular, but most were polygonal. This agreed with our earlier studies (Moriarty 1975, 1976 a, b), in which we reported a decrease in the percentage of ovoid LH cells during proestrus and estrus (Figs. 1 and 2). Blake also found that, at 2200 proestrus, the overall number of granules per cell profile had decreased and the remaining granules were small. At 0200, during estrus, there were a number of degranulated LH cells with a few very small granules. The nuclei were also irregularly shaped after the proestrous LH surge.

Recently, Yoshimura et al. (1984) also described changes in the gonadotrope population throughout the estrous cycle. They reported optical density measurements of staining that showed that the percentage of intensely stained LH cells increased at 1200 proestrus and decreased at 1200 estrus. This agrees well with data presented in Table 1. During early proestrus, the cells contained an accumulation of small secretory granules near the blood vessels, whereas at estrus the cells were degranulated. Thus the studies of Yoshimura et al. agreed with those in our laboratory (1975, 1976) and the laboratory of Blake (1980) that showed changes

Table 1. Counts of dispersed cells stained for LH or FSH (Pre-Embedding)

Stage of Cycle (10)	LH cells (%)	FSH cells (%)
Diestrus II	9.4% ±1[b]	10.6±0.7[b]
Proestrus	13.7% ±1[a]	10.8% ±1[b]
Estrus	7.0% ±0.8	7.7% ±1

Cells were plated for 24 h and then fixed in 2.5% glutaraldehyde. After a buffer wash, they were stained for LH or FSH with the avidin-biotin-complex (ABC) technique (Childs 1984). At least three coverslips six–eight fields/coverslip) were analyzed per experiment. The data are averages of three separate experiment. LH cells from proestrus rats are most numerous and stain most intensely.
[a] Significantly different from diestrus and estrus.
[b] Significantly different from estrus.

in granule distribution and number with the different stages of the cycle. The changes correlate with presurge and postsurge storage levels of LH.

In addition, Yoshimura et al. (1984) showed changes in the size of granules throughout the cycle. During proestrus, the smaller granules predominated (80–160 nm). As the rats entered estrus, granules larger than 160 nm began to appear and the diameter of granules in diestrus was in the range 120–220 nm. These data agree with our early studies that evaluated the cells by type on the basis of the different sizes of secretion granule. In 1976, we reported that there were more of the type I cells (cells with both large and small granules) during diestrus (Fig. 3) and more of the type II cells (cells with a homogeneous population of small granules) during proestrus and estrus (Fig. 1; Moriarty 1976a).

The studies by Yoshimura et al. (1984) correlated granule volume and number with pituitary and serum LH and FSH and showed that the granule volume was highest on the morning of proestrus and lowest after the LH surge. The population of small granules was also more frequent just before the LH surge, and less frequent throughout the remaining stages. These authors suggested that the small granules are a select population reserved for LH secretion. Blake, in contrast, reported that there was an increase in the number of small granules following the LH surge (1980). Thus, further work is needed to differentiate the functions of the granule subpopulations.

2.3 Changes in Gonadotropes Following Gonadectomy

As described in Sect. 1.1, pioneering studies by Farquhar and Rinehart (1954) described the various stages in the formation of gonadectomy cells. They showed that the number of secretory granules is reduced and that the rer is dilated markedly. The presence of dense regions in the rer suggested that it might condense and package the hormone and take on functions of the Golgi complex (Farquhar

1971). This suggestion was later supported by Pelletier (1974), who showed that following exposure to tritiated fucose, (a terminal sugar that is normally added in the Golgi region), label is seen first in the Golgi region in normal gonadotropes. In contrast, most of the label is in the rer in gonadectomy cells.

The first ultrastructural immunocytochemical studies of gonadectomy cells were reported by Tixier-Vidal et al. (1975). Stored gonadotropins were detected in both dilated rer and secretion granules. All castration cells identified contained both gonadotropins. As stated earlier, the filigreed cell described by Farquhar and Rinehart (1954) was stained for LH-beta.

Garner and Blake (1981) correlated serum and pituitary LH levels with the development and function of immunocytochemically stained ovariectomy cells. They reported a 22-fold increase in serum LH and a 5-fold increase in pituitary LH by 92 days after surgery. The cells expanded as described in the previous studies; in addition they contained more secretion granules per cell profile and an irregularly shaped nucleus. In contrast to the previous studies (Tixier-Vidal et al. 1975), there was no LH staining in cells that resembled filigreed cells.

In 1976, Kurosumi et al. described a morphometric study of castration cells showing an increase in overall gonadotrope cell volume and volume of rer. In later studies, mitotic figures were found in normal pituitary cells (Kurosumi 1979), yet none were seen in castration cells. Thus, the question of the origin of the gonadectomy cell arose.

In a more recent study, Inoue and Kurosumi (1981) used immunocytochemistry to identify the LH cells, and showed that the volume of the castration cells increased more than 6.5 times after 3 months. They also observed mitotic cells that were stained for LH as early as 2 weeks after surgery. They then combined autoradiographic studies with immunocytochemistry and determined that, one day after castration, there was a ten fold increase both in the number of cells labeled with tritiated thymidine and also of mitotic LH cells. There were no further increases with additional time after surgery.

Smith and Keefer (1982) recently reported that there was an increase in the number of mitotic figures after ovariectomy, and that the majority of these dividing cells were ovariectomy cells. They also described some small dividing cells that did not stain well for any of the hormones. These may be equivalent to the immature gonadotropes that were described in the recent studies by Kihara (1984). In this report, Kihara described changes in subpopulations of gonadotropes at different times after castration and noted that 18 months after surgery there was a striking increase in the percentage of small, poorly granulated gonadotropes. The signet ring cells also disappeared by 12 months after surgery. The serum LH levels remained high 6–18 months after castration, in spite of the remarkable change in the morphology of the cell population. Perhaps the "long-term" gonadectomy cells are more like those in human pituitary gonadotropic adenomas.

In 1983, Sakuma et al. compared the effect of castration on immature and mature rat pituitaries and found that the total number of mitotic cells increased 1.4 times in immature animals, but that there was no increase in numbers in the mature animals. Analyses of individual cell populations showed that following gonadectomy the mitotic index of LH cells rose in both groups (five–six times); however, the mitotic activity of prolactin cells (and their total numbers) was reduced

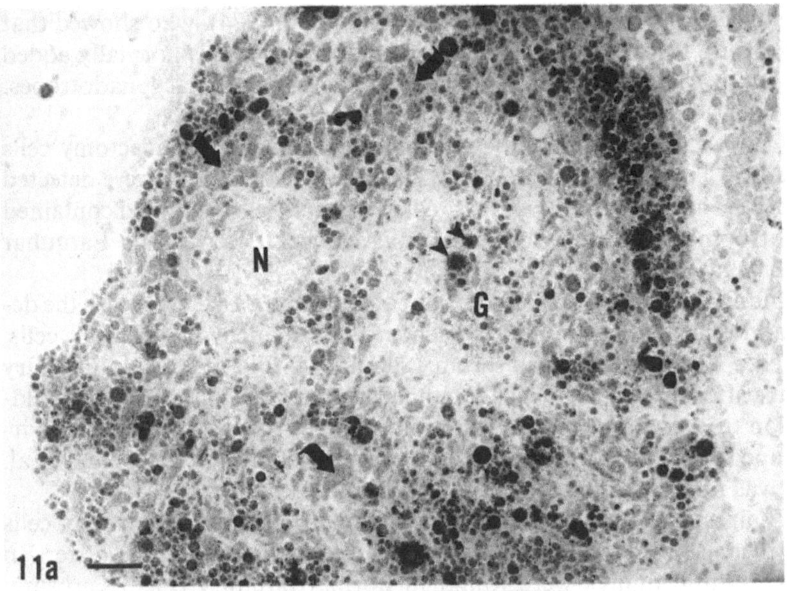
N
G
11a

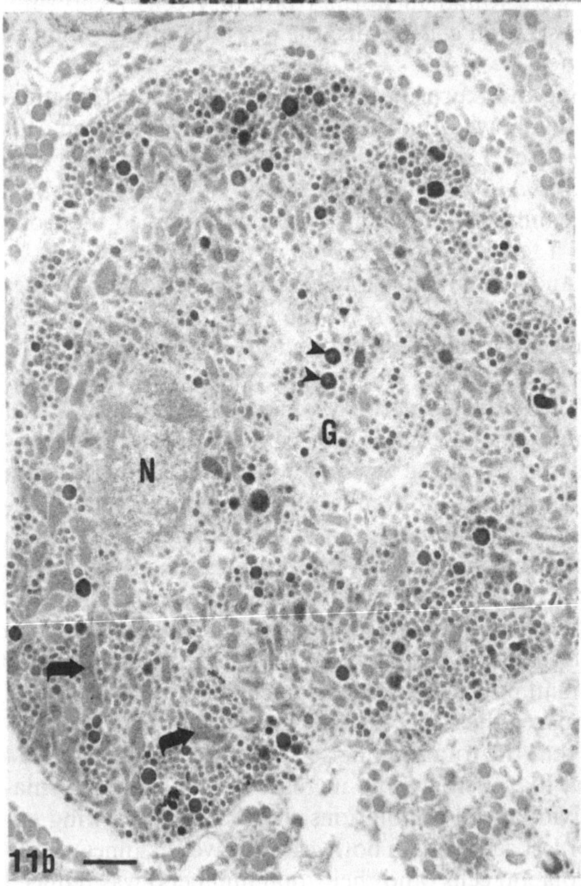
G
N
11b

by one-third in immature rats and by two-thirds in adult rats. This latter reduction would partly account for the lack of change in the overall numbers of mitotic cells in the mature animals.

In 1982, we reported morphometric immunocytochemical studies of the development of castration cells during the first 24 h after surgery. Staining for LH and FSH varied in intensity in most of the castration cells. Strong staining for LH was seen in both rer and the secretion granules. The staining for FSH was weaker than that seen in normal cells (Fig. 11). The percentages of each cell population were increased to 14%–15% as soon as 12 h after surgery (Fig. 5). In addition, the percentage of gonadotropes that stored both hormones increased. The analysis showed that 91% of the castration cells were multihormonal, compared with 70% in the normal rat (Fig. 5). This again supports the hypothesis that gonadotropes may shift storage patterns following GnRH stimulation. However, because the overall percentages of LH and FSH gonadotropes were increased to 15%, we could not rule out a possible de novo origin for the additional multihormonal gonadotropes. This was supported further by the recent study by Inoue and Kurosumi (1981); reviewed on p 71.

The development of castration cells included an increase in area, as determined by morphometry. The LH cell population showed a significant increase in area from an average of 133 μm^2 to 170 μm^2 24 h after castration, and the FSH cell population showed an increase in area from 158 μm^2 to 208 μm^2 in the same time period. Note that the average area of FSH cells in normal male rats is larger than that of LH cells. This correlated with the earlier studies by Denef et al. (1978), who showed more FSH release than LH release from larger-cell fractions. For more information see page 90.

By 1984 we had shown two physiological states where the gonadotrope population shifted its storage patterns to cells that were predominantly multihormonal. These included neonatal development (Childs et al. 1981) and gonadectomy (Fig. 5; Childs et al. 1982 b). In both cases, GnRH stimulation presumably mediated the change. During this same period, we also studied the effect of adrenalectomy on the development of castration cells, in order to learn what mechanisms were behind the temporary postponement of the rise in LH and FSH that ensues if rats are adrenalectomized and castrated simultaneously (Schwartz and Justo 1977; Lorenzen et al. 1980).

In these studies, we found a striking effect of adrenalectomy or sham adrenalectomy on both the percentages of cells staining for LH and on the area of gonadotropes (Figs. 12 and 13; Childs et al. 1983 b). The areas were reduced below those seen in the castrated rat, and LH cell percentages were reduced below normal. In addition, much to our surprise, we found that the joint surgeries also resulted in a dramatic reduction in the percentage of gonadotropes that stored both hormones (Fig. 5). The percentages of multihormonal gonadotropes were suppressed to below normal levels (30% of all gonadotropes) by either adrenalec-

Fig. 11 a, b. Developing castration cell (24 h after surgery) from male rat showing stain for **a** LH and **b** FSH on granules and rer (*arrows*). *Arrowheads* indicate stained serially sectioned granules in the Golgi complex region (*G*). *N*, nucleus; *Bar*, × 1 μm. Stain as in Fig. 1, × 12000

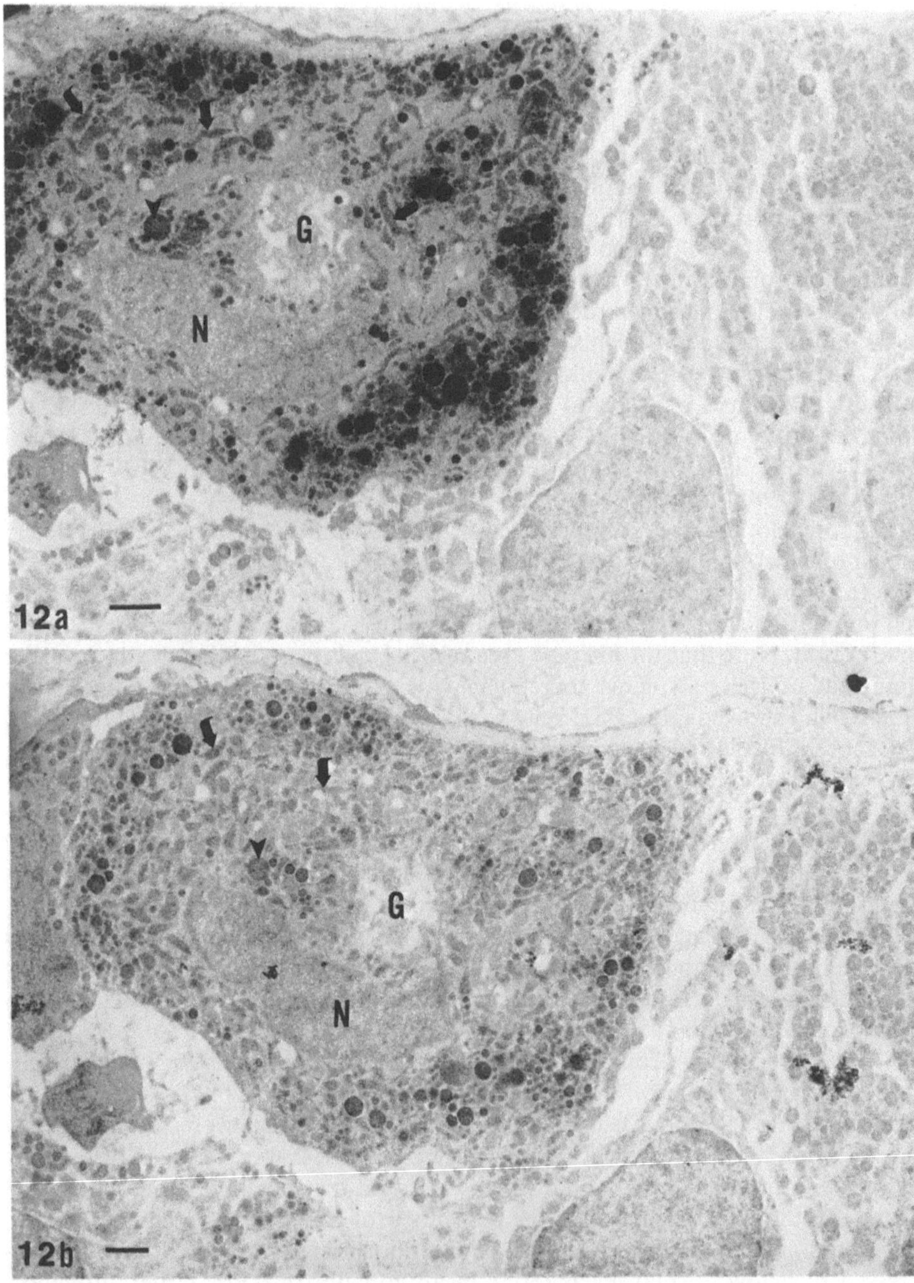

Fig. 12 a, b. Developing castration cell from rat that was adrenalectomized and castrated 12 h previously. The cells are not enlarged as are cells in rats that were castrated only. Some rer is expanded (*arrows*), and the cells are stained intensely for **a** LH and **b** FSH. This lack of expansion in area and the buildup in LH storage per cell fits with the lower serum levels seen in these animals (Childs et al. 1983 b). Granules noted by *arrowheads* in the Golgi region (*G*) are serially sectioned. *N*, nucleus; *Bar*, 1 μm. Stain as in Fig. 1, × 11 000

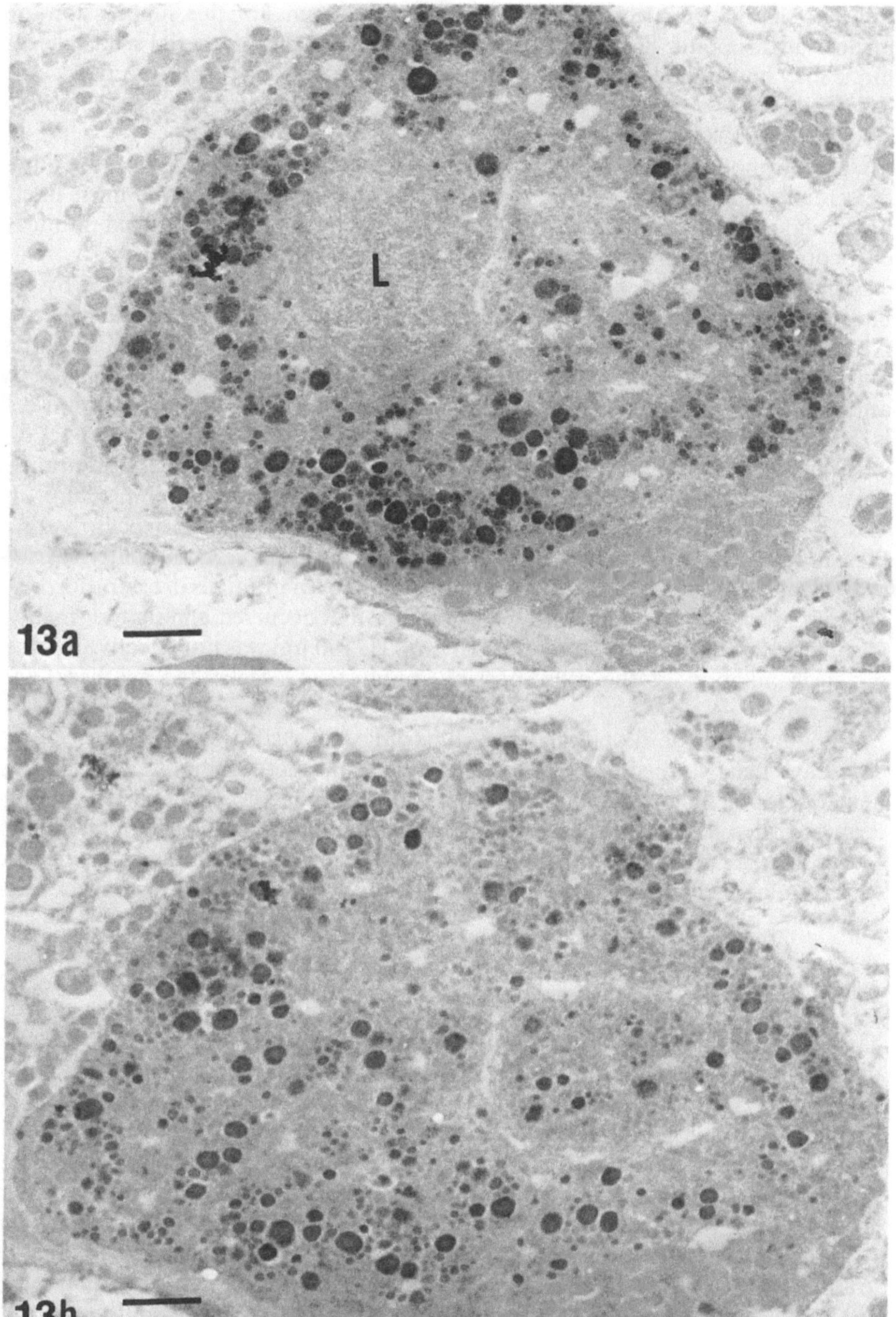

Fig. 13 a, b. Another cell from the group shown in Fig. 12, in which there is no expansion of area or rer. Development of this castration cell is completely retarded by the simultaneous adrenalectomy. Stain for **a** LH (*L*) and **b** FSH is weaker than normal. *Bar*, 1 μm. Stain as in Fig. 1, × 12000 tomy or sham adrenalectomy (Fig. 5). The mechanisms for this suppression are unclear at this point; however, it is important to note that the serial section technique may have failed to detect dissociated storage of gonadotropins in the same cell (Childs 1985). Thus, the apparent reduction might be partly due to a sampling error (Childs 1984).

The hormonal basis for the suppression of activity in gonadotropes by adrenalectomy or sham adrenalectomy is unclear. However, one of the most likely hormonal candidates might be endorphin secretion from the stimulated pituitary corticotrope. Further studies of this penomenon are needed.

Several laboratories have shown that the gonadotropes hypertrophied by gonadectomy can secrete gonadotropins in response to GnRH. In an early study, Arimura et al. (1976) reported that castration cells secreted LH and FSH after GnRH stimulation, but that their secretion and hypertrophy could be blocked by antibodies to GnRH. This indicated that the development of the castration cell is dependent upon high circulating levels of GnRH.

Shiino (1982) reported that GnRH stimulated numerous exocytosis profiles in castration cells that correlated with a GnRH-mediated elevation in serum LH and FSH. Kurosumi and Inoue (1980) also described numerous exocytosis profiles and pits that contained dense granules in castration cells. Recently, Yoshimura et al. (1984) showed a reduction in the number of granules in castration cells after an injection of 100 ng GnRH. The major loss occurred among the smallest granules (120–160 nm). A larger dose of GnRH (100 µg) was not effective in causing degranulation of the cells, possibly because of accelerated synthesis and packaging.

In summary, the development and cytology of castration and ovariectomy cells have been studied extensively at the electron microscopic level. Intriguing new developments include the finding that this cell type does not persist in long-term castrates. There appears to be further differentiation and development after 1 year in order to sustain the high serum LH and FSH. Further work is needed to determine if the new form of gonadotrope is monohormonal or multihormonal.

2.4 The Effect of GnRH Stimulation on Gonadotropes

The studies of GnRH stimulation of gonadotropes can be divided into two groups. The first group of investigators have concentrated on the ultrastructural morphology and staining of the GnRH-stimulated cells. The other group have examined the binding and internalization of labeled GnRH. One purpose of this section will be to correlate the two types of studies to provide a complete view of the cytologic changes in gonadotropes following GnRH stimulation.

Early in vivo studies reported effects of GnRH stimulation on gonadotrope cytology. In 1971, Pelletier et al. showed that hypothalamic extracts stimulated granule extrusion from gonadotropes (identified by morphology). Rennels et al. (1971) and Shiino et al. (1972) also reported the extrusion of secretion granules from pituitary gonadotropes 1–15 min after a single injection of purified GnRH. Their work was confirmed in 1973 by Mendoza et al., who used synthetic GnRH.

The effect of GnRH on the synthesis and release of gonadotropins following portal vessel infusion was correlated with ultrastructural morphology in 1975 by Luborsky-Moore et al. They showed that 1–3 min after LHRH infusion the male rat gonadotropes exhibited an increase in the number of exocytosis profiles, with several granules being extruded from the same region (compound exocytosis). The rer was enlarged at 15 min after infusion, but the number of exocytosis profiles had decreased. The Golgi complex began to fill with newly packaged granules during the first 15 min. By 30 min after infusion, the rer was no longer as dilated and the number of granules in the Golgi complex had increased. By 60 min after stimulation, the cells were very well-granulated. Thus the initial loss in granules correlated with the rise in serum LH that occurred within minutes after stimulation. The cells then appeared to increase synthetic and storage activity in order to replenish granule stores.

In 1977, Hopkins and Gregory showed that the ferritin-labeled GnRH bound to the surfaces of cells that resembled gonadotropes and that it entered the cell within 10 min. The stain could then be found in lysosomes or the Golgi complex. However, no major changes in ultrastructural morphology of the gonadotrope were reported.

The following year, Römmler et al. (1978) reported an in vivo study of GnRH-stimulated gonadotropes in female rats that correlated electron microscopic observations with serum and pituitary levels of LH following injections of GnRH. These investigators noted an initial depletion of secretion granules in gonadotropes 3–15 min after GnRH injection, which coincided with the rapid rise in serum LH and the loss in pituitary LH. The cells exhibited numerous exocytosis profiles; some were single, others were compound in that several granules extruded into the same extracellular space. The compound exocytosis profiles formed pouches and tubules filled with extruded granular material. The space in the pouches was continuous with the extracellular space, and it invaginated deeply into the cytoplasm.

The changes in the other organelles were similar to those seen by Luborsky-Moore et al. (1975), except that the regranulation process was slower in the female. The rer was dilated and the Golgi complex filled with immature granules 3–15 min after stimulation. This was followed by an accumulation of secretion granules in 60 min that supported the second rise in LH. Thus, after a second injection of GnRH 2 h later, the serum LH rose to a peak that was twice the level of the previous peak.

Almost 3 h after the GnRH injection, the number of granules returned to control levels and the investigators noted an increase in the frequency of organelles that resembled lysosomes. They concluded that this represented the phenomenon of crinophagy described by Farquhar (1971), a process in which the level of intracellular hormone is controlled by lysosomal degradation of secretory granules.

In 1979, Garner and Blake reported an in vivo study of GnRH stimulation of LH cells in phenobarbitol-blocked rats that were continuously infused with 50 ng GnRH per minute. In contrast to studies reported by Römmler et al. (1978), pituitary LH content had not decreased 15 min after the *continuous* GnRH exposure, and there was no decrease in the number of secretory granules. The first decrease in granule number was seen after 2 h of continuous infusion, when the pituitary LH content was also decreased. Serum LH remained high during this period, but fell after 2–3 h of infusion. There was a marked decrease in pituitary LH, in the number of stained granules, and in serum LH after 5 h of infusion of GnRH, indicating that the cells had become refractory to further stimulation.

It is difficult to separate effectively the effects of exogenous and endogenous GnRH stimulation in vivo. The use of phenobarbitol-blocked rats (Garner and Blake 1979) may represent one method. Another is seen in the use of hypogonadal mice that lack hypothalamic GnRH. In studies by McDowell et al. (1982), the gonadotropes from such mice were analyzed 5 min after the injection of 40 ng of GnRH. There were no changes in the density of granules or in the numbers of exocytosis profiles in GnRH-stimulated cells. However, there was an increase in the number of membrane-bound vacuoles filled with flocculent material.

After 20 days of GnRH stimulation, there was an increase in the frequency of LH cells and LH-containing granules. The gonadoropes were hypertrophied and the lipid droplets continued to accumulate. This correlated with the 63-fold increase in pituitary FSH and the 2-fold increase in pituitary LH. Increases in testicular weight (5-fold) indicated that the gonadotropes were able to release gonadotropins in response to LHRH.

Another animal model is the inbred dark mink (Tung et al. 1984). In a recent study, we showed that the infertile animals with dark coats had normal percentages of gonadotropes. They were also responsive to injections of GnRH, in that they showed increases in gonadotrope cell area, staining intensity, and LH secretion. As in the hypogonadal mouse, the fertility defect may be at the hypothalamic level.

Collectively, the ultrastructural studies of the 1970s agreed that gonadotropes responded to GnRH stimulation within minutes by secreting granular material into single or multiple exocytosis profiles. This was followed by enlargement of the rer and more frequent packaging of granules by the Golgi complex. The granule population is thereby replenished, and initially storage is greater. This is seen at the light microscopic level by an increase in the intensity of the stain (Yoshimura et al. 1984) after GnRH stimulation. Normal storage patterns are recovered within several hours, unless the GnRH is continuously infused (Garner and Blake 1979) or there is another pulsatile GnRH exposure (Römmler et al. 1978). This recovery may be due to the lytic digestion of excess secretion granules (Farquhar 1971).

In the 1980s, another series of studies described the process that followed GnRH binding to gonadotropes with the use of labeled GnRH analogs. In 1981, Hopkins et al. used ferritin-labeled analogs to show the redistribution of GnRH on the surfaces of gonadotropes. An octapeptide GnRH analog that does not release LH by itself was induced to release the hormone by antisera that caused aggregation and internalization of the labeled GnRH analog. After 15 min of exposure, the labeled GnRH was found in multivesicular bodies or in the Golgi complex region. Internalization of the labeled peptide did not correlate with cytologic changes in the stimulated cells, however.

Pelletier et al. (1982) used autoradiography to trace the internalization of iodinated GnRH analog. The label was mostly membrane-bound 3 min after injection. At 10–30 min, the label was found in the Golgi complex, lysosomes, and secretory granules. Label was also seen over rer, especially the dilated rer found in castration cells. Another study by Duello et al. (1983) showed a similar shift in the label with time. These workers used radioactive lysine-containing analogs that were stabilized by cross-linkages with glutaraldehyde fixative. They found significant labeling of the Golgi complex and lysosomes with time. However, the grain density over rer or secretory granules was low and of uncertain significance.

A more recent autoradiographic study by Hazum et al. (1985) described the internalization pathway of a photoaffinity probe for GnRH that could be covalently linked to the receptor by photolysis. In addition to label on other organelles described by the previous studies, these investigators described significant labeling of secretion granules.

Another type of cytochemical probe was used by Jennes et al. (1983), who applied ferritin- or gold-labeled GnRH analogs to trace the ligand into the cell. After 5 min, the label was mostly associated with the plasma membrane. It shifted to vesicles and lysosomes after 15 min with small amounts also evident in the Golgi complex or the rer. Little labeling was seen in the secretory granules throughout the time of incubation.

Thus, while the pathway of internalization appeared similar, the final distribution of label differs slightly with each study. Specifically, the distribution of label in the secretory granules was relatively high only in the study by Hazum et al. (1985). This may reflect the increased stabilization of the photoaffinity probe by photolysis.

It is also clear from the above groups of studies that GnRH is internalized by gonadotropes in a pattern that resembles classical "receptor-mediated endocytosis." However, is the endocytosis coupled to exocytosis of gonadotropins? In studies by Conn et al. (1981), it was shown that endocytosis was not required for LH release. The major requirement seems to be the aggregation of receptors.

The studies of GnRH internalization provided little information about gonadotrope secretory responses to stimulation, perhaps because a description of the cytology of the gonadotrope itself was beyond the scope of the reports. However, in some cases, the biological activity of the labeled peptide may have been reduced, so that secretory activity was slowed. A direct correlation of changes in serum or media levels of gonadotropins with cytochemical studies of GnRH uptake was needed.

Our studies of the effect of GnRH stimulation on gonadotropes began in 1982, when we developed avidin-biotin-complex (ABC) stains for a biotin-conjugated analog of GnRH (Childs et al. 1983 a) and then used immunocytochemistry to colocalize LH or FSH (Childs et al. 1983 c). Secretory activity from the stimulated cell population was also evaluated by radioimmunoassay to confirm that the biotinylated GnRH was fully potent at the time of the cytochemical test. There was no loss in binding affinity or potency in the GnRH-mediated release of LH or FSH when the biotinylated analog was compared to the unlabeled analog. Furthermore, significant release of LH and FSH was detected in a time course study that matched the periods used to study the receptor-mediated endocytosis of GnRH.

Figure 14 shows GnRH-stimulated gonadotropes that were stained for biotinylated GnRH. The staining pattern matched that seen in the previous studies.

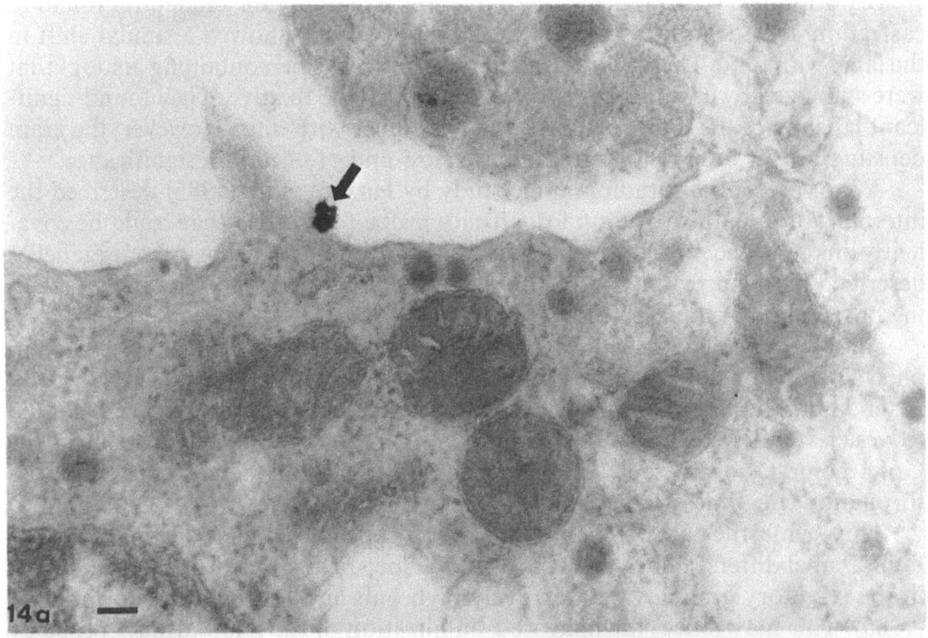

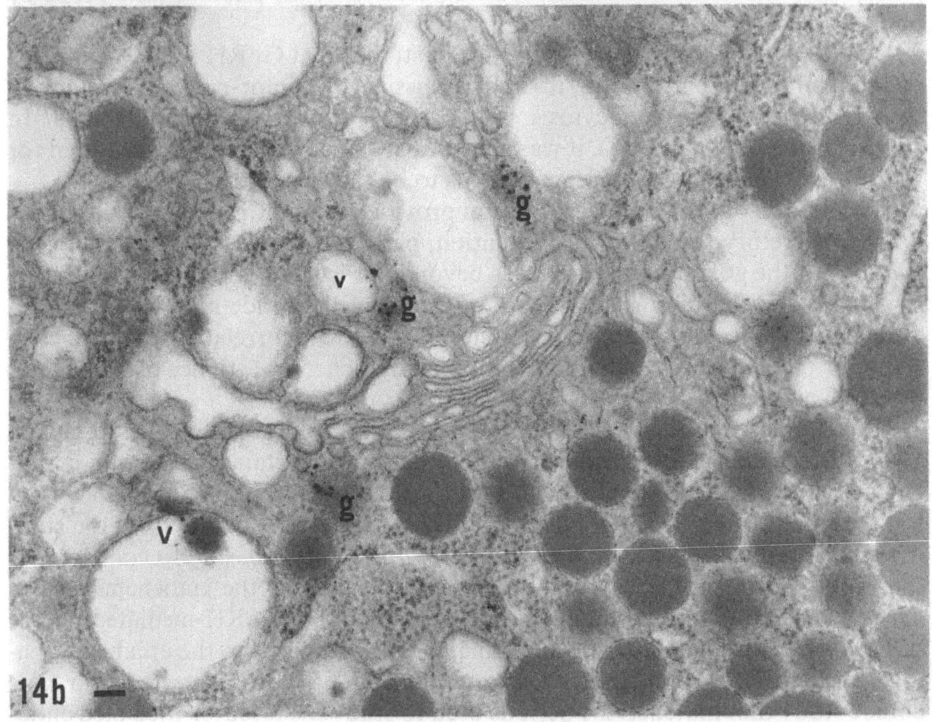

Fig. 14 a–d. Pituitary cells from female rats in suspension stimulated for 1–30 min with biotinylated GnRH and then stained with ABC-peroxidase. In **a**, a patch of stain (*arrow*) is seen on the membrane near a microvillus 1 min after stimulation. The cell is a small gona-

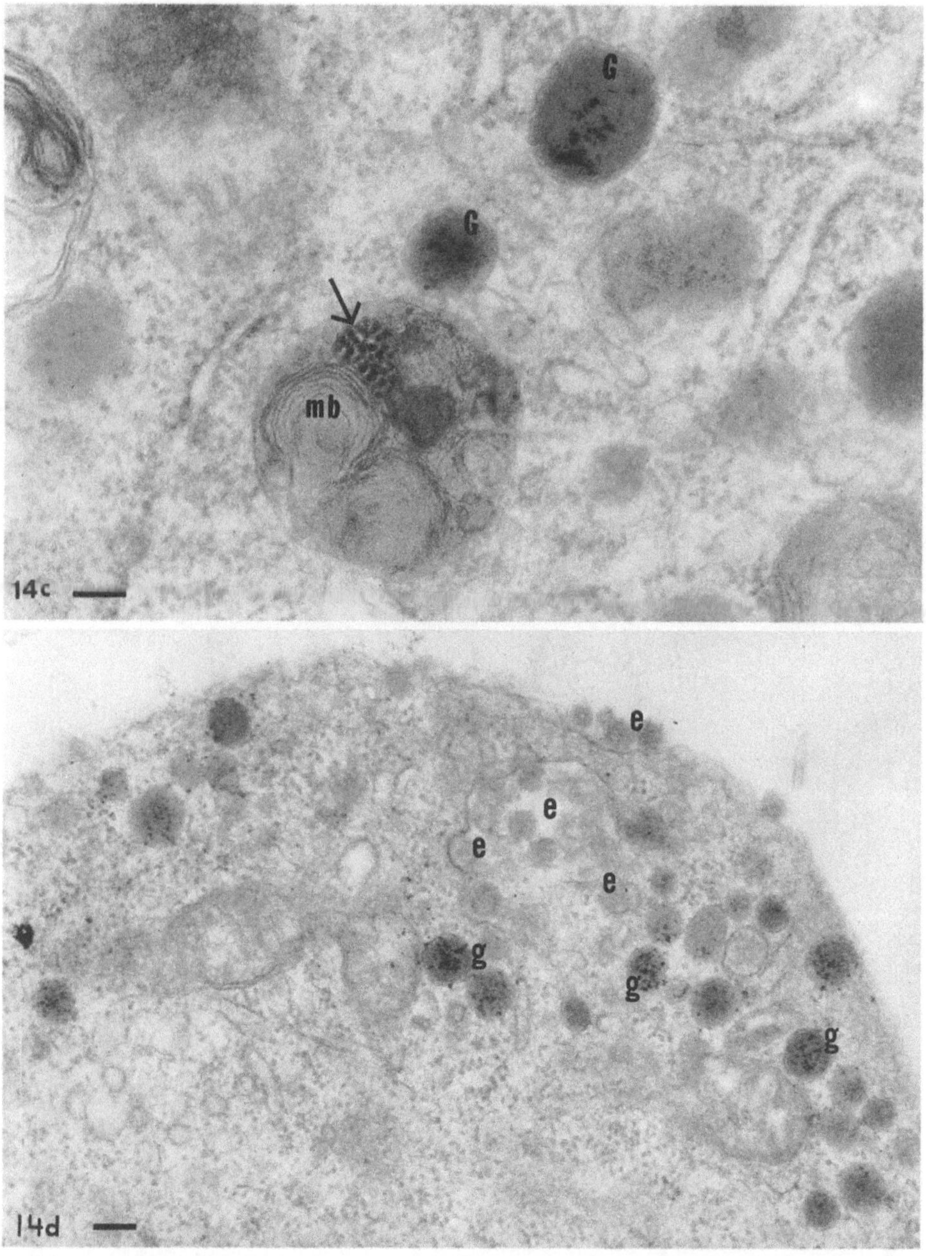

dotrope, from elutriation fraction 2. In **b**, stain is seen in granules (*g*) and condensing vacuoles (*V*) in the Golgi complex 3 min after stimulation (v = vesicle). In **c**, stain (*arrow*) for biotinylated GnRH is seen in a multivesicular body (*mb*) 5 min after stimulation *G*, Golgi complex. In **d**, stain for biotinylated GnRH in a subpopulation of granules in the cytoplasm (*g*) 5 min after stimulation. Nearby is a compound exocytosis profile (*e*) containing extruded granules, which is a characteristic response seen after stimulation by GnRH. **a** × 81900; **b** 62650; **c** 104000; **d** 82500; *Bar*, 0.1 µm. (Figures taken from Childs et al. 1986a)

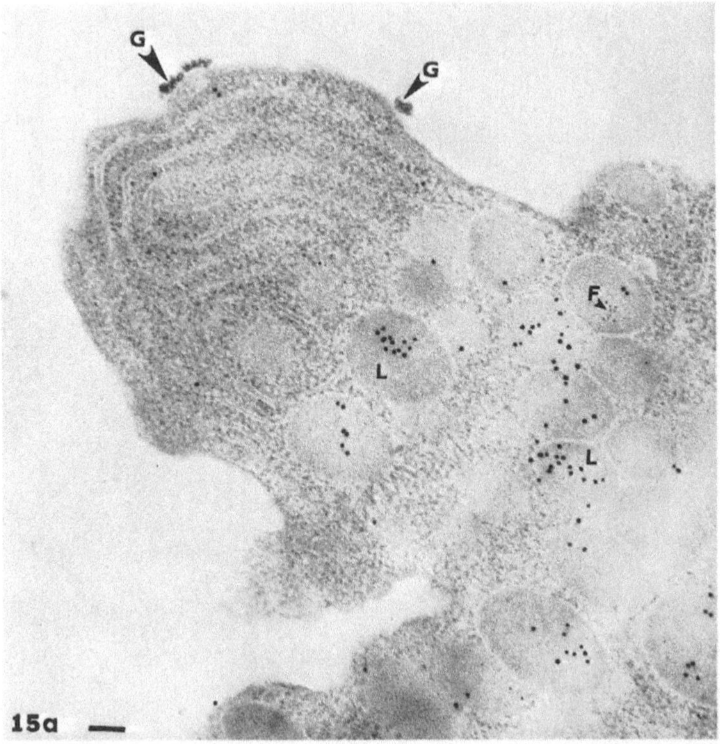
G
G
F
L
L
15a

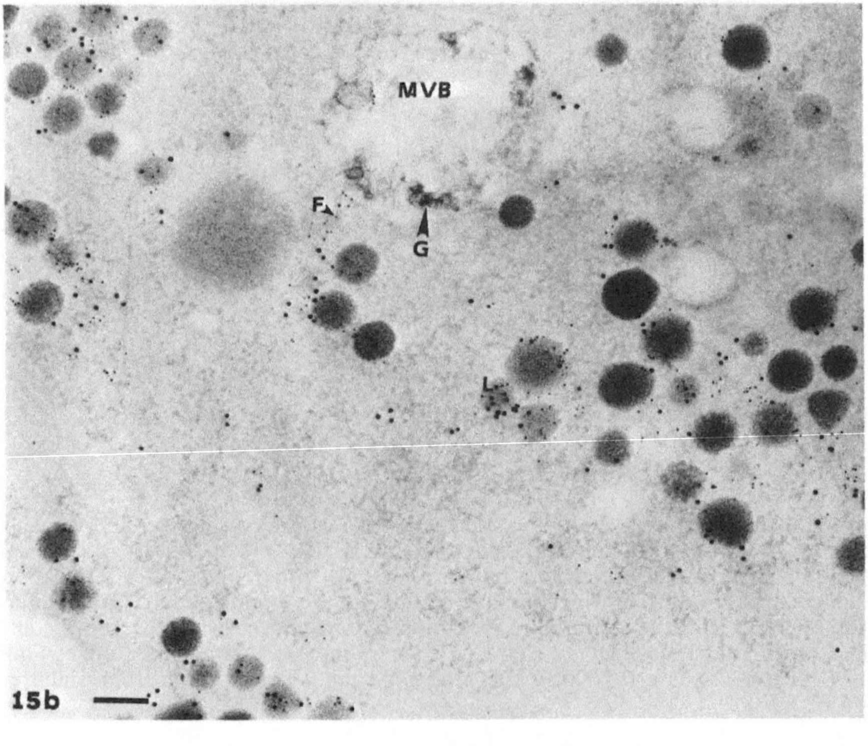
MVB
F
G
L
15b

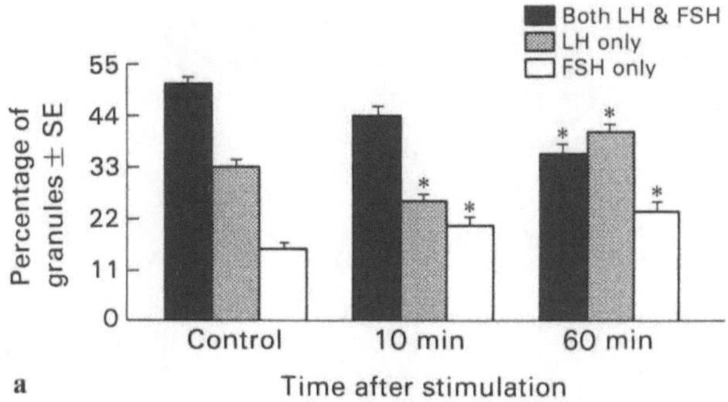

a

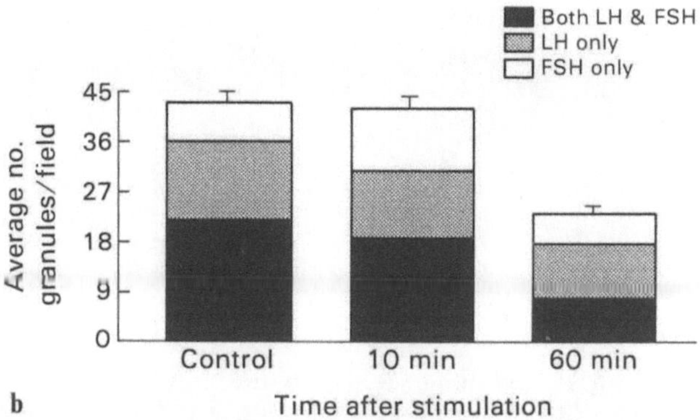

b

Fig. 16 a, b. a Gold stains and GnRH stimulation. After stimulation with GnRH, the percentage of multihormonal granules decreases, whereas the proportion of monohormonal LH and FSH granules increases. *, significantly different from control (unstimulated). The increase in the proportion of monohormonal LH granules is striking 60 min after stimulation. In **b**, the overall number of granules/field also drops by 50% during the first 60 min after stimulation

At 1 min after stimulation, the stain was diffuse on the plasma membrane (Fig. 14 a). After 3–5 min, the stain was in patches, vesicles, and receptosomes (Fig. 14 b). Stain was also seen in the Golgi complex, in vacuoles and immature granules (Figs. 14 b, c). With time after stimulation (15 min), a subpopulation of secretory granules was labeled (Fig. 14 d) (Childs et al., 1986 b).

As the gonadotropes were stimulated by the biotinylated GnRH, they responded in a manner similar to that described by Römmler et al. (1978). Compound or single exocytosis profiles were seen (Fig. 14 d). The initial sites of stimulation were characterized by the presence of cellular processes, or blebs. Often, the exocytosis profiles were most numerous at the base of the bleb. The bleb itself was filled with secretory granules, vesicles, and rer (Fig. 15 a).

Fig. 15 a, b. Triple stain for biotinylated GnRH (*G*), LH (*L*), and FSH (*F*), performed by combining ABC-peroxidase and immunogold stains on freshly dispersed cells from female rats. The GnRH is **a** in patches on a cellular process or **b** in a multivesicular body (*MVB*). Nearby are secretory granules containing LH (10 nm) and FSH (5 nm); some are monohormonal and some are multihormonal. Magnification. *Bar*, 1 μm. × 78 000; **b** 120 000

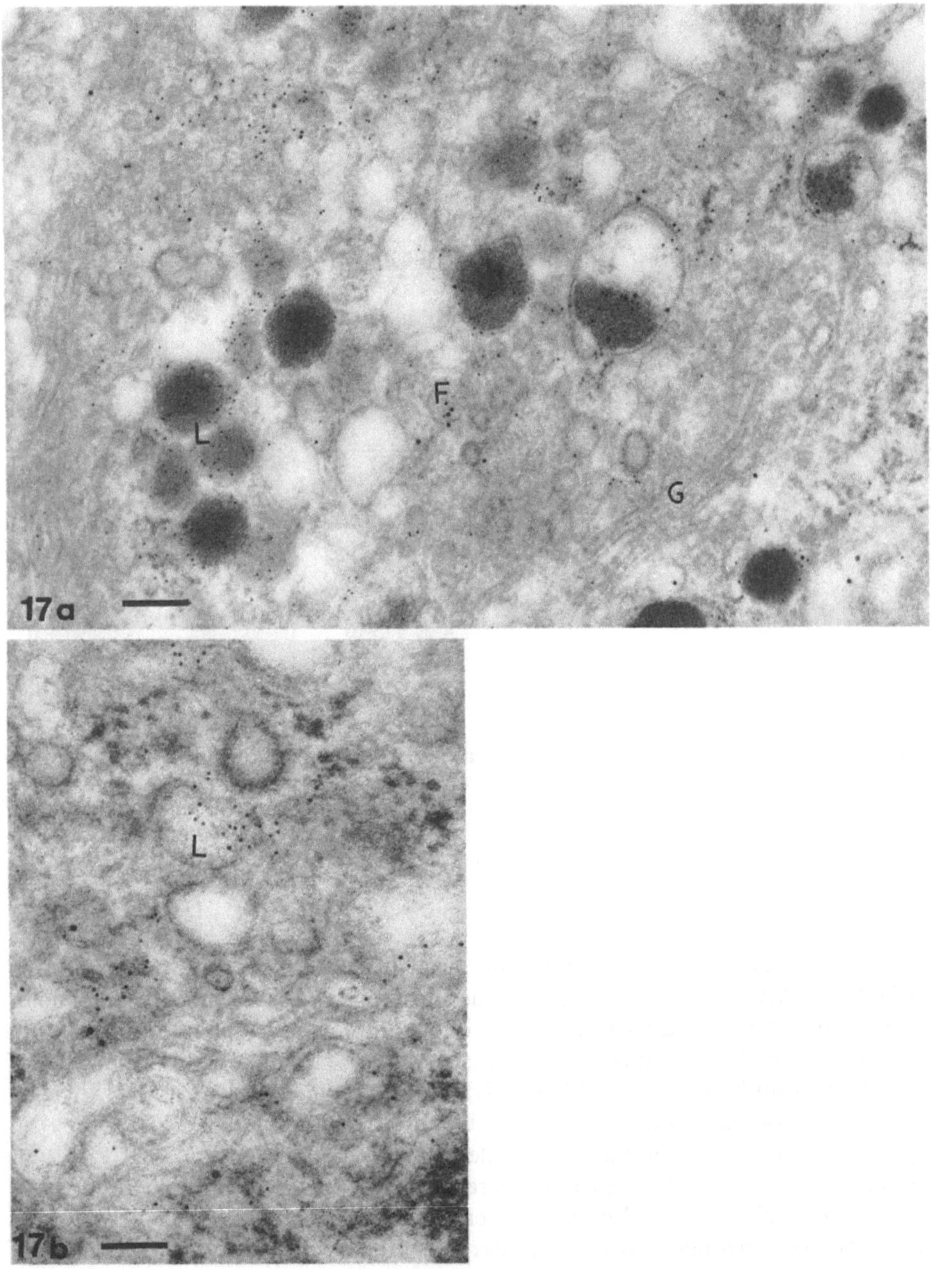

Fig. 17 a, b. Analysis of two Golgi regions of GnRH-stimulated dispersed cells from female rats shows monohormonal vacuoles and granules, indicating that the sorting of LH and FSH packages occurs in the Golgi complex (*G*). *L*, 5 nm gold; *F*, 10 nm gold; bar, 0.1 μm. ×125000

The most recent phase of our study has involved the development of double colloidal gold stains for gonadotropins on secretory granules (Childs et al. 1986a). Figure 15 shows fields that were stained for biotinylated GnRH, followed by immunogold stains for LH and FSH. The analysis of granule storage patterns demonstrated that the LH and FSH were stored separately or together. Furthermore, the proportion of multihormonal granules decreased with time after GnRH stimulation (Fig. 16a). The percentage of monohormonal LH granules decreased initially, while the proportion of monohormonal FSH granules increased.

Finally, after 30–60 min of stimulation by GnRH, our counts showed that the overall number of granules was reduced to 50% of control levels (Fig. 16b). We next studied Golgi regions to learn where monohormonal packages were produced. Figure 17 shows a Golgi region from a GnRH-stimulated gonadotrope in which monohormonal packaging was evident in immature granules and vacuoles.

Parallel in vitro LM studies were designed to determine whether GnRH had an effect on storage in the overall gonadotrope cell population (Childs 1985). To address this question, we analyzed 3-day monolayers stimulated for 10–240 min with [D-Lys^6] GnRH (Childs 1985) and then double-stained for LH and FSH. There was a gradual decrease in the percentage of monohormonal gonadotropes. The percentages of multihormonal gonadotropes were increased after 60 min of stimulation (Fig. 18). Because the percentages of total gonadotropes did not change, the shift indicated that the monohormonal gonadotropes might have the capacity to produce the other hormone when stimulated by GnRH.

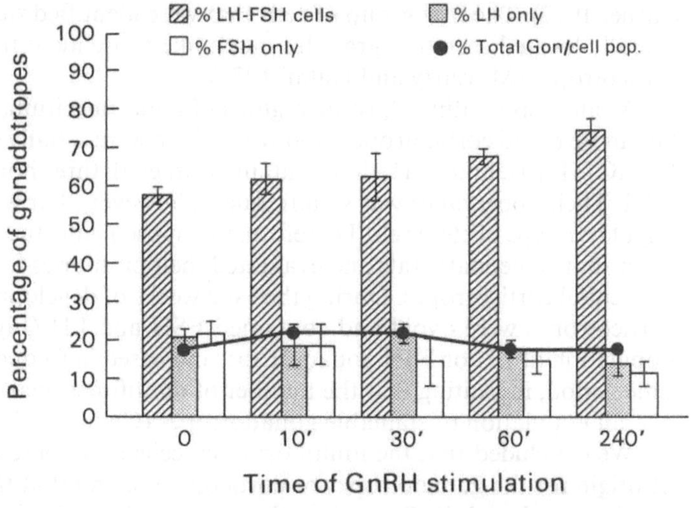

Fig. 18. Effect of GnRH stimulation on gonadotropin storage in female rats. Dispersed pituitary cells from female rats were grown in monolayer culture and stimulated with [D-Lys^6] GnRH for up to 4 h. Media levels of FSH and LH were assayed to prove GnRH-mediated stimulation of secretion (Childs 1985). Cells were double-stained for LH and FSH (Childs 1984) and their storage patterns analyzed. There was a significant increase in the percentage of multihormonal gonadotropes after GnRH stimulation and a decrease in monohormonal LH and FSH gonadotropes. The *line graph* shows no change in the overall percentage of gonadotropes in the cell population

These studies also showed separate storage of gonadotropins in different regions of the same cell. This is supported at the electron microscopic level by the finding that there are monohormonal secretory granules (Figs. 15–18; Childs et al. 1986 a). Thus, the multihormonal gonadotrope has a mechanism for selective packaging of one of the gonadotropin molecules. It may also sequester the granules in a separate region. Together, these features may provide one mechanism for the control of nonparallel release of gonadotropins from the same cell.

Finally, we agree that the effect of GnRH on gonadotrope ultrastructure is striking. The gonadotropes exhibit the cytologic effects of GnRH stimulation that were described by the early studies (Luborsky-Moore et al. 1975; Römmler 1978). In addition, our latest cytochemical studies suggest that the population of new granules that are formed after a pulse of GnRH are mostly monohormonal (Figs. 16 a, b). In addition, we confirm the work of Hazum et al. (1985), who demonstrated significant localization of labeled GnRH on secretory granules (Fig. 14). These findings show that GnRH may be involved in modulation of granule content, or that it may be recycled with its receptor to the surface. (Childs et al. 1986 b).

2.5 The Gonadotrope Is a Multifunctional Cell Storing Hormones Other than LH and FSH

Other hormones have been found in gonadotropin cells, usually in a subpopulation. In 1977, we reported the localization of ACTH with anti-ACTH (17–39) or anti-ACTH (25–39) antisera in cells that also contained FSH (Moriarty and Garner 1977). The first group of cells that were identified were stellate, with peripherally located secretion granules much like those identified in early studies as corticotropes (Moriarty and Halmi 1972).

As analysis continued, some regions of adult rat pituitaries appeared to contain more ovoid corticotropes, and when these were serially sectioned and stained for ACTH, LH, and FSH, they contained either all three hormones or ACTH and FSH. Their appearance was so infrequent, however, that we concluded that they might be a type of stem cell. Therefore, in a subsequent study (Childs et al. 1982 a) we examined neonatal rats and evaluated their pituitaries for the presence of multipotential corticotropes. During the 1st 2 weeks of development, over 60% of the corticotropes were ovoid and contained FSH and LH (Figs. 6, 9, and 10). The population of LH or FSH gonadotropes increased in frequently during this same time period, indicating that the number of multifunctional cells had added to the overall population of stainable gonadotropes (Childs et al. 1982 a).

We concluded that the multihormonal cells represented a common stem cell of origin for the gonadotropes or corticotropes, and that they persist in residual numbers in the adult. Recently, other workers have described round cells in the adult rat that resemble gonadotropes and stain for ACTH (Newman et al. 1983; Kurosumi et al. 1983, 1984). Studies reporting ACTH immunoreactivity in gonadotropes have also been confirmed in rats by Yoshimura and Nogami (1980) and in porcine pituitaries by Dacheux (1981). Unpublished work by K. Braas in our laboratory has demonstrated both endorphin and enkephalin sequences in a subpopulation of gonadotropes. In this regard, it is important to note a recent study by Knepel et al. (1985), who demonstrated GnRH-mediated corelease of LH, FSH, and a peptide derived from the proenkephalin precursor (dynorphin). Since corticotropin-releasing factor (CRF) had no effect on release of this peptide, the authors concluded that it must be stored in gonadotropes and released with gonadotropins.

Another set of antigens found in a gonadotrope subpopulation is constituted by renin (Naruse et al. 1981) and angiotensin II (A-II; Steele et al. 1982; Deschepper et al. 1985). In 1981, renin was localized in the same cells that store LH and, more recently, A-II was also found in gonadotropes. The significance of the finding may relate to paracrine interactions between gonadotropes and other cell types, since A-II is known to stimulate the secretion of both prolactin (Denef and Andries 1983) and ACTH.

The significance of another group of hormones found in gonadotropes may be related to receptor-mediated endocytosis. In early studies, GnRH was localized in gonadotrope secretory granules by immunocytochemistry (Sternberger and Petrali 1975; Sternberger et al. 1978; Dacheux 1981; Bauer et al. 1981). Bauer et al. (1981) and Dacheux (1981) showed that serially sectioned granules contain both GnRH and gonadotropins. This site may reflect the GnRH that is bound to granules during the process of receptor-mediated exocytosis (Fig. 14; Hazum et al. 1985; Childs et al. 1986b). If so, the findings indicate that the GnRH remains immunoreactive in the secretory granule. Sternberger et al. also showed that the granules themselves contained GnRH binding sites (1978).

Serotonin is known to modulate gonadotrope function, and therefore it is significant that Nunez et al. (1981) reported uptake of serotonin by bat pituitaries in cells that were labeled by immunocytochemistry for LH. The label is found scattered in the cytoplasm over secretion granules and also in the nucleus. These authors also described seasonal fluctuation in the labeling that correlates with state of activity and hibernation.

The E-36 neuropeptide has also been found in a subpopulation of LH cells in female rat pituitaries (Monaghan et al. 1984). Only a subpopulation of the granules in these LH cells were stained. Interestingly, no colocalization of E-36 was found in the male rat LH gonadotropes. While its significance is uncertain in neuronal and non-neuronal tissues, its presence in a subpopulation of female rat gonadotropes may be related to heterologous functions in the population (Denef et al. 1978).

2.6 Structural and Functional Sorting of Gonadotropes

Cytophysiological and electrophysiological studies of gonadotropes have been hampered for many years because they are part of a mixed cell population and are themselves a heterogeneous group (Denef et al. 1978; Childs et al. 1983d). Therefore, there have been many attempts to develop techniques to obtain enriched fractions of these cells in order to study their secretory, storage, and electrophysiologic properties.

Denef et al. (1978) and Naor et al. (1982) concluded that unit gravity sedimentation was of limited value in the purification of gonadotropes from adult rats.

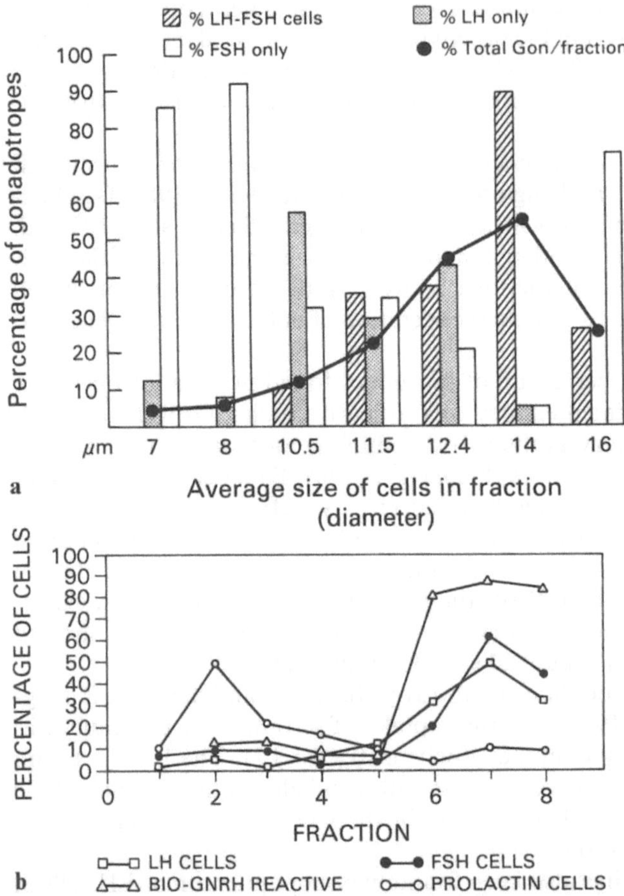

Fig. 19. **a** Gonadotropin storage in female rats – studies of separated fractions. Double stains for LH and FSH were applied to pituitary cells from female rats separated by centrifugal elutriation. This *graph* shows the fractions listed by average size (diameter) of the cells in each. The *line graph* shows the total percentage of all gonadotropes in that fraction. Gonadotrope-enriched fractions contain mostly multihormonal cells, whereas the small-cell fractions contain monohormonal cells. **b** Elutriation fractions in female rats. Studies of GnRH reactivity in these elutriation fractions show that it is highest in the gonadotrope-enriched fractions and lowest in the small-cell fractions. The percentages of LH, FSH, and prolactin cells were calculated after the application of single stains for each hormone to individual coverslips. Therefore, they do not reflect the percentage of total gonadotropes, as do the values in **a** (*line graph*). Note that prolactin cell enrichment is in the small-cell fractions (Taken from Childs et al. 1986b)

Enriched fractions obtained from juvenile female rats were seen in the larger cell fractions. However, there appears to be considerable spread along the gradient when cells from the adult rats are used.

In 1982 and 1983, two studies used fluorescence-activated cell-sorting methods on gonadotropes that were bound to rhodamine-labeled GnRH (MacInnes et al. 1983) or on rhodamine microspheres that were bound to either anti-LH or GnRH (Thorner et al. 1982). The cells were then sorted by their fluorescence activity, and fractions containing enriched gonadotropes were obtained. MacInnes et al. (1983) reported an enrichment in gonadotropes to 80% when immature rates were used. Thorner et al. 1982 reported an enrichment to 52% with the use of adult rat gonadotropes bound to anti-LH or GnRH fluorescent beads.

Another technique useful for sorting gonadotropes by size is centrifugal elutriation. Recently, we have been successful in the application of this method to pituitary cells from both male and female rats. In the large-cell fractions there is an enrichment as high as 88% of GnRH-reactive cells (Fig. 19 b; Childs et al. 1986 b). In parallel studies of gonadotropin storage, we showed that the enriched fraction was mainly multihormonal (Fig. 19 a; Hyde et al. 1982; Childs et al. 1983 d).

The elutriation method also separates functionally distinct populations of gonadotropes, since the smaller cells tend to be monohormonal (Fig. 19 a; Childs et al. 1983 d). Thus, the different subtypes can be studied separately. However, only the larger cells are enriched, since the smaller gonadotropes are found with the numerous prolactin and growth hormone cells. More recently, we evaluated ovariectomy cells after elutriation and found that the hypertrophy of the gonadotropes allows enrichment to almost 90% in the larger-cell fractions. A population of small gonadotropes remain, however.

Perhaps future studies can combine elutriation with fluorescence technology to first separate the different subtypes of gonadotropes and then sort the smaller cells by their binding to fluorescent analogs. In that way, further structural–functional correlations can be made over the entire gonadotrope population.

2.7 Summary and Conclusions

Gonadotropes are a diverse group of cells that have been programmed to respond to GnRH, steroids, and other neuromodulators to release gonadotropins in a finely controlled pattern and to maintain cyclicity and function of the reproductive system. The ultrastructural analyses of the cells have provided important clues to the mechanisms behind this action. In addition, the studies have produced some surprising results. Some of the new findings that require further exploration are summarized as follows:

1. The cells may maintain nonparallel release of gonadotropins by mechanisms that include storage in separate cells and in separate regions in the same cell (Childs 1985). This is supported by findings that show the presence of gonadotropins in separate granule profiles (Figs. 15–18) in the same cell.
2. The initial stimulation by GnRH appears to increase the rate of packaging of gonadotropins (Luborsky-Moore et al. 1975; Römmler et al. 1975). The granule package is smaller (Yoshimura et al. 1984), and most of the new granules appear to be monohormonal (Figs. 16 and 17).
3. The monohormonal gonadotropes may be multipotential, since stimulation by GnRH in vivo (Childs et al. 1981, 1982 b) or in vitro (Childs 1985) results in the appearance of a higher proportion of multihormonal gonadotropes. Studies with probes for messenger RNA (mRNA) for the other gonadotropin are needed to support these findings.

4. Gonadotropes may play a role as paracrine stimulators of other pituitary cells during development (Bégeot et al. 1983; Denef and Andries 1983). The hormones that support this function may include alpha subunit or A-II, both of which are found in gonadotropes.
5. The gonadotropes in turn may be influenced by other pituitary cells, including corticotropes that produce endorphin. Ultrastructural studies show that states that stimulate corticotrope function inhibit gonadotrope expansion, especially that of LH cells (Childs et al. 1983b).
6. Pars tuberalis cells, especially those from ovine pituitaries, contain a rich store of gonadotropes that are closely associated and may be directly innervated by fibers that contain GnRH and other neuromodulators (Gross 1984).
7. Gonadotropes may originate from a stem cell that also gives rise to opiocortin peptide-producing cells (Figs. 4 and 8; Childs et al. 1982a). Studies with probes for mRNA are needed to support the synthesis of gonadotropins by corticotropes.
8. Gonadotropes internalize the hormones that affect them, such as GnRH (Fig. 4; Pelletier et al. 1982; Duello et al. 1983; Jennes et al. 1983; Hazum et al. 1984) and serotonin (Nunez et al. 1981). The pathway ends with labeling in organelles in a pattern that suggests a subcellular regulatory role for the neuromodulating hormone. These may include the Golgi complex and rer. Other labeled organelles, such as multivesicular bodies, the Golgi complex, or the secretory granules, may be involved in receptor recycling and regulation of receptors.
9. Gonadotropes show patching and capping of GnRH receptors, followed by internalization of the peptide (Pelletier et al. 1982; Duello et al. 1983; Jennes et al. 1983; Hazum et al. 1984). At the site of the cap, there appears to be active secretion of the gonadotropins as seen by morphological, cytophysiological, and immunocytochemical methods (Fig. 14; Childs et al. 1983c). This supports the hypothesis that there may be functional differentiation of that part of the membrane bound by GnRH that might allow more rapid release of LH and FSH. Since internalization of GnRH is not required for LH release (Conn et al. 1981), the aggregation of receptors themselves may stimulate the necessary molecular changes in the membrane. Such functional-structural polarization may be very useful to an ovoid cell that normally sits with only one face on a capillary.

Acknowledgements. The author wishes to acknowledge the generous gift of RIA Kits for rLH-β, rFSHβ from NIADD (National Institute Arthritic, Digestive Diseases), Dr. A. Parlowe, Hormone Distribution Office, and of anti-bLHβ from Dr. J.G. Pierce. She also thanks Dr. Laura Garner-Render, Dr. Charlene Hon, Dayle G. Ellison, Linda P. Foster, and Geda Unabia for excellent technical assistance. She is grateful for the typing skills of Ms. Phyllis Fletcher.

Note Added in Proof

For more information about changes in area or percentages of gonadotropes after long term castration or ovariectomy please consult a recent paper by Ibrahim (1986) and Childs (1986b).

References

Aguado LI, Hancke JL, Rodriguez S, Rodriguez EM (1982) Changes in the luteinizing hormone content of the rat pars tuberalis during the estrous cycle and after lesions in the preoptic area. Neuroendocrinology 35:178–185

Arimura A, Matsuo H, Baba Y, Debeljuk L, Sandow J, Schally AV (1972) Stimulation of release of LH by synthetic LH-RH in vivo. I. A comparative study of natural and synthetic hormones. Endocrinology 90:163–168

Arimura A, Debeljuk L, Schally AV (1973) Stimulation of FSH release in vivo by prolonged infusion of synthetic LH-RH. Endocrinology 91:529–532

Arimura A, Shiino M, De La Cruz KG, Rennels EG, Schally AV (1976) Effect of active and passive immunization with luteinizing hormone-releasing hormone on serum luteinizing hormone and follicle stimulating hormone levels and the ultrastructure of the pituitary gonadotrophs in castrated male rats. Endocrinology 99:291–303

Asa SL, Kovacs K, Bilbao JM (1983) The pars tuberalis of the human pituitary. Virchows Arch [A] 399:49–59

Baker BL (1977) Cellular composition of the human pituitary pars tuberalis as revealed by immunocytochemistry. Cell Tissue Res 182:151–163

Baker BL, Yu YY (1975) Immunocytochemical analysis of cells in the pars tuberalis as revealed by immunocytochemistry. Cell Tissue Res 156:443–449

Baker BL, Karsch FJ, Hoffman DL, Beckman WC (1977) The presence of gonadotropic and thyrotropic cells in the pituitary pars tuberalis of the monkey (*Macaca mulatta*). Biol Reprod 17:232–240

Barnes BG (1962) Electron microscope studies on the secretory cytology of the mouse anterior pituitary. Endocrinology 71:618–628

Batten TFC, Hopkins CR (1978) Discrimination of LH, FSH, TSH, and ACTH in dissociated porcine anterior pituitary cells by light and electron microscope immunocytochemistry. Cell Tissue Res 192:107–120

Bauer TW, Moriarty CM, Childs GV (1981) Studies of immunoreactive gonadotropin releasing hormone (GnRH) in the rat anterior pituitary. J Histochem Cytochem 29:1171–1178

Bégeot M, Heming FJ, Dubois PM, Combarnous Y, Dubois MP, Albert ML (1984) Induction of pituitary lactotrope differentiation by luteinizing hormone α-subunit. Science 226:566–568

Blake CA (1980) Correlative study of changes in the morphology of the LH gonadotroph and anterior pituitary gland LH secretion during the 4-day rat estrous cycle. Biol Reprod 23:1097–1108

Bugnon C, Bloch B, Fellmann D (1976) Études cyto-immunologiques des cellules gonadotropes du foetus humaine. Bull Assoc Anat 169:259–268

Bugnon C, Bloch B, Fellmann D (1977) Cytoimmunological study of the ontogenesis of the gonadotropic hypothalamo-pituitary axis in the human fetus. J Steroid Biochem 8:565–575

Childs GV (1984) Application of dual pre-embedding stains for gonadotropins to pituitary cell monolayers with avidin-biotin (ABC) and peroxidase-antiperoxidase (PAP) complexes: light microscopic studies. Stain Technol 58:281–289

Childs GV (1985) Shifts in gonadotropin storage in cultured gonadotropes following GnRH stimulation, in vitro. Peptides 6:103–107

Childs GV, Ellison DG, Garner LL (1980) An immunocytochemists' view of gonadotropin storage in the adult male rat: cytochemical and morphological heterogeneity in serially sectioned gonadotropes. Am J Anat 158:397–409

Childs GV, Ellison DG, Foster LP, Ramaley JA (1981) Postnatal maturation of gonadotropes in the male rat pituitary. Endocrinology 109:1683–1696

Childs GV, Ellison DG, Ramaley JA (1982 a) Storage of anterior lobe adrenocorticotropin in corticotropes and a subpopulation of gonadotropes during the stress-nonresponsive period in the neonatal male rat. Endocrinology 110:1676–1692

Childs GV, Ellison DG, Lorenzen JR, Collins TJ, Schwartz NB (1982b) Immunocytochemical studies of gonadotropin storage in developing castration cells. Endocrinology 111:1318–1328

Childs GV, Naor Z, Hazum E, Tibolt R, Westlund KN, Hancock MB (1983a) Localization of biotinylated gonadotropin releasing hormone with avidin-biotin peroxidase complexes. J Histochem Cytochem 31:1422–1425

Childs GV, Ellison DG, Lorenzen JR, Collins TJ, Schwartz NB (1983b) Retarded development of castration cells after adrenalectomy or sham adrenalectomy. Endocrinology 113:166–177

Childs GV, Naor Z, Hazum E, Tibolt R, Westlund KN, Hancock MB (1983c) Cytochemical characterization of pituitary target cells for biotinylated gonadotropin releasing hormone. Peptides 4:549–555

Childs GV, Hyde C, Naor Z, Catt K (1983d) Heterogeneous luteinizing hormone and follicle stimulating hormone storage patterns in subtypes of gonadotropes separated by centrifugal elutriation. Endocrinology 113:2120–2128

Childs GV, Unabia G, Tibolt R (1985) How the fixative and embedding protocol affects the specificity and efficiency of immunocytochemical stains for gonadotropin subunits. Am J Anat 174:409–417

Childs GV, Ellison DG, Unabia G (1986a) Cytochemical studies of pituitary hormones with PAP, ABC, and immunogold techniques: Evolution of technology to best fit the antigen. Am J Anat 175:307–330

Childs GV, Hazum E, Amsterdam A, Limor R, Naor Z (1986b) Cytochemical evidence for different routes of gonadotropin-releasing hormone processing by large gonadotropes and granulosa cells. Endocrinology 119

Conn PM, Smith RG, Rogers DC (1981) Stimulation of pituitary gonadotropin release does not require internalization of gonadotropin releasing hormone. J Biol Chem 256:1098–1101

Currie RW, Faiman C, Thliveris JA (1981) An immunocytochemical and routine electron microscopic study of LH and FSH cells in the human fetal pituitary. Am J Anat 161:281–297

Dacheux F (1978) Ultrastructural localization of gonadotrophic hormones in the porcine pituitary using the immunoperoxidase technique. Cell Tissue Res 191:219–232

Dacheux F (1980) Ultrastructural localization of LH and FSH in the porcine pituitary. In: Jutisz J, McKerns KW (eds) Synthesis and release of adenohypophyseal hormones. Plenum, New York, pp 187–195

Dacheux F (1981) Ultrastructural localization of gonadotropin-releasing hormone in the porcine gonadotropic cells. Cell Tissue Res 216:143–150

Dacheux F (1984) Functional differentiation of the anterior pituitary cells in the fetal pig. Cell Tissue Res 235:623–633

Dacheux F, Dubois MP (1976) Ultrastructural localization of prolactin, growth hormone and luteinizing hormone by immunocytochemical techniques in the ovine pituitary. Cell Tissue Res 174:245–260

Dacheux F, Dubois MP (1978) LH-producing cells in the ovine pituitary. An electron microscopic immunocytochemical study. Cell Tissue Res 188:449–463

Debeljuk LA, Arimura A, Shiino M, Rennels EG, Schally AV (1973) Effects of chronic treatment with LH/FSH-RH in hypophysectomized pituitary-grafted male rats. Endocrinology 92:921–930

Denef C, Andries M (1983) Evidence for paracrine interaction between gonadotrophs and lactotrophs in pituitary cell aggregates. Endocrinology 112:813–822

Denef C, Hautekeete E, DeWolf A, Vanderschuren B (1978) Pituitary basophils from immature male and female rats: distribution of gonadotrophs and thyrotrophs as studied by unit gravity sedimentation. Endocrinology 103:724–738

Deschepper CF, Seidler CD, Steele MK, Ganong WF (1985) Further studies on the localization of angiotensin-II-like immunoreactivity in the anterior pituitary gland of the male rat, comparing various antisera to pituitary hormones and their specificity. Neuroendocrinology 40:471–475

Dubois PM, Bégeot M, Dubois MP (1975) Dissociation entre les sous-unités de LH dans l'antéhypophyse foetale humaine. Annu Endocrinol Paris 36:321–322

Dubois PM, Bégeot MP, Herbert DC (1978) Immunocytological localization of LH, FSH and TSH and their subunits in the pituitary of normal and anencephalic human foetus. Cell Tissue Res 191:249–265

Duello TM, Nett TM, Farquhar MG (1983) Fate of a gonadotropin-releasing hormone agonist internalized by rat pituitary gonadotrophs. Endocrinology 112:1–10

Dussault JH, Walker P, Dubois JD, Labrie F (1977) The development of the hypothalamic-pituitary axis in the neonatal rat: sexual maturation in male and female rats as assessed by hypothalamic LHRH and pituitary and serum LH and FSH concentrations. Can J Physiol Pharmacol 55:1091–1130

Farquhar MG (1971) Processing of secretory products by cells of the anterior pituitary gland. Mem Soc Endocrinol 19:79–124

Farquhar MG, Rinehart JF (1954) Electron microscopic studies of the anterior pituitary gland of castrate rats. Endocrinology 54:516–541

Friend JN, Judge DM, Sherman BM, Santen RJ (1976) FSH-secreting pituitary adenomas: stimulation and suppression studies in two patients. J Clin Endocrinol Metab 43:650–657

Garner LL, Blake CA (1979) Morphological correlates for LHRH self-priming and anterior pituitary gland refractoriness to LHRH in proestrous rats: an immunocytochemical study. Biol Rep 20:1055–1066

Garner LL, Blake CA (1981) Ultrastructural, immunocytochemical study of the LH-secreting cell of the rat anterior pituitary gland: changes occurring after ovariectomy. Biol Rep 24:461–474

Girod C, Dubois MP, Trouillas J (1980) Immunohistochemical study of the pars tuberalis of the adenohypophysis in the monkey, *Macaca irus*. Cell Tissue Res 210:191–203

Girod C, Dubois MP, Trouillas J (1981) Immunohistochemical localization of FSH and LH in the pars distalis of vervet (*Cercopithecus aethiops*) and baboon (*Papio hamadryas*) pituitaries. Cell Tissue Res 217:245–257

Gross DS (1978) Effect of castration and steroid replacement on immunoreactive luteinizing hormone cells in the pars tuberalis of the rat. Endocrinology 103:583–588

Gross DS (1983) Hormone production in the hypophysial pars tuberalis of intact and hypophysectomized rats. Endocrinology 112:733–744

Gross DS (1984) The mammalian hypophysial pars tuberalis: a comparative immunocytochemical study. Gen Comp Endocrinol 56:283–298

Gross DS, Page RB (1979) Luteinizing hormone and follicle-stimulating hormone production in the pars tuberalis of hypophysectomized rats. Am J Anat 156:285–291

Gross DS, Turgeon JL, Waring DW (1984) The ovine pars tuberalis: a naturally occurring source of partially purified gonadotropes which secrete luteinizing hormone in vitro. Endocrinology 114:2084–2091

Grotjan HE, Leveque NW, Berkowitz AS, Keel BA (1984) Quantitative studies of LH subunits released by rat anterior pituitary cells in primary culture. Mol Cell Endocrinol 35:121–129

Hazum E, Koch Y, Liscovitch M, Amsterdam A (1985) Intracellular pathways of receptor-bound GnRH agonist in pituitary gonadotropes. Cell Tissue Res 239:3–8

Herbert DC (1976) Immunocytochemical evidence that luteinizing hormone (LH) and follicle-stimulating hormone (FSH) are present in the same cell type in the Rhesus monkey pituitary gland. Endocrinology 98:1554–1557

Herbert DC (1978) Identification of the LH- and TSH-secreting cells in the pituitary gland of the Rhesus monkey. Cell Tiss Res 190:151–161

Herbert DC, Silverman AY (1983) Topographical distribution of the gonadotrophs, mammotrophs, somatotrophs, and thyrotrophs in the pituitary gland of the baboon (*Papio cynocephalus*). Cell Tissue Res 230:233–238

Hopkins CR, Gregory H (1977) Topographical localization of the receptors for luteinizing hormone-releasing hormone on the surface of dissociated pituitary cells. J Cell Biol 75:528–540

Hopkins CR, Semoff S, Gregory H (1981) Regulation of gonadotropin secretion in the anterior pituitary. Philos Trans R Soc Lond [Biol] 296:73–81

Horvath E, Kovacs K (1984) Gonadotroph adenomas of the human pituitary: sex-related fine-structural dichotomy. Am J Pathol 117:429–440

Hyde CL, Childs G, Wahl LM, Naor Z, Catt K (1982) Preparation of gonadotropin-enriched cell populations from adult rat pituitary cells by centrifugal elutriation. Endocrinology 111:1421–1423

Ibrahim SN, Moussa SM, Childs GV (1986) Morphometric studies of rat anterior pituitary cells after gonadectomy: correlation of changes in gonadotropes with the serum levels of gonadotropins. Endocrinology 119

Inoue K, Kurosumi K (1981) Mode of proliferation of gonadotrophic cells of the anterior pituitary after castration–immunocytochemical and autoradiographic studies. Arch Histo Jpn 44:71–85

Jennes L, Stumpf WE, Conn PW (1983) Intracellular pathways of electron-opaque gonadotropin-releasing hormone derivatives bound by cultured gonadotropes. Endocrinology 113:1683–1689

Kihara H (1984) Compensatory development of immature LH cells after long-term gonadectomy: an immunocytochemical electron microscopic and cell count study. Endocrinol Jpn 31:395–406

Knepel W, Schwaninger M, Dohler KD (1985) Co-release of dynorphin-like immunoreactivity, luteinizing hormone, and follicle-stimulating hormone from rat adenohypophysis in vitro. Endocrinology 117:481–487

Kovacs K, Horvath E (1975) Gonadotrophs following removal of the ovaries: a fine structural study. Endokrinologie 66:1–8

Kovacs K, Horvath E, Van Loon GR, Rewcastle NB, Ezrin C, Rosenbloom AA (1978) Pituitary adenomas associated with elevated blood follicle-stimulating hormone levels: a histologic, immunocytologic and electron microscopic study of two cases. Fertil Steril 29:622–628

Kovacs K, Horvath C, Rewcastle NB, Ezrin C (1980) Gonadotroph cell adenoma of the pituitary in a woman with long-standing hypogonadism. Arch Gynecol 229:57–65

Kurosumi K (1979) Formation and release of secretory granules during mitosis in the anterior pituitary gland. Arch Histol Jpn 42:481–486

Kurosumi K, Inoue K (1980) Surface pits of typical gonadotrophs and castration cells of the rat anterior pituitary suggestive of exocytosis and micropinocytosis. Arch Histo Jpn 43:373–382

Kurosumi K, Oota Y (1968) Electron microscopy of two types of gonadotrophs in the anterior pituitary glands of persistent estrous and diestrous rats. Z Zellforsch 85:34–46

Kurosumi K, Kawrai Y, Yukitake Y, Inoue K (1976) Electron microscopic morphometry of the rat castration cells. Gunma Symp Endocrinol 13:221–236

Kurosumi K, Inoue K, Taniguchi Y (1983) Evidence for two types of ACTH-secreting cells of the rat anterior pituitary: ultrastructural, morphometric and immunocytochemical studies. Biomed Res 4:117–134

Kurosumi K, Taniguchi Y, Inoue K (1984) Electron microscopic immunocytochemistry by the protein A-Gold method of two types of rat anterior pituitary ACTH cells. Arch Histol Jpn 47:279–291

Legait H (1969) Recherches expérimentales sur le lobe tubéral de l'hypophyse chez le rat impubère. C R Soc Biol 163:714–715

Lorenzen JR, Schlepphorst C, Schwartz NB (1980) The interaction of castration and adrenalectomy on pituitary responses to loss of target gland negative feedback in the male rat. Endocrinology 1061:592–603

Luborsky-Moore JL, Poliakoff SJ, Worthington WC (1975) Ultrastructural observation of anterior pituitary gonadotrophs following hypophysial portal vessel infusion of luteinizing hormone-releasing hormone. Am J Anat 144:549–555

MacInnes DG, Green DK, Harmar A, Nairn EG, Fink G (1983) Neuropeptide-receptor interactions studied with the aid of fluorescence-activated cell sorting and video intensification microscopy. J Neurosci Methods 7:309–316

McDowell IFW, Morris JF, Charlton HM, Fink G (1982) Effects of luteinizing hormone-releasing hormone on the gonadotrophs of hypogonadal (hpg) mice. J Endocrinol 95:331–340
Mendoza D, Arimura A, Schally AV (1973) Ultrastructural and light microscopic observations of rat pituitary LH-containing gonadotrophs following injection of synthetic LH-RH. Endocrinology 92:1153–1160
Monaghan P, Sappino AP, Roberts JDB, Knight MA, Ellis J, Nakajima T (1984) Monoclonal antibody LICR-LON-E36 recognizes gonadotrophs in female rat pituitaries. J Histochem Cytochem 32:1048–1054
Moriarty GC (1973) Adenohypophysis: ultrastructural cytochemistry, a review. J Histochem Cytochem 20:855–892
Moriarty GC (1975) Electron microscopic-immunocytochemical studies of rat pituitary gonadotrophs: a sex difference in morphology and cytochemistry of LH cells. Endocrinology 97:1215–1225
Moriarty GC (1976a) Ultrastructural-immunocytochemical studies of rat pituitary gonadotrophs in cycling female rats. Gunma Symp Endocrinol 13:207–219
Moriarty GC (1976b) Immunocytochemistry of the pituitary glycoprotein hormones. J Histochem Cytochem 24:846–863
Moriarty GC, Garner LL (1977) Immunocytochemical studies of cells in the rat adenohypophysis containing both ACTH and FSH. Nature 265:356–358
Moriarty GC, Halmi NS (1972) Electron microscopic study of the adrenocorticotropin producing cell with the use of unlabeled antibody and the soluble peroxidase-antiperoxidase complex. J Histochem Cytochem 20:590–603
Nakane PK (1970) Classifications of anterior pituitary cell types with immunoenzyme histochemistry. J Histochem Cytochem 18:9–20
Nakane PK (1971) Application of peroxidase-labelled antibodies to the intracellular localization of hormones. Acta Endocrinol [Suppl] (Copenh) 153:190–203
Naor Z, Childs GV, Leifer AM, Clayton RN, Amsterdam A, Catt KJ (1982) Gonadotropin-releasing hormone binding and activation of enriched populations of pituitary gonadotrophs. Mol Cell Endocrinol 25:85–97
Naruse K, Takii Y, Inagami T (1981) Immunohistochemical localization of renin in luteinizing hormone-producing cells of rat pituitary. Proc Natl Acad Sci USA 78:7579–7583
Newman GR, Jasani B, Williams ED (1983) Metal compound intensification of the electron density of diaminobenzidine. J Histochem Cytochem 31:1430–1434
Nunez EA, Gershon MD, Silverman AJ (1981) Uptake of 5-hydroxytryptamine by gonadotrophs of the bat's pituitary: a combined immunocytochemical radioautographic analysis. J Histochem Cytochem 29:1336–1346
Odronneau P, Petrusz P (1980) Immunocytochemical demonstration of anterior pituitary hormones in the pars tuberalis of long-term hypophysectomized rats. Am J Anat 158:491–506
Osamura RY, Watanabe K (1978) An immunohistochemical study of epithelial cells in the posterior lobe and pars tuberalis of the human adult pituitary gland. Cell Tissue Res 194:513–524
Pearson AK, Licht P (1982) Morphology and immunocytochemistry of the turtle pituitary gland with special reference to the pars tuberalis. Cell Tissue Res 222:81–100
Pelletier G (1974) Autoradiographic studies of synthesis and intracellular migration of glycoproteins in the rat anterior pituitary gland. J Cell Biol 62:185–192
Pelletier G, Peillon F, Vila-Porcile E (1971) An ultrastructural study of sites of granule extrusion in the anterior pituitary of the rat. Z Zellforsch 115:501–507
Pelletier G, Leclerc R, Labrie F (1976) Identification of gonadotropic cells in the human pituitary by immunoperoxidase technique. Mol Cell Endocrinol 6:123–128
Pelletier G, Robert F, Hardy J (1978) Identification of human anterior pituitary cells by immunoelectron microscopy. J Clin Endocrinol Metabol 46:534–542
Pelletier G, Dubé D, Guy J, Séquin C, Lefebvre FA (1982) Binding and internalization of a luteinizing hormone-releasing hormone agonist by rat gonadotrophic cells. A radioautographic study. Endocrinology 111:1068–1076

Phifer RF, Midgley AR, Spicer SS (1973) Immunohistologic and histologic evidence that follicle-stimulating hormone and luteinizing hormone are present in the same cell types in the human pars distalis. J Clin Endocrinol 17:125–141
Pierce JG, Parsons TF (1981) Glycoprotein hormones: structure and function. Annu Rev Biochem 50:465–501
Purves HD, Griesbach WE T (1954) The site of follicle-stimulating and luteinizing hormone production in the rat pituitary. Endocrinology 55:785–793
Rennels EG, Bogdanove EM, Arimura A, Saito M, Schally AV (1971) Ultrastructural observations of rat pituitary gonadotrophs following injection of purified porcine LH-RH. Endocrinology 88:1318–1326
Römmler A, Seinsch W, Hasan AS, Haase F (1978) Ultrastructure of rat pituitary LH gonadotrophs in relation to serum and pituitary LH levels following repeated LH-RH stimulation. Cell Tissue Res 190:135–149
Sakuma S, Shirasawa N, Yoshimura F (1983) A histometrical study of immunohistochemically identified mitotic adenohypophysial cells in immature and mature castrated rats. J Endocrinol 100:322–328
Schally AV, Arimura A, Kastin AJ (1971) Hypothalamic regulatory hormones. Science, N.Y. 179:341–350
Schwartz NB, Justo S (1977) Acute changes in serum gonadotrophins and steroids following orchidectomy in the rat: role of the adrenal gland. Endocrinology 100:1550–1567
Sétáló G, Nakane PK (1972) Studies on the functional differentiation of cells in fetal anterior pituitary gland of rats with peroxidase-labeled antibody method. Anat Rec 172:403–404
Sétáló G, Nakane PK (1976) Functional differentiation of the fetal anterior pituitary cells in the rat. Endocrinol Exp (Bratis) 10:155–166
Shiino M (1982) Ultrastructural evidence of gonadotrophin release from castration cells following injection of LHRH in the rat. Cell Tissue Res 222:213–222
Shiino MA, Arimura A, Schally AV, Rennels EG (1972) Ultrastructural observations of granule extrusion from rat anterior pituitary cells after injection of LH-releasing hormone. Z Zellforsch Mikroskop Anat 128:152–161
Shirasawa N, Kihara H, Yoshimura F (1985) Fine structural and immunohistochemical studies of goat adenohypophysial cells. Cell Tissue Res 240:315–321
Smith PF, Keefer DA (1982) Immunocytochemical and ultrastructural identification of mitotic cells in the pituitary gland of ovariectomized rats. J Reprod Fertil 66:383–388
Steele MK, Brownfield MS, Ganong WF (1982) Immunocytochemical localization of angiotensin immunoreactivity in gonadotropes and lactotropes of the rat anterior pituitary gland. Neuroendocrinology 35:155–158
Sternberger LA, Petrali JP (1975) Quantitative immunocytochemistry of pituitary receptors for luteinizing hormone-releasing hormone. Cell Tissue Res 162:141–176
Sternberger LA, Hardy PH Jr, Cuculis JJ, Meyer HG (1970) The unlabeled antibody enzyme method of immunohistochemistry: preparation and properties of soluble antigen-antibody complex (horseradish peroxidase-antiperoxidase) and its use in identification of spirochetes. J Histochem Cytochem 18:315–333
Sternberger LA, Petrali JP, Joseph SA, Meyer HC, Mills KR (1978) Specificity of the immunocytochemical luteinizing hormone-releasing hormone receptor reaction. Endocrinology 102:63–73
Thorner MD, Borges JL, Cronin MJ, Keefer DA, Hellmann P, Lewis D, Dabney LG, Quesenberry PJ (1982) Fluorescence-activated cell sorting of functional anterior pituitary cells. Endocrinology 110:1831–1832
Tixier-Vidal A (1972) Ultrastructure and immunocytology of cellular gonadotropes in culture. In: Pituitary glycoprotein hormones. INSERM, pp 49–64
Tixier-Vidal A, Tougard C, Kerdelhué B, Jutisz M (1975) Light and electron microscopic studies on immunocytochemical localization of gonadotropic hormones in the rat pituitary gland with antisera against ovine FSH, LH, LHα, and LHβ. Annu NY Acad Sci 254:433–460
Tougard C, Kerdelhué MB, Tixier-Vidal A, Jutisz MM (1971) Localisation par cyto-immunoenzymologie de la LH de ses sous-unités, α at β, et de la FSH dans l'adenohypophyse de la ratte castrée. C R Seances Acis Sci III 273:897–900

Tougard C, Kerdelhué B, Tixier-Vidal A, Jutisz M (1973) Light and electron microscope localization of binding sites of antibodies against ovine luteinizing hormone and its two subunits in rat adenohypophysis using peroxidase-labeled antibody technique. J Cell Biol 58:503–521
Tougard C, Picart R, Tixier-Vidal A (1977) Cytogenesis of immunoreactive gonadotropic cells in the fetal rat pituitary at light and electron microscope levels. Dev Biol 58:148–163
Tougard C, Picart R, Tixier-Vidal A (1980) Immunocytochemical localization of glycoprotein hormones in the rat anterior pituitary. J Histochem Cytochem 28:101–114
Trouillas J, Girod C, Sassolas G, Claustrat B, Lheritier M, Dubois MP, Goutelle A (1981) Human pituitary gonadotropin adenoma: histological, immunocytochemical, and ultrastructural and hormonal studies in eight cases. J Pathol 135:315–336
Tung KSK, Legrande E, Ellis, Childs GV, Dufau M (1984) The dark mink: a model of male infertility. Endocrinology 114:922–929
Yoshimura F, Nogami H (1980) Immunohistochemical characterization of pituitary stellate cells in rats. Endocrinol Jpn 27:43–51
Yoshimura F, Nakamura F, Nogami H, Suzuke K (1984) In: Ochiai K, Arai Y, Shioda T, Takahashi M (eds) Endocrine correlates of reproduction. Jpn Sci Soc Press, Tokyo. Springer, Berlin Heidelberg New York, pp 41–58

Ultrastructure of Anterior Pituitary Cells

K. KUROSUMI and K. INOUE

Contents

1 Introduction

The anterior pituitary gland together with the exocrine pancreas was the target for electron microscopists who wanted to investigate the ultrastructural background of biological phenomena occurring in the cell in order to elaborate secretion material (Farquhar and Wellings 1957). These authors clearly ascribed the origin of secretory granules the Golgi apparatus, and one of them – Farquhar (1961) – demonstrated the mechanism of release of the secretory material out of the cell, which was later called "exocytosis." The first author who described this mechanism of secretion release in the anterior pituitary was Ichikawa (1959), but since his paper was written in Japanese, it was unknown to cytologists outside Japan.

During the two decades between 1960 and 1980, the interest of endocrinological morphologists was concentrated on the determination of the function of cells in the anterior pituitary, i.e., which cell produces which hormone. This trend was accelerated by the advancement of immunocytochemistry, especially that technique in electron microscopy which may be called immunoelectron microscopy.

Department of Morphology, Institute of Endocrinology. Gunma University, Maebashi 371, Japan

Current Topics in Neuroendocrinology, Vol. 7

Although this trend prevailed in the field of anterior pituitary research, enzyme cytochemistry, autoradiography, the freeze-fracture method, freeze-substitution fixation, and other recent advances in morphological techniques together with immunoelectron microscopy have enhanced our knowledge of the biology of anterior pituitary cells, especially concerning their physiological activities in respect to hormone production.

This article deals with two major topics: the ultrastructural contribution to the cell physiology of endocrine activities of the anterior pituitary cells, and the morphological identification of the producer cells of each hormone in the anterior pituitary of experimental mammals, especially rats.

2 Ultrastructure of Cell Organelles

2.1 Endoplasmic Reticulum

The endoplasmic reticulum (ER) of the anterior pituitary cells is mostly rough-surfaced, because they are typical protein-producing cells. As is well known, the rough ER is a complex of saccular elements often called cisternae, on whose outer surface facing the cytoplasmic matrix there are a great number of attached ribonucleoprotein granules called ribosomes (Fig. 1 a). These are often randomly attached to the cisternal surface but are sometimes arranged in small clusters or spirals.

The cisternae of the rough ER of the anterior pituitary cells are either flattened or dilated. In the former case the ER appears to be two membranes because the cavity of the cisternae is collapsed and very narrow. These double membranes are sometimes solitary, as in a section of the surface of collapsed sacs, but they are frequently arranged parallel to each other and called "parallel lamellae." If

Fig. 1. a High magnification view of the rough endoplasmic reticulum (*ER*) of a rat anterior pituitary cell. Ribosomes are attached to the outer surface of ER membranes, which are shown as trilaminar unit membranes. Flocculent filamentous substance within the ER cavity is attached to the inner surface of the ER membrane and thought to be the newly synthesized hormone protein. × 180000. **b** Low magnification of a gonadotropic cell whose cisternae of rough ER are dilated considerably at 2 months after castration. × 4000. **c** Dilated cisternae of rough *ER* of a gonadotropic cell fuse with one another and the cytoplasm containing secretory granules becomes islet-like fragments at 2 months after castration. × 6000. **d** A typical signet-ring cell derived from a gonadotropic cell changed markedly 3 months after castration. A large vacuole (*ER*) situated at the center of the cell is formed by fusion of dilated cisternae of rough ER. The nucleus (*N*) and the Golgi apparatus (*G*) are pushed to the periphery of the cell. × 3000. **e** Dilated cisternae (*ER*) of the castration cell contain a substance which reacted moderately with methenamine silver. This method stained the secretory granules intensely. Glycoprtein in the cisternal cavity is also stained weakly, because its concentration is lower than that in the secretory granules. × 10000. **f** A thyroidectomy cell which may be an altered thyrotropic cell about 10 days after thyroidectomy. Cisternae of the rough ER are dilated and contain round dense granules (intracisternal granules). Lysosomes are also numerous in the cytoplasm (*arrowheads*). × 14000

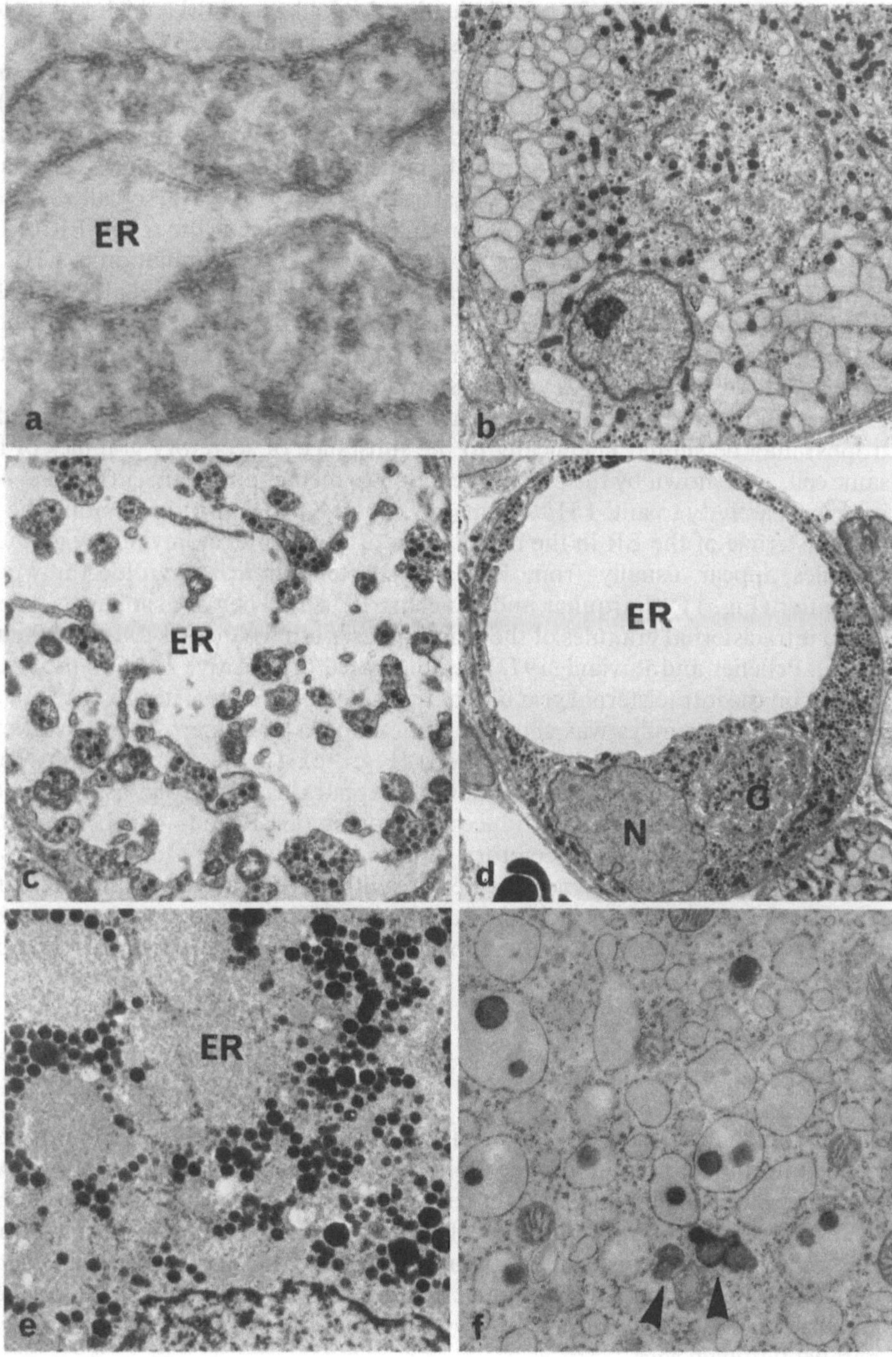
ER
a
b
ER
c
ER
G
N
d
ER
e
f

the lamellae are curved to form a circle, they look like a whorl, which was called Nebenkern by classical electron microscopists (Haguenau 1958).

It is well known that the secretory protein is synthesized at the ribosomes and transferred to the cavity of the ER, where it is stored for a while. The accumulation of such a newly synthesized material in the cisternal cavity may cause dilatation of the cisternae, which is extremely noticeable when the pituitary cells have been stimulated, as for example after castration (Fig. 1 b, c). The so-called signet-ring cells are the result of the fusion of dilated cisternae of the rough ER of the gonadotrophic cells of the anterior pituitary owing to the stimulation of LHRH, whose secretion was enhanced by castration (Fig. 1 d) (Kurosumi et al. 1976). Hypothalamic hormones such as LHRH may increase the rate of synthesis of hormones in the corresponding cell; the newly synthesized hormone is accumulated in the cavity of ER and dilates each cisterna of the ER.

The content of dilated cisternae in the castration cells (hypertrophied gonadotrophs) may be similar chemically to the substance of secretory granules of the same cell, as is shown by the positive reaction to methenamine silver (Fig. 1 e) and immunoreactivity to anti-LHβ serum (Inoue and Kurosumi 1981). Within the dilated cisternae of the ER in the thyrotrophs after thyroidectomy, many globular granules appear usually from 7–10 days after surgical thyroidectomy and thereafter (Fig. 1 f) (Farquhar and Rinehart 1954 b). They are similar to the so-called intracisternal granules of the guinea pig exocrine pancreas found by Palade (1956). Pelletier and Puviani (1973) demonstrated the positive reaction for glycoprotein on the intracisternal granules of thyroidectomy cells. The presence of hormone in these granules was shown by immunoelectron microscopy using anti-TSHβ serum (unpublished observation) (Fig. 9 c). In the thyroidectomy cells, morphological changes occur at the Golgi apparatus and Golgi-associated endoplasmic reticulum which produces lysosomes (GERL) as will be described later; the transportation of secretory substance through the Golgi apparatus or GERL is hindered, and hence the newly synthesized material at the ER is not smoothly transferred into the Golgi apparatus. This hindrance to the passage of secretory substance may cause the formation of intracisternal granules (Kurosumi and Baba 1969).

2.2 Golgi Apparatus and GERL

The granulated anterior pituitary cells which are thought to produce protein or peptide hormones contain a large Golgi apparatus, usually situated in the juxtanuclear region near the center of the cell. The overall shape of the Golgi apparatus of the anterior pituitary cell is almost spherical. The periphery of the apparatus is formed by a shell consisting of parallel flat sacs, whose cavities are almost collapsed. As they resemble the annual rings of a tree, they are also called Golgi lamellae or Golgi stacks. In pituitary cells the Golgi stacks consist of five or six parallel flattened sacs.

The outer (convex) one or two sacs are called cis cisternae (Ehrenreich et al. 1973) and show a strong affinity to osmium, as demonstrated by the fact that they contain black precipitate when treated with osmium tetroxide solution for a relatively long period. The reverse side of the Golgi stack that faces the center of the Golgi area, i.e., the inner or concave side, is called the trans side (Ehrenreich et al. 1973). Electron microscopic histochemistry demonstrates the activity of thiamine pyrophosphatase (TPPase) in two to three such sacs on the trans side (Inoue and Kurosumi 1977; Kurosumi and Inoue 1983).

In many cells in the anterior pituitary, an additional membrane structure which may be also a flattened sac is frequently observed on the trans side of the Golgi apparatus. They are often separated, at a short distance from the Golgi stack. The membrane is usually straight or slightly curved, keeping an almost uniform distance between both walls. It is seen as slightly thicker than the membrane of the Golgi stack and often accompanied by a fuzzy substance coating the outside surface. Because it looks thick and rigid, it is often called "rigid lamellae" (Claude 1970); the structure is considered a part of the GERL named by Novikoff (1964) (Fig. 2a).

As Novikoff and his coworkers (Novikoff et al. 1977) stated, GERL is positive to acid phosphatase activity as well as lysosomes; they thought that the GERL is the production site of lysosomes. The rigid lamellae or GERL of anterior pituitary cells are always observed inside the Golgi stack and often closely associated with a cisterna of the rough ER, although no direct communication with the ER has been observed.

The end of the GERL is always a little swollen and covered with a bristle-like coating which is thought to be clathrin basket. This feature suggests that the end of GERL may be pinched off and thus forms a small coated vesicle, which may be the primary lysosome conveying lytic enzymes to the phagosomes or autophagic vacuoles.

The site of the formation of secretory granules was the object of long debate among cytologists studying the glandular cells. In the early days of electron microscopic cytology of the pituitary gland, Farquhar and Wellings (1957) and Ichikawa (1959) demonstrated that the secretory granules of the anterior pituitary cells are formed in the Golgi apparatus. At that time the presence of a specific membrane structure related to the Golgi apparatus, later named GERL, was not known. Pelletier and Novikoff (1972) argued that the secretory granules of anterior pituitary cells (especially gonadotrophs) may arise from GERL but not from the Golgi apparatus. We have demonstrated, however, that the secretory granules may arise from the trans Golgi stack which is positive to TPPase activity (Fig. 2b) and simultaneously from GERL which is acid phosphatase positive (Fig. 2c) (Inoue and Kurosumi 1977; Kurosumi and Inoue 1983).

As both the GERL and the Golgi stacks may produce the same kinds of secretory granules, the so-called GERL is evidently a part of the Golgi apparatus. This idea was supported by many authors from laboratories other than ours (Fujita and Okamoto 1979; Goldfischer 1982). As the heterogeneity of secretory granules was recently proved in rat corticotrophs (Tanaka and Kurosumi 1986), it is possible that the granules formed from GERL may be different from those formed from Golgi stacks, but the evidence of sorting the secretory substance in GERL or some of the Golgi cisternae has not yet been obtained.

As the site of the formation of secretory granules is known to be the Golgi apparatus and/or GERL, the secretory substance must be transported from the site

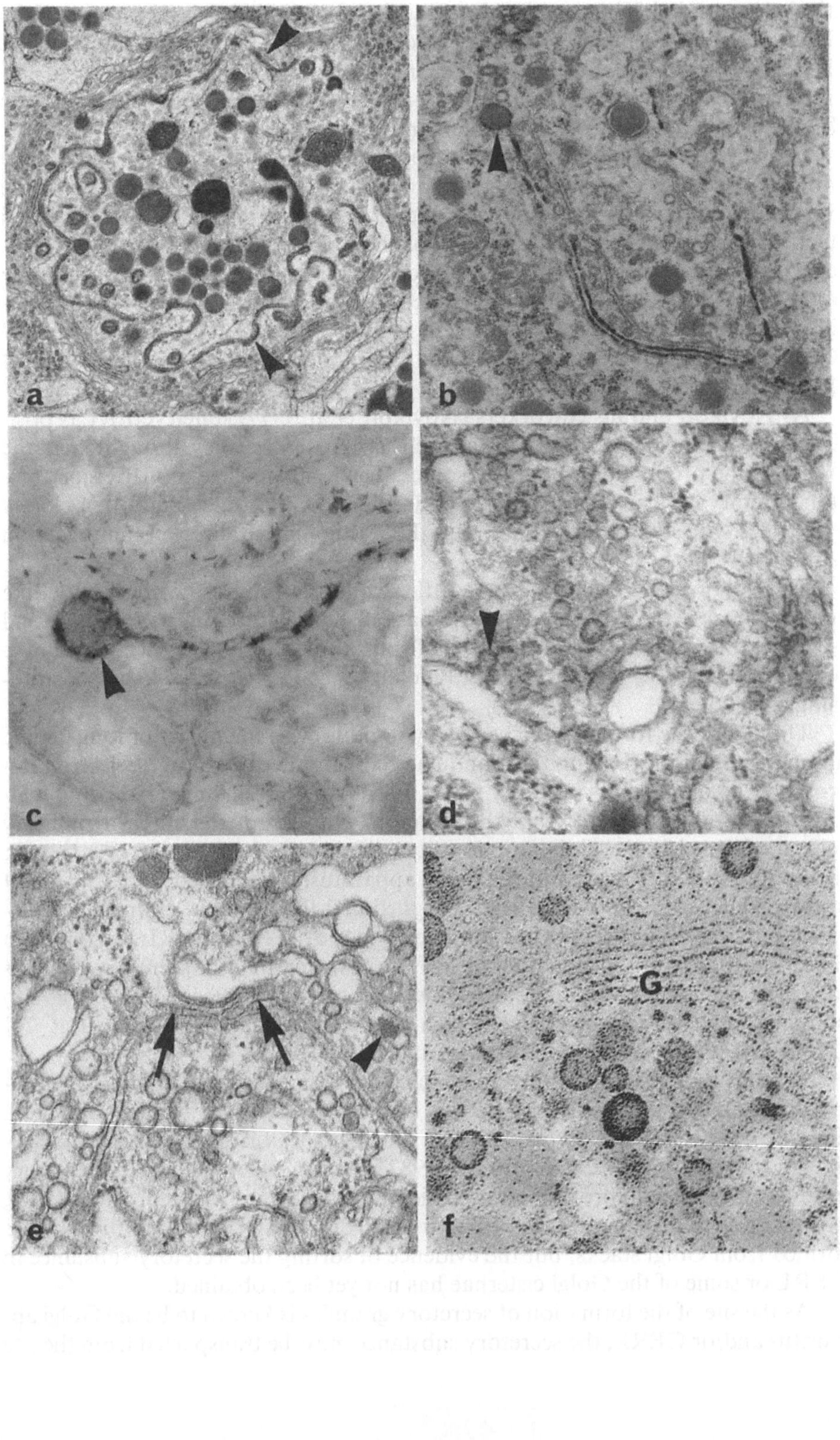
a
b
c
d
G
e
f

of its synthesis, i.e., the rough ER, into the Golgi apparatus or GERL. Zeigel and Dalton (1962) demonstrated a possible mechanism of transportation of secretory material from the ER to the Golgi apparatus in various kinds of protein-secreting cells. Blebbing refers to the formation of blebs (vesicles) containing the newly synthesized secretory material obtained as a result of the pinching off of a small projection extended from the cisternae of the rough ER facing the Golgi apparatus (transitional elements) (Fig. 2d). Such transport vesicles may attach themselves to and fuse with the Golgi cisternae or GERL, and convey the substance into the site of secretory granule formation. The direct communication between an ER and Golgi stacks is sometimes observed both in the neurosecretory cells (Kawabata 1966) and the anterior pituitary cells (Fig. 2e), and is thought to be one of the routes for transportation of secretory substance.

The secretory protein transported from the ER into the Golgi cisternae may be condensed and become small immature granules visible by electron microscopy within the cavity of Golgi and Golgi-related sacs. Neutra and Leblond (1966) demonstrated that the sugar moiety of glycoprotein to be secreted is incorporated into the secretory protein at the Golgi apparatus in mucus-secreting cells using ^{3}H-glucose, but in thyroidectomy cells autoradiographic localization of ^{3}H-fucose was shown both in the ER and the Golgi apparatus (Pelletier and Puviani 1973). Staining with periodic acid methenamine silver is stronger in the cavities of Golgi saccules on the trans side than on the cis side (Fig. 2f). Nucleoside diphosphatase and TPPase activity, which is strongly positive in the trans cisternae, may be correlated with the binding of saccharides and sulfates to the secretory protein in glycoprotein-secreting (basophile) cells of the anterior pituitary.

The sugar moiety of the glycoprotein hormones of the anterior pituitary is well demonstrated by the concanavalin A-iron dextran method (Inoue and Kurosumi 1980). The simple protein hormones such as growth hormone and prolactin pro-

Fig. 2. a An enlarged Golgi area of a gonadotropic cell following administration of cholchicine after castration. Stacked Golgi saccules enclose the cytoplasmic area containing secretory granules and lysosomes. A membrane structure probably coincident with GERL (*arrowheads*) can be seen inside the Golgi area, is filled with a dense material, and formes coated vesicles. × 16000. **b** Electron microscopic histochemistry showing the activity of TPPase in some cisternae on the tran side. The GERL is negative to TPPase. A secretory granule (*arrowhead*) is continuous with the TPPase-positive Golgi cisterna, and the granule is surrounded by dark reaction product for TPPase activity. This feature may suggest the formation of secretory granules at the Golgi stack. × 15000. **c** A secretory granule continuous to GERL (*arrowhead*), both of which are positive to acid phosphatase activity. It suggests that some of the secretory granules may arise from GERL, which has a positive histochemical reaction to acid phosphatase. × 47000. **d** The Golgi area and surroundings of an anterior pituitary cell. A cisterna of rough ER sends out a small projection (*arrowhead*), which may be pinched off and turn into a vesicle carrying newly synthesized protein from ER to the cavity of Golgi apparatus. × 40000. **e** A cisterna of ER with ribosomes attached to the outer surface on one side, but on the other side a long extension (*arrows*) reaches to the Golgi stack, which are probably continuous to each other. A small immature granule is formed within a sac of Golgi apparatus (*arrowhead*). × 36000. **f** The result of EM histochemistry of carbohydrate moiety of the secretory material of a gonadotropic cell after castration. The dark substance indicates that the Golgi cisternae (*G*) reacted with methenamine silver and is intensive at trans cisterna as compared with cis cisterna. Secretory granules are also positive. × 36000 duced by acidophile cells are also positive for carbohydrate in the secretory granules, as detected by the concanavalin A-iron dextran method. As the reaction is strong along the periphery of the granule, it is possible that the carbohydrates related to membrane particles of the limiting membrane of secretory granules may be reacted, but not the hormones themselves. However, the incorporation of carbohydrates into the protein related to the receptors to be localized in the cell surface membrane may also occur at the trans cisternae of the Golgi apparatus.

2.3 Discharge of Secretory Substance

Mature secretory granules may leave the Golgi apparatus and move toward the cell surface for discharge. For the movement of secretory granules, some of the cytoskeleton such as microfilaments and microtubules may participate as suggested by Lacy et al. (1968) for the secretion of insulin from the beta cells of the pancreatic islet. No direct evidence of participation by the cytoskeleton in the transportation and discharge of secretory granules has been reported in the anterior pituitary.

Secretory granules approach the cell surface and attach themselves to the inside surface of the plasma membrane. The fusion of the limiting membrane of the secretory granule with the plasma membrane of the cell surface occurs at the contact point of the two membranes (Fig. 3a), and then the granule may slip out (Fig. 3b). This mechanism of extrusion of secretory granules is now widely referred to as exocytosis but was first observed in the anterior pituitary by Ichikawa (1959), who described it as merocrine secretion. The same feature of granule discharge was reported by Farquhar (1961) with more convincing illustrations, but she did not propose any special terms designating this mechanism of granule discharge. Kurosumi (1961) published a new classification of mechanisms of secretion discharge and classified each type of secretion discharge with noncommittal numerals rather than words. In this classification, the mode of extrusion now known as exocytosis was type IV. For the sake of explanation, however, the author described it as opening mechanism, but later used the term "eruptocrine secretion"

Fig. 3. a A secretory granule of GH cell just attached to the surface plasma membrane showing the fusion of two membranes at the attachment site (*arrowhead*). × 98000. **b** Fission of fused membrane occurs at the location of granule attachment to the surface plasma membrane (*arrows*). The granule of secretory substance is ready to be extruded. × 98000. **c** Multigranular exocytosis (*arrows*) of an ACTH cell after injection of CRF. × 73000. **d** Exocytosis of secretory granules (*arrows*) associated with endocytosis (*arrowhead*) at the surface of a typical gonadotropic cell a few days after castration. Secretory granules during exocytosis are slightly dissolved. × 31000. **e** A similar exocytosis associated with endocytosis in a gonadotropic cell a few days after castration. One large pit which may be the sac of secretory granule just after opening for exocytosis already contains no material (*crossed arrow*), suggesting the liquefaction of secretory substance and its rapid release. Another pit contains remnants of secretory granule (*arrow*) and a small pit suggestive of endocytosis is attached (*arrowhead*) to the exocytotic pit. × 90000. **f** Four round bodies gathered at the Golgi area of a GH cell: two contain moderate dark substance, small granules, and a myelin figure, suggesting that they are lysosomes (*L*), and the other two contain small vesicles and are considered to be multivesicular bodies (*M*). × 30000

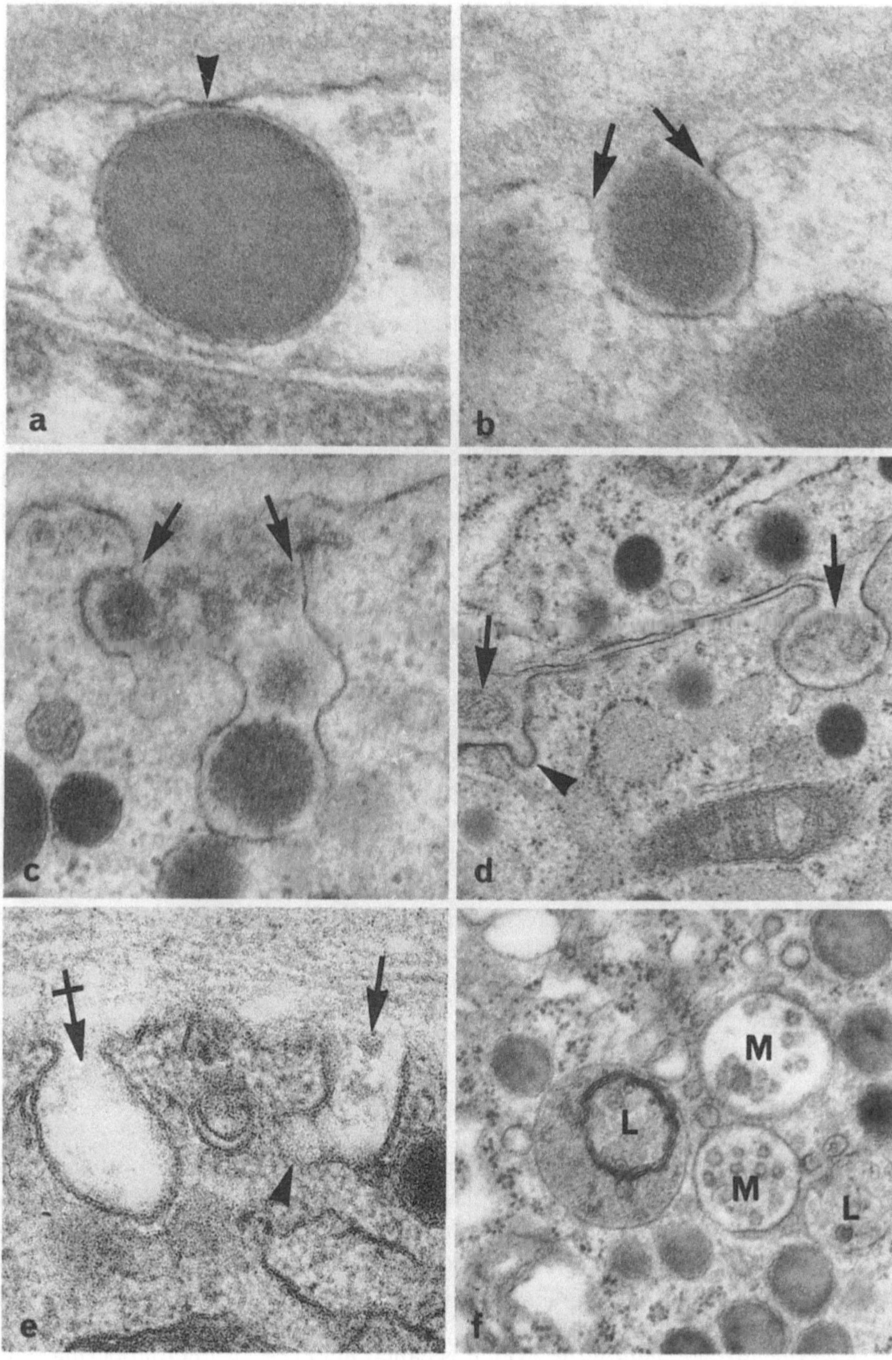
a
b
c
d
e
M
L
M
L
f

(Kurosumi 1965) in accordance with the proposal of the late Professor Seki. The most popular term – exocytosis – was proposed by de Duve (1963) as the antonym of endocytosis, which was also proposed by him as a general term covering phagocytosis and pinocytosis.

The participation of calcium ions in the granule discharge was suggested by Douglas (1968) and Kanno (1980). Morphological evidence of calcium involvement in exocytosis was provided by Fujita et al. (1983) through the use of calcium ionophore A23187 applied to the rat anterior pituitary tissue in vitro. Application of this agent caused a massive extrusion of secretory granules, which was called "multigranular exocytosis" by the authors (Fujita et al. 1983; Kurihara and Fujita 1984). In vivo application of a hypothalamic releasing factor such as corticotropin-releasing factor (CRF) may cause granule extrusion by a mechanism similar to the so-called multigranular exocytosis (Fig. 3c).

Osamura et al. (1982) studied prolactin (PRL) cells stimulated by castration and estrogen treatment, using immunocytochemical procedures by the method of preembedding peroxidase-labeled antibody, and postulated the secretion mechanism of PRL through two different pathways: the long pathway which is the ordinary mechanism of protein secretion, i.e., ER → Golgi → secretory granule → exocytosis, and the short pathway seen in the accelerated secretion which is the direct discharge from the ER to the extracellular space. They showed that the ER cisternae including immunoreactive PRL extended vertical projections toward the plasma membrane and almost contacted it.

If such a direct pathway from the ER to the extracellular milieu exists in every kind of anterior pituitary cell, how can we explain the occurrence of strong dilatation of the ER after castration, thyroidectomy, and adrenalectomy? If there is such a direct way, an extraordinarily enhanced production of secretory substance would flow out it and the cisternae would not dilate to form the signet-ring cell even in an extreme case. Therefore, it must be considered that the PRL cell is an exceptional case which is provided with a direct pathway for secretion discharge other than the exocytosis of secretory granules, as suggested by the fact that this cell type does not show strong dilatation of the ER after stimulation of hormone synthesis. But this possibility may become more convincing if more direct morphological evidence can be obtained.

2.4 Retrieval and Recycling of Membrane Material

Granule discharge by exocytosis is accompanied by the relocation of the membrane covering the discharging granules to the cell surface. During the period of recovery following exocytosis, membrane must be removed from the cell surface to compensate for the membrane added by exocytosis, in order to keep the cell surface area constant. The removal of a part of the cell surface membrane may take place through the shedding of a small part of the cytoplasm covered by a piece of the surface membrane, which is called apocrine or microapocrine secretion. This is the case observed in the mammary gland (Kurosumi et al. 1968). Another mechanism of membrane removal is the retrieval of a part of the membrane through the pinching off of the surface membrane invaginated into the cytoplasm.

This is known as micropinocytosis (a form of endocytosis) and is observed in the anterior pituitary (Fig. 3 d, e).

Farquhar et al. (1975) proved membrane retrieval by demonstrating the uptake of horseradish peroxidase (HRP) by dissociated anterior pituitary cells incubated in a medium containing HRP, and stimulated exocytosis by adding cyclic adenosine monophosphate (AMP). After the stimulation much more HRP was found within the cell than in an unstimulated control. The HRP was seen in vesicles and tubules near the cell surface membrane, and also found in the lysosomes and the innermost sacs of the Golgi apparatus.

We observed that the small surface pits and small vesicles suggesting micropinocytosis are markedly increased at the cell surface of gonadotrophs a few days after castration (Kurosumi and Inoue 1980). On the lateral surface of an opening vacuole containing fragments of the secretory granule in the course of exocytosis, a small pit with a bristle-like coating was associated (Fig. 3 d, e). This feature may indicate the occurrence of micropinocytosis on the wall of a discharging secretory granule.

Ishimura et al. (1980) observed the images of exocytosis and endocytosis of anterior pituitary cells by the freeze-fracture method. According to them, the P face of the plasma membrane is devoid of membrane particles at the small elevation covering a secretory granule coming into contact with the plasma membrane. The limiting membrane of the secretory granule is also particle-free at the part facing the plasma membrane. Therefore, the fusion of the two membranes may occur at the particle-free part. In contrast, the depressed region of the cell surface formed from the secretory granule membrane after exocytosis is rich in membrane particles. Membrane retrieval by endocytosis commences in the particle-rich depressed region of the plasma membrane. When observed in thin sections, the endocytotic pit for membrane retrieval is always covered with a bristle-like coating which is now known to be a basket-like structure of clathrin (Aggeler and Werb 1982). It is speculated that the clathrin molecules forming the coating on the cytoplasmic surface of the plasma membrane may be attached either directly or indirectly to the membrane particles gathered at the deep part of the surface depression formed after exocytosis.

The fate of the retrieved membrane material was investigated by the use of cationized ferritin (CF) as a marker of the surface plasma membrane (Farquhar 1978). Within 15–60 min after adding CF into the incubation medium of dissociated pituitary cells, CF was found within stacked Golgi cisternae, especially – and most remarkably – in the trans-most cisterna, as well as within forming secretory granules, rigid lamellae corresponding to GERL, and lysosomes. The CF molecules are increasingly deposited within the secretory granules of PRL cells ranging from the simplest type to the more complicated irregular type, as time elapses after application of CF. The internalization of plasma membrane material not only compensates for the membrane added during exocytosis, but the membrane substance comes to the lysosome where it is digested or comes to the Golgi stacks where it undergoes recycling, taking part of the membrane surrounding the new secretory granules.

2.5 Lysosomes and Multivesicular Bodies

Every cell contains lysosomes in greater or lesser number, which are believed to be produced from GERL or the trans-most cisterna of the Golgi apparatus and are positive to acid phosphatase activity as shown by electron microscopic histochemistry. Acid phosphatase (AcPase) is a marker enzyme of the lysosome, which may contain more than a dozen different kinds of lytic enzyme with acidic optimal pH. These enzymes may be synthesized at the rough ER and are then transferred to the Golgi apparatus and/or GERL, just as in the synthesis and transport of proteinous hormones in the anterior pituitary cells.

The multivesicular bodies have a unique structure, which is a vacuole containing many small vesicles (Fig. 3 f). Friend and Farquhar (1967) observed the multivesicular body of the vas deferens epithelial cells and showed the positive reaction to the AcPase activity on the matrix of the multivesicular body, but not in the small inside vesicles. As the multivesicular body is positive to the marker enzyme for lysosomes, it seems to be one of the variety of lysosomes.

The lysosomes and multivesicular bodies of the anterior pituitary cells may act in two ways: they may act in the normal physiological course of hormone production, but they may also increase in number and size in certain pathological conditions, such as after the removal of the target organ, e.g., castration and thyroidectomy. An example of the first activity of lysosomes was demonstrated by Smith and Farquhar (1966) in the PRL cells after weaning of young. The overproduction of PRL owing to lactation is suddenly followed by cessation of lactation. Therefore, the PRL-containing secretory granules are no longer useful for the mother rat. The secretory granules may be taken into lysosomes or multivesicular bodies, and digested there. This mechanism is one of autophagy but the substance which is digested is the secretory material (hormone in this case) and therefore this mechanism is called crinophagy (Farquhar 1977). We observed the increase of crinophagy in growth hormone (GH) cells after adrenalectomy or CRF treatment. Dense round granules supposed to be secretory granules were contained in the multivesicular bodies or in a large dense body probably corresponding to the lysosome. The idea that the granules are really secretory granules containing GH

Fig. 4. a The result of immunoelectron microscopy using anti-GH serum which is labeled with colloidal gold. Secretory granules and a dark substance within a multivesicular body (*arrow*) are positive to immunoreaction, indicating the presence of GH. The result shows the dark substance in the multivesicular body is nothing but a secretory granule in the couse of crinophagy. × 51 000. **b** A secretory granule positive to immunoelectron microscopic detection of GH is contained in a lysosome (*arrow*) also showing crinophagy. × 41 000. **c** A complex of peculiar membrane structures filled with a dark substance found near the Golgi apparatus of a thyroidectomy cell. × 14000. **d** Dark substance filling the peculiar membrane structures of a thyroidectomy cell is positive to acid phosphatase, as shown by electron microscopic histochemistry. This structure may correspond to GERL, producing a number of primary lysosomes. × 16000. **e** Mitotic cell division of PRL cell of a normal female pituitary. Chromosomes are now dividing, suggestive of the anaphase. A secretory granule indicated by an *arrow* is shown in Fig. 4f in a higher magnification. × 6000. **f** A secretory granule is expelled from the cell by exocytosis (*arrow*). Chromosome (*CH*) is seen in the upper part of the micrograph, showing that the exocytosis of secretory granule may occur during mitosis. × 33000

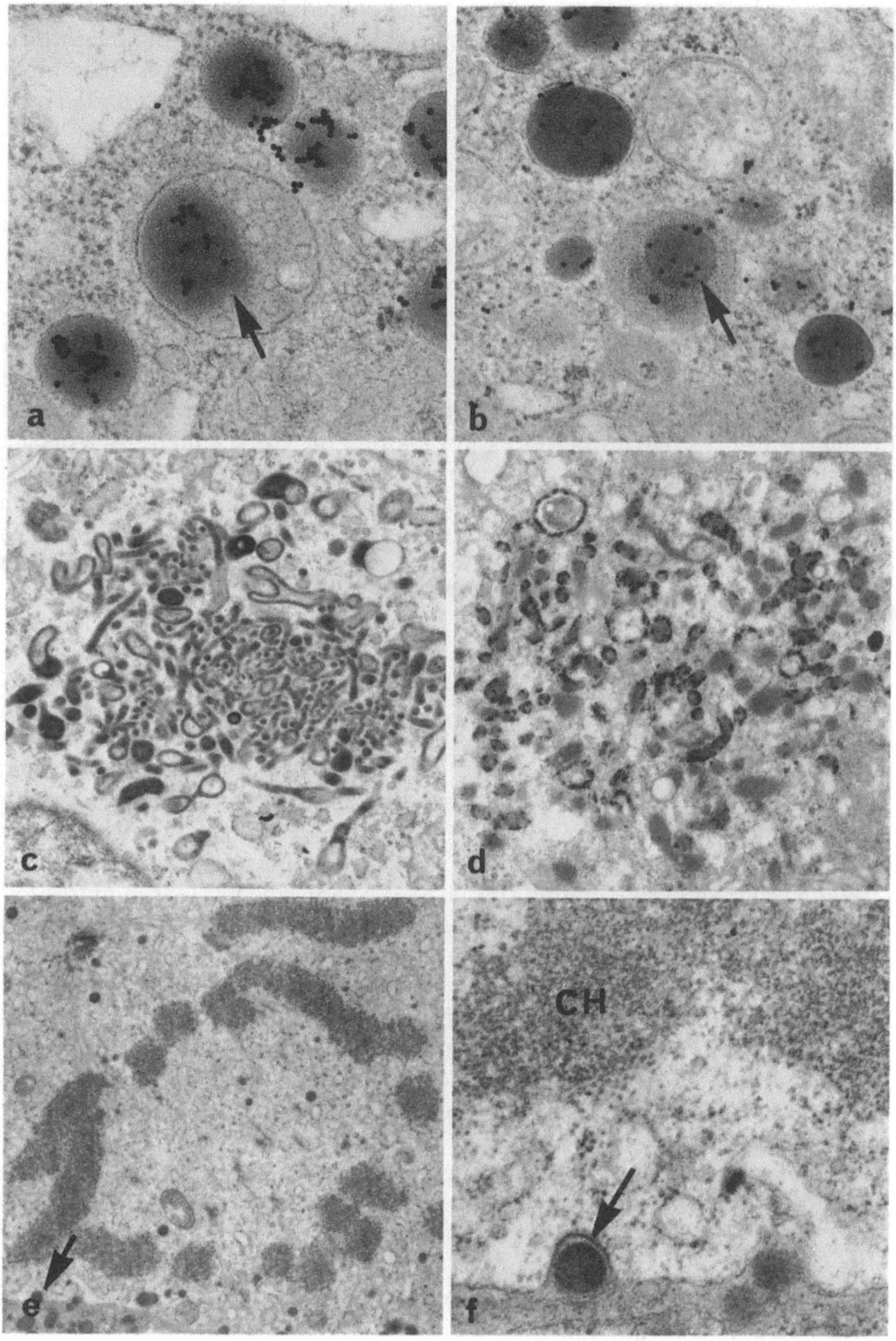
a
b
c
d
e
CH
f

was proved by immunocytochemistry, using colloidal gold after anti-GH antibody had been applied to the section. The gold particles indicating the localization of GH were seen on the granules either within the multivesicular bodies (Fig. 4a) or within the dense bodies (lysosomes) (Fig. 4b) as well as on the intact secretory granules of the same cell.

The second function of lysosomes is concerned with pathological degeneration or necrosis of cells. Extraordinary stimulation may cause the hypertrophy and hyperfunction of cells such as those after the extirpation of target organs that is normally stimulated by the hormone secreted from the cells which are changed by the experiment. After prolonged stimulation the cells which were first hypertrophied become degenerate, probably owing to the overload. The thyroid-stimulating cells (TSH cells) showed about two weeks after thyroidectomy a degenerating change in the Golgi apparatus and ER as well as an increase in lysosomes. The cisternae of the rough ER dilate to an extreme degree and contain round dark granules (intracisternal granules). Near or within the Golgi apparatus, a peculiar heap of membrane-bound structure is observed, whose interior is filled with dark substance (Fig. 4c). The cytochemical reaction for AcPase activity is positive in the dark material within this membrane-bound structure (Fig. 4d), which may correpond to GERL from which the primary lysosomes may arise (Kurosumi 1972).

2.6 Mitosis of Anterior Pituitary Cells

It is well known that the anterior pituitary cells may increase in number as well as in size after the removal of the target organ, such as the gonads. The hypertrophy of the gonadotropic cells was clearly shown after castration, owing mainly to the strong dilatation of the rough ER (Kurosumi et al. 1976). The proliferation of gonadotrope was also shown by the use of immunohistochemistry at light microscopic level and cell counting (Inoue et al. 1985). The increase in mitoses was also shown in gonadotrope after castration (Inoue et al. 1985). As the luteinizing hormone (LH)-immunoreactive cells undergo mitosis after castration, the proliferation of anterior pituitary cells following the removal of target organs is evident. Heidenhain (1911) assumed that fully differentiated cells such as endocrine cells of the anterior pituitary could not divide any more, but this is not the case in the anterior pituitary.

Electron microscopy clearly showed the production of secretory granules at the Golgi apparatus as well as their release out of the cells during mitosis (Kurosumi 1971, 1979) (Fig. 4e, f). The great increase in the number of gonadotropic cells after castration may well be caused by mitotic cell division.

3 Identification of Cells Containing Each Pituitary Hormone by Immunoelectron Microscopy

The function of each cell type of anterior pituitary was suggested partly by experimental cytological techniques such as extirpation of target organs, but the precise hormone producers could not be identified until the discovery of the technique of immunocytochemistry. Such a technique applied to morphology was first performed in light microscopy, i.e., immunofluorescence microscopy and immunoenzyme histochemistry. The latter technique can also be applied to electron microscopy (Nakane 1970).

Electron microscopic application of immunocytochemistry is called immunoelectron microscopy (IEM) and there are three groups of IEM technique: preembedding, postembedding and nonembedding methods. In the preembedding method, the immunological reaction takes place before embedding for ultramicrotomy. Only immunoenzyme cytochemistry can be used for the preembedding method. Although mild fixation with aldehydes is usually carried out before immunological treatment, postfixation with osmium is often applied to the specimen after immunocytochemical reaction. The postfixation makes the preservation of cell surface and intracellular membranes generally good, but treatment by freezing and thawing or with certain detergents such as triton X or saponin slightly destroys the cell membrane or membranes of secretory granules in order for antibody molecules to penetrate into the cell and into the secretory granules. Owing to such treatments destroying the membrane of secretory granules, the antigens (hormones in our case) may diffuse out into the cytoplasmic matrix, and antigen-antibody reaction may occur in the cytoplasmic matrix where no antigen exists in vivo. The result is the so-called false positive reaction which is often seen in the preembedding method.

The postembedding method is a technique in which the antigen-antibody reaction takes place on the surface of a thin section which was already processed through fixation, either chemical or physical (freezing), then dehydrated and embedded in resin. The first antibody applied to the section is the specific antibody against each of the anterior pituitary hormones or their fragments, and the second antibody or protein A specifically binds to the primary antigen. The second antibody or protein A is labeled with either enzyme (horseradish peroxidase) or heavy metal compounds (ferritin or colloidal gold). The reaction products are usually seen only on the secretory granules but not on the cavity of the Golgi apparatus and the rough ER where the antigen (hormone) must be present. This result is the so-called false negative reaction. It may be minimized, however, by the selection of an embedding medium such as a low-temperature polymerizing resin and the use of fixation such as rapid freeze substitution.

The third method is nonembedding IEM, i.e., immunocytochemistry applied on the thin sections made by cryomicrotomy without embedding. This method may be sensitive and the reaction on the material in the cavity of the Golgi apparatus and ER may easily occur, but the results of this method have not been reported as far as the anterior pituitary is concerned because the technique of sectioning is not easy and the demonstration of ultrastructure is not satisfactory.

In this review the technique of postembedding IEM, especially the forms using ferritin or colloidal gold, is preferred because its results are the most reliable and the reaction is very clear, as shown with particles of heavy metal compounds.

There are two major categories of anterior pituitary cells: cells with secretory granules (granulated cells) and nongranulated (agranular) cells. The latter are the chromophobe cells in light microscopy and are called folliculostellate cells because of their ultratructural morphology, i.e., they are often either star-shaped or surrounding a follicle. They do not contain secretory granules and are thus not thought to produce hormone.

The granulated cells are subclassified into two groups: cells producing simple protein hormones, corresponding to the acidophiles in light microscopy, and cells producing glycoprotein hormones, corresponding to the basophiles. Several years ago the theory of "one cell, one hormone" prevailed, but recent advances in immunocytochemistry have demonstrated that some cells produce more than two different kinds of hormones, and also that several morphologically different types of cells may produce the same hormone. Therefore, the "one cell, one hormone" theory no longer survives.

3.1 Growth-Hormone-Producing Cell

Comparison of light and electron micrographs indicates that the cells containing large round granules about 350 nm in average diameter correspond to one type of acidophile. As this type of cell could not be observed in the anterior pituitary of the dwarf mouse, it is thought to produce GH (Kurosumi 1974). Two types of acidophile cells were noticed in early studies with electron microscopes: one containing round secretory granules was thought to contain GH, and the other containing large irregularly shaped granules was thought to produce PRL (Hedinger and Farquhar 1957).

When the electron microscopic fields of sections from the rat anterior pituitary were carefully observed, it was believed that some cells containing granules of about the same size as that of typical GH cells might also contain some small round granules intermingled with the typical large granules (Kurosumi 1968). Yoshimura et al. (1973, 1974) described three types of rat pituitary acidophile cells: the first contains large secretory granules about 350 nm in diameter corresponding to the classical GH cells first described by Hedinger and Farquhar (1957); the second contains a mixture of large (350 nm) and small (150 nm) granules which was noticed by us as mentioned above (Kurosumi 1968); and the third contains only small granules (130 nm). However, since the authors did not use the technique of immunocytochemistry, these three types could not be identified as true GH cells. Yoshimura et al. (1974) considered that the first two types might secrete GH but the third type might produce adrenocorticotropic hormone (ACTH) instead of GH.

Immunoelectron microscopic techniques have developed greatly in recent times, and identification anti-GH serum has made it possible to identify GH cells under the electron microscope. Three morphologically different types of cells reactive to anti-GH serum were observed and classified as types I, II, and III ac-

→

Fig. 5. a Part of a type I GH cell with large granules. Colloidal gold particles bound to anti-GH are almost exclusively observed on the secretory granules. × 11 000. **b** A type II GH cell immunoreactive with anti-GH serum contains mixed large and small secretory granules. × 11 000. **c** A type III GH cell containing a large number of secretory granules immunoreactive with colloidal gold bound to anti-GH serum. Almost all secretory granules are small in size. × 11 000

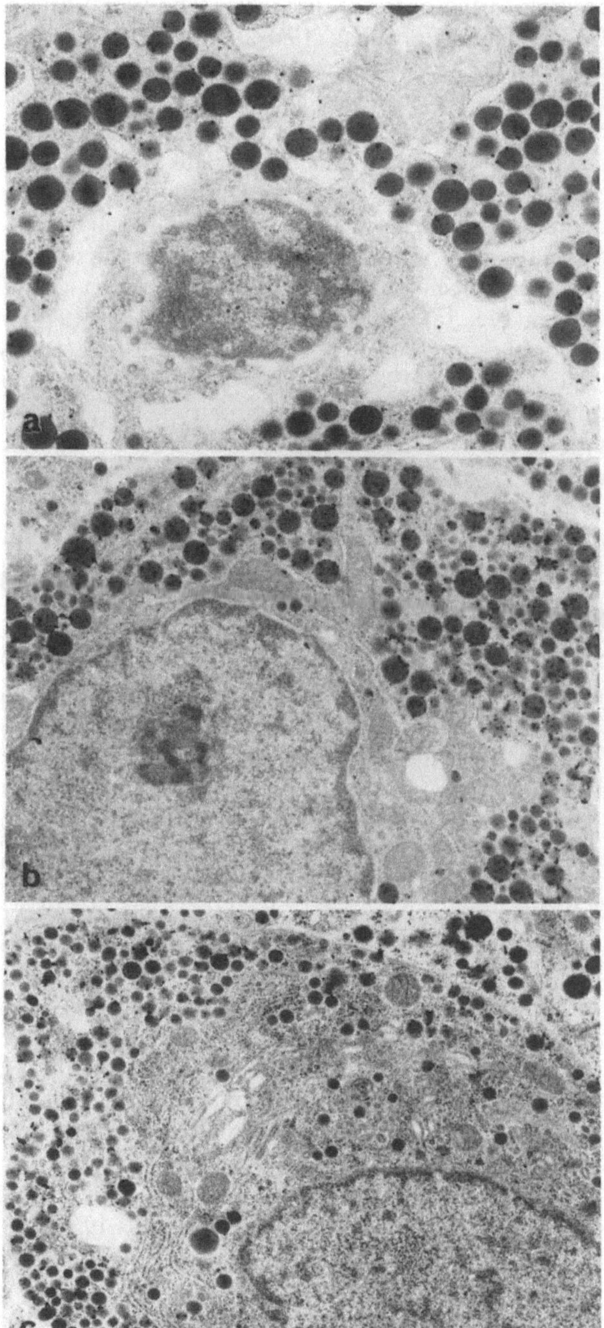
a
b
c

cording to the order of discovery. The type I GH cell, originally discovered by Hedinger and Farquhar (1957), contains large spherical granules about 350 nm in diameter (Fig. 5a). This cell contains not only secretory granules but also well-developed cell organelles. The rough ER consists of parallel lamellae of cisternae, mainly arranged along the surface of the cell. The Golgi apparatus is also well developed and situated near the nucleus. Mitochondria are round and well developed. The matrix of the mitochondria is usually seen clear. The type II GH cell contains both large (350 nm) and small (150 nm) secretory granules. These two populations of granules of different sizes are completely intermingled (Fig. 5b). The type III cell contains only small secretory granules about 150–200 nm in diameter (Fig. 5c). Although Yoshimura et al. (1974) argued that the small granule acidophiles might secrete ACTH but not GH, the type III GH cell found by us (Kurosumi et al. 1984) was clearly shown by immunoelectron microscopy to contain GH. Neither the granules of type III GH cells nor the small granules of type II GH cells reacted to anti-ACTH serum, as seen by immunoelectron microscopy.

Immunoreactive GH cells in adult male rats are mostly type I (about 67%) and type II (25%), while type III cells total only about 6%. Adult female rats contain slightly more type II cells (53%) than type I cells (40%), but almost the same percentage of type III cells as the male. In neonatal rats the rate of occurrence of type I and II cells is almost the same as in adults, but the percentage of type III cells is higher in both males (16%) and females (24%). It is possible that type III cells are an immature form of GH cells which may develop into type I cells. Type II cells may intermediate between type III and I cells, being situated at the halfway point in the development of GH cells.

3.2 Prolactin-Producing Cell

Light microscopy as well as electron microscopy of rather low magnification can recognize two morphologically different types of acidophile, one of which is the GH cell mentioned above; the other is the PRL cell. Using light microscopy, PRL cells of rodents are known to stain strongly with azocarmine or erythrosine. Using electron microscopy, Hedinger and Farquhar (1957) differentiated PRL cells from GH cells, the former containing irregularly shaped or oval (polymorphic) secretory granules larger than those of GH cells. Most PRL cell granules have a maximum diameter of about 700 nm. Such cells are very frequently observed in the female anterior pituitary and are thought to produce PRL since they are found exclusively in the graft transplanted under the renal capsule, where only PRL-producing cells can survive (Rennels 1962).

Fig. 6. a Part of a typical (classical) PRL cell which is also called type I PRL cell. The secretory granules are large and irregular in shape, and shown immunolabeling with colloidal gold bound to anti-PRL serum. ×25000. **b** Part of a type II PRL cell containing variable-sized round secretory granules. Immunolabeling is shown on these granules. × 25000. **c** Part of a type III PRL cell with immunoreactive small secretory granules arranged along the cell periphery. × 20000

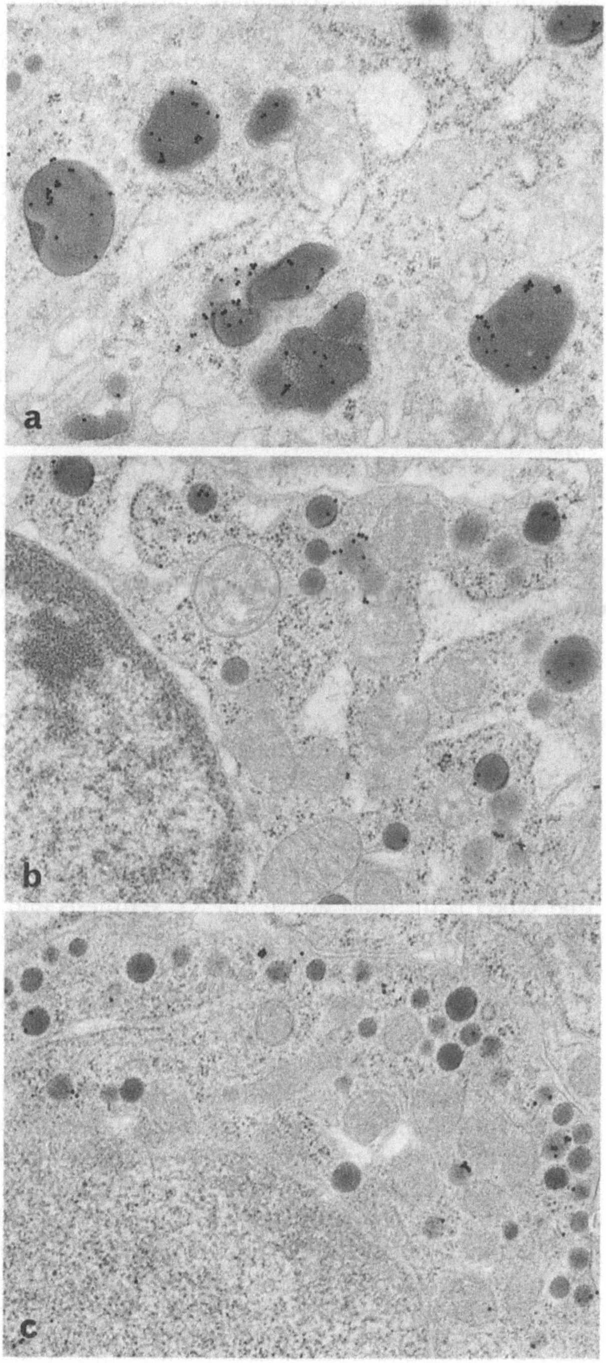
a
b
c

Probable PRL cells containing only round secretory granules were reported in many different species of mammals, such as the bovine (Dacheux and Dubois 1976), the guinea pig (Beauvillain et al. 1977), the ovine (Parry et al. 1978), the porcine (Dacheux 1980), the rat (Nogami and Yoshimura 1980), and the mouse (Harigaya et al. 1983).

IEM using anti-PRL serum reveals atypical PRL cells with round secretory granules together with the classical typical PRL cells containing polymorphic large secretory granules, which were called type I PRL cells (Fig. 6a). The atypical PRL cells found by Nogami and Yoshimura (1980), which may be called type II cells, contain round secretory granules about 150–250 nm in diameter (Fig. 6b). Besides these round-granule PRL cells, we can observe PRL immunoreactive cells containing small round granules about 100 nm in diameter. This cell type is called the type III PRL cell and is the least common of the different types of immunoreactive PRL cells of the adult rat anterior pituitary, but accounts for about 90% of PRL cells in neonatal infants (Fig. 6c).

The order of the numerals used for the names of cell types described here represents the order of discovery. This is the rule for our nomenclature of anterior pituitary cells. Type I is the earliest-found cell type in each hormone producer of the anterior pituitary. This nomenclature is different from that used by Nogami (1984) and Harigaya et al. (1983). The type II cell, which is the same as in their nomenclature, contains round granules of variable sizes and is frequently observed in adult male rats, but is rare in the female anterior pituitary, in which type I accounts for more than 90% of PRL cells. In the adult male pituitary, type II cells total 56.9%, type I cells 36.5%, and type III cells 6.1%. No type III cells are found in the adult female pituitary (Koyama 1985).

Both type I and II cells contain well-developed organelles such as the rough ER and the Golgi apparatus, but these are usually more developed in type I cells than in type II cells. In the former cell type, the cisternae of the rough ER form a thick stack of parallel lamellae on one side of the cell. Therefore, it is conjectured that type I is the most active in PRL synthesis and secretion. Type II PRL cells are often crescent-shaped, embracing a gonadotropic cell. Both type I and II cells frequently show the features of secretory granule extrusion by the mechanism of exocytosis, but the PRL cells of type III do not show granule extrusion.

Type III cells are usually small and oval in shape, and contain a rather large nucleus, so that the nucleocytoplasmic ratio is very large. A small amount of cytoplasm contains moderately developed cell organelles and small secretory granules about 100 nm in diameter arranged along the cell periphery.

Nogami (1984) used a technique of superimposition of an ultrathin section stained with ordinary electron staining and the adjacent thick section stained with an immunohistochemical method of peroxidase-labeled antibody against PRL for light microscopy. He demonstrated that type I (classical PRL cells) increased in both in male and female rats after estrogen (E2) treatment. Testosterone (TP) administration increased type II cells in males. Nogami (1984) applied steroids to the intact rats, and Koyama (1985) tried the same experiments, applying E2 and TP to castrated rats of both sexes. Koyama's method of immunocytochemistry is direct IEM with the colloidal gold-antibody method. The results of his E2 administration were almost the same as those obtained by Nogami (1984), namely that E2 stimulated PRL cells and increased type I cells (classical PRL cells) remarkably, but that the effects of TP are complex. The administration of TP to the ovariectomized female rats increased type I cells, as did the administration of E2, though to a lesser extent. The effect of TP on castrated male rats is not so remarkable, but TP inhibited the decrease of type I cells due to orchidectomy. The similarity of TP to E2 in the stimulation of the PRL cells of spayed rats may be the result of conversion of TP into estrogen by aromatase existing in organs other than the ovary (Koyama 1985).

Observation of various stages of development indicated that the type III cells of our classification (Koyama and Kurosumi 1986) (small-granule PRL cells) are undoubtedly at an immature stage, which may develop into type II cells with medium-sized round granules and type I cells with polymorphic large granules as the effects of endogenous or exogenous estrogens. Like estrogen, thyroxine also has the effect of stimulating the PRL cells, as demonstrated by Shiino and Yamauchi (1985). Thyroidectomy performed on neonatal rats causes infrequent small round secretory granules to appear in the PRL cells, but administration of thyroxine restores the size and number of PRL granules.

The coexistence of GH and PRL either in the same cell or in the same granule has been postulated by some authors. Ishikawa et al. (1983) studied the human pituitary adenoma secreting both GH and PRL, i.e., a mixed tumor. They used superimposition of two adjacent sections of the pituitary adenoma cells, one stained for GH and the other for PRL by the enzyme-antibody method for electron microscopy. As a result, they found some secretory granules containing both GH and PRL. With a more advanced IEM technique, i.e., the double labeling method using protein A-gold particles of different sizes, Fumagalli and Zanini (1985) demonstrated the coexistence of GH and PRL in the same cells and in the same granules of the anterior pituitary of the cow. Our study using a similar double labeling technique of colloidal gold-antibody method on the rat pituitary shows that it is difficult to prove the coexistence of two hormones in the same cell or in the same granule. Sometimes two different-sized particles, one indicating GH and the other PRL, can be seen in the same granule, but the mixing of different-sized gold particles does not prevail against the nonspecific reaction such as the occurrence of particles on the nucleus, mitochondria, or extracellular space.

3.3 Pro-opiomelanocortin-Producing Cell

The recent progress of molecular biology has clarified the molecular structure of the precursor of coticotropin (ACTH), which is called pro-opiomelanocortin (POMC) because this large protein may give rise to the opioid substance endorphin and to melanophore-stimulating hormone (MSH), besides ACTH. Now immunocytochemical morphology, especially at the electron microscopic level, demonstrates that the anterior pituitary cells thought to produce ACTH also produce α- and β-endorphins, α-MSH and other peptide fragments of POMC (Pelletier et al. 1977; Guy et al. 1980; Dacheux 1981; Cantin et al. 1983; Tanaka and Kurosumi 1986).

Since the earliest research into the anterior pituitary with the electron microscope (Rinehart and Farquhar 1953), the identification of pituitary cells which might produce ACTH has been one of the most difficult problems of the cytophysiology of the pituitary gland, and there have been a great number of repeated debates. In 1957, Farquhar ascribed the secretion of ACTH to the follicular cells which contain no secretory granules. But after about a quarter century, it is evident that this type of cell, often called a folliculostellate cell, contains neither ACTH nor any other hormones of the anterior pituitary but S-100 protein, which is said to be specific to the neuroglia (Shirasawa et al. 1983). Rennels (1964) carefully searched with the electron microscope for any changes in the anterior pituitary after the scalding of an animal, which had been known to cause a marked discharge of ACTH stored in the anterior pituitary. But he failed to find a marked loss of secretory granules from any type of granulated pituitary cells, and the most obvious change he found was an accumulation of lipid droplets in the nongranular follicular cells that Farquhar (1957) had assumed to be the ACTH producer.

Siperstein and Allison (1965) carried out an electron microscopic study by which the morphological changes in the pituitary cells after adrenalectomy were clearly demonstrated. The so-called adrenalectomy cells which appeared in the anterior pituitary of a bilaterally adrenalectomized rat were cells with an irregular outline which had a tendency to extend cytoplasmic projections or to encompass a neighboring cell partially or wholly. They contained secretory granules about 200 nm in diameter. The authors, however, could not detect the adrenalectomy-type cell or any other cell types which might be morphologically transitional from it in the normal rat anterior pituitary.

Kurosumi and Kobayashi (1966) demonstrated a cell type not previously described in the normal rat pituitary, and showed the structural transition of this cell to a cell which undergoes a marked change after adrenalectomy. It is irrgular, polyhedral, elongate, or retort-like in shape, and has several processes often reaching to the wall of the blood capillaries. The secretory granules are 150–200 nm in diameter and their electron density varies from dense granules to empty-like vesicles, often provided with a dense core. Although this cell type undergoes a striking change in structure after adrenalectomy, it does not show morphological changes after either thyroidectomy or gonadectomy (Kurosumi and Oota 1966). Therefore, we considered that it is the cell responsible for ACTH secretion.

Rennels and Shiino (1968) observed the combined effects of unilateral adrenalectomy and treatment with cyanoketone on the rat anterior pituitary. Their results on presumptive corticotrops were consistent with our previous findings (Kurosumi and Kobayashi 1966; Kurosumi and Oota 1966). Yamada and Yamashita (1967) as well as Costoff (1973) agreed with our concept concerning the morphology of ACTH-producing cells.

On the other hand, Nakayama et al. (1969) reported another type of anterior pituitary cell as the producer of ACTH in rats. This second type of presumptive ACTH cell is characterized by an irregular stellate shape extending may long slender proceses with a low electron density of the cytoplasm, and by the secretory granules arranged along the cell periphery. Many successive researchers agreed with the hypothesis of Nakayama et al. (1969) and argues that the cell type reported by them is the only producer of ACTH in the rat anterior pituitary (Siperstein and Miller 1970; Pelletier and Racadot 1971; Moriarty and Halmi 1972; Bowie et al. 1973).

The cell found by Kurosumi and Kobayashi (1966) was called the type I ACTH cell and the cell described by Nakayama et al. (1969) was called the type II ACTH cell, because the former cell was discovered earlier. However, some authors who supported the theory of Nakayama et al. (1969) and Siperstein and Miller (1970) proved that the type II ACTH cells really contain ACTH by using immunocytochemical techniques. Moriarty and Halmi (1972) used the electron microscopic peroxidase-antibody method and Bowie et al. (1973) utilized the comparative method of light microscopic immunohistochemical findings and the electron microscopy of the adjacent section with the usual electron staining. Although the type II cells were found later than the type I ACTH cells, the immunocytochemical demonstration of ACTH succeeded earlier in type II cells. Additionally, type II cells have a characteristic stellate shape with long processes. This unique morphology makes the identification of this cell type easy in both light and electron microscopy, as well as with positive immunocytochemical reaction. This cell can thus reasonably be called the typical ACTH cell, and type I cells were then called atypical ACTH cells (Kurosumi et al. 1984).

Kurosumi et al. (1983, 1984) demonstrated, however, that both the typical (type II) and atypical (type I) cells are reactive to anti-ACTH serum using the peroxidase-antibody and ferritin-antibody methods (Kurosumi et al. 1983) as well as the protein A-gold method (Kurosumi et al. 1984) (Fig. 7a, b). Cells of an intermediate type showing the characteristics of both types are occasionally observed. Morphometry of the ACTH cells after adrenalectomy indicated that both type I and type II cells underwent hypertrophy of the ER and Golgi apparatus, and showed a decrease in secretory granules, although the peak of the alteration is different among type I and type II cells (Kurosumi et al. 1983).

Not only the serum against ACTH but also antibodies against various fragments of POMC, such as α-MSH, CLIP, β-LPH, β-endorphin (Fig. 7c), and α-endorphin, were obtained and IEM was performed using these various antibodies. The common results are that both type I and type II cells (and the intermediate type on some occasions) show reactions for these antibodies for various fragments of POMC. Therefore, types I and II ACTH cells as demonstrated by electron microscopy are to be called types I and II POMC cells.

Among various antibodies, only α-MSH showed a unique reaction, i.e., type I cells react well to anti-α-MSH serum, but type II cells hardly react to it. Only a few of the secretory granules arranged along the periphery of type II cells may react to the antibody for α-MSH, although almost all granules may react with anti-ACTH (Fig. 7d). Therefore, light microscopy shows type I ACTH cells to be reactive with α-MSH, but type II cells are negative to α-MSH, although they are positive to ACTH (Tanaka and Kurosumi 1986). This is a clear method for distinguishing type II (typical POMC cells) from type I (atypical POMC cells) as observed by light microscopy, because the former reacts only to anti-ACTH but the latter reacts to both anti-α-MSH and anti-ACTH sera. IEM for peptides derived from POMC other than α-MSH showed no difference between type I and type II cells.

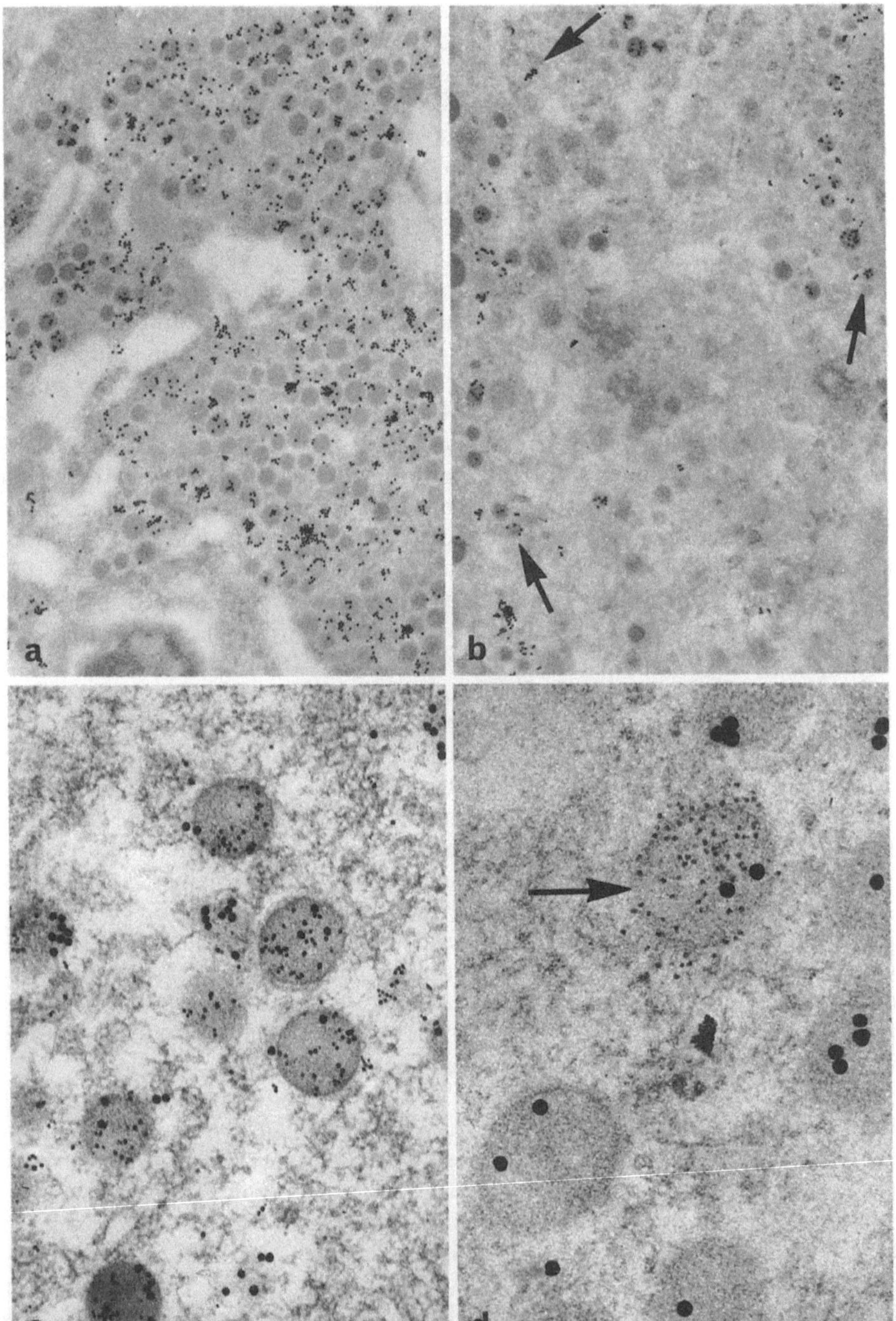
a
b
c
d

Yoshimura and Nogami (1981) classified ACTH-containing cells of the rat anterior pituitary into four types as a result of comparative light and electron microscopy of the adjacent thick and thin sections, the former of which was used for immunohistochemistry. They differentiated the so-called Siperstein (stellate) cell from the Moriarty (spindle-shaped) cell, named types 1 and 2, respectively. However, the difference between a stellate shape and a spindle shape in cell outline is not so important and no further difference in ultrastructure was reported between these two types. Similarly, type 3 and type 4 cells bear a resemblance to each other and look rather like the type I ACTH cells of our classification. Because Yoshimura and Nogami (1981) did not use simple osmium fixation but double fixation with glutaraldehyde and osmium, the secretory granules of their ACTH cells appeared all dense and solid, but no cored granules characteristic of our type I ACTH cell could be observed. Therefore, there is a possibility that the type I ACTH cell of our classification may be included in either or both of types 3 and 4 of Yoshimura and Nogami (1981). Direct demonstration of antigens under the electron microscope seems better than the indirect method of comparative thin and thick sections, the latter of which was processed by the light microscopic immunohistochemical procedure. As the peroxidase, ferritin and colloidal gold methods for IEM were all able to demonstrate the immunoreactivity of the secretory granules of these two different types of presumptive ACTH cells to the antibodies against ACTH and further related peptides of POMC (Kurosumi et al. 1983, 1984; Tanaka and Kurosumi 1986), the two types and the additional intermediate type (three types altogether) shown by IEM are undoubtedly cells engaged in production of POMC (ACTH) residing in the anterior pituitary.

3.4 Gonadotropin-Producing Cell

The hormones that stimulate gonads (ovary in the female and testis in the male) are called inclusively gonadotropins (gonadotropic hormones or GTH). In most vertebrates there are two GTH: follicle-stimulating hormone (FSH) and luteinizing hormone (LH). Both FSH and LH are glycoprotein and consist of two subunits – alpha and beta chains. The alpha chain is common among FSH, LH, and also TSH (thyroid-stimulating hormone), but the beta chain varies. Therefore, antibodies against beta-chain antigens must be used to differentiate between FSH and LH in immunocytochemical studies.

Fig. 7. a Part of a type I ACTH cell showing a great number of secretory granules filling almost the entire cytoplasm. The IEM technique is protein A-gold method using anti-ACTH serum. × 20000. **b** Part of type II ACTH cell. Immunoreactive secretory granules labeled with protein A-gold bound to anti-ACTH are arranged along the cell periphery (*arrows*). Almost no granules are observed in the deeper part of the cytoplasm. × 20000. **c** Double labeling with different-sized gold particles, i.e., large particles for ACTH and small particles for β-endorphin. Most granules in both types of ACTH cells are labeled with ACTH and β-endorphin antibodies. × 70000. **d** Double labeling with ferritin and colloidal gold, i.e., ferritin (very small particles) for α-MSH and colloidal gold (large particles) for ACTH. In type II ACTH cells only a few granules showed coexistence of α-MSH and ACTH (*arrow*), and most granules are only labeled with anti-ACTH and show no reaction for α-MSH. × 120000

The first research with the electron microscope concerning gonadotropic cells (GTH cells) of the rat anterior pituitary was carried out by Farquhar and Rinehart (1954a) and observed morphological changes in GTH cells after castration. They suggested two different types of GTH cells in the castrated rat pituitary but did not make clear which cell secretes FSH or LH.

Barnes (1962) studied the mouse anterior pituitary in the female (various stages of the estrous cycle, pregnancy, lactation, and ovariectomy) as well as in the male (normal and castrated). She identified five different types of anterior pituitary secretory cells, each of which was ascribed to the producer of each of five pituitary hormones except ACTH. Her work was characterized by the differentiation of producer cells of FSH and LH, which were called β- and γ-cells respectively in the first report (Barnes 1962) but FSH cells and LH cells in a later report (Barnes 1963). The reported size of secretory granules is greater in FSH cells than LH cells: 150–200 nm for the former, 100–200 nm for the latter. Furthermore, the granules of FSH cells are relatively low in electron density as compared with the granules of LH cells. Barnes (1962, 1963) reported that both LH and FSH cells underwent hypertrophy after castration and markedly changed parallel to each other according to the estrous cycle.

Kurosumi and Oota (1968) applied the identification criteria for FSH and LH cells originally proposed for the mouse by Barnes (1962, 1963) to the rat anterior pituitary. Comparison between two types of GTH cells in experimentally induced persistent estrous and diestrous rats led to a conclusion supporting Barnes' concept of GTH cell identification (Kurosumi and Oota 1968). The FSH cells of the rat are large round cells containing two types of secretory granules, one small and dense, about 200 nm in diameter, and the other large and less dense, about 500–700 nm in diameter (Fig. 8a). The FSH cells in the male rat contain many large granules and some small granules, but those in the female rat contain only a few large granules. The LH cells contain only small granules, most of which are accumulated in the cytoplasm of the cell pole facing the blood capillary (Fig. 8b). This cell type is rare in the male but is observed with moderate frequency in the female pituitary.

Immunocytochemistry, an advanced technique, demonstrated the existence of LH and FSH in the same pituitary cell (Nakane 1970; Phifer et al. 1973). Because the two kinds of gonadotropins are contained in the same cell type as mentioned above, the nomenclature of FSH cell and LH cell was no longer sustainable, and therefore we called these cells type I gonadotrope and type II gonadotrope, respectively (Kurosumi 1974). Then, similarly, Childs (Moriarty) et al. (1980) gave

Fig. 8. a Ordinary electron micrograph of a type I GTH cell containing large and small secretory granules, taken from an adult male rat. × 8000. **b** Ordinary electron micrograph of a type II GTH cell from an adult female rat, containing only small secretory granules accumulated in the cytoplasm of the vascular pole of the cell. × 7000. **c** Immunoelectron micrograph of ferritin-antibody method using anti-LHβ serum, made after rapid freeze and acetone substitution. Ferritin particles mostly label small secretory granules and only a few are seen on the large granules. × 100000. **d** Immunoelectron micrograph using anti-FSHβ serum with the same technique as for **c**. Ferritin particles indicating FSHβ are mostly shown on the large granules and very few or no particles are seen on the small granules. × 100000

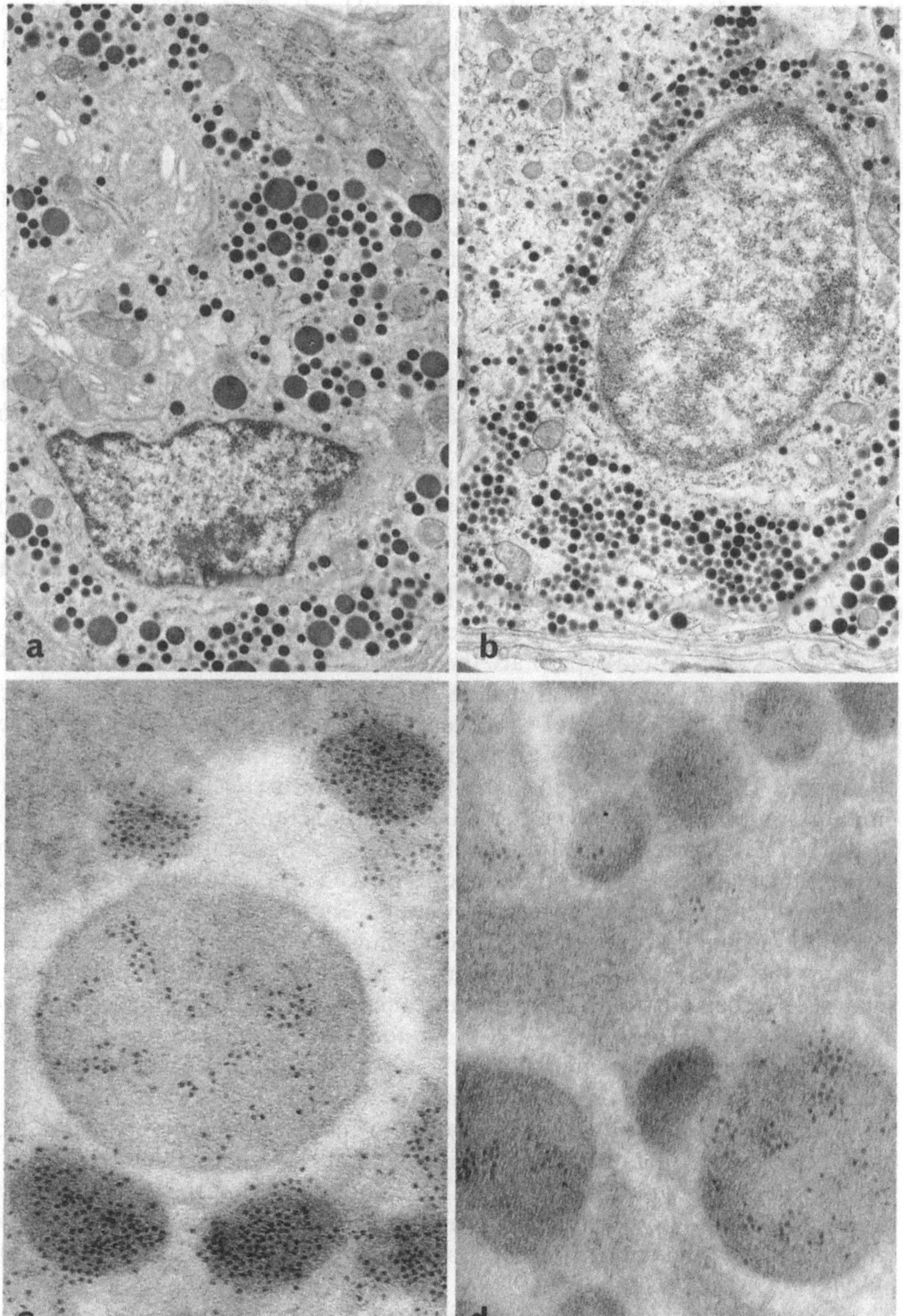
a
b
c
d

the names type I to FSH cells and type II to LH cells. Tixier-Vidal and her coworkers (Tougard et al. 1973, 1980; Tixier-Vidal et al. 1975) named these cells type A and type B, instead of type I and type II, respectively. Both Tixier-Vidal's group and that of Childs (Moriarty 1976; Moriarty and Garner 1977; Childs et al. 1980) described the third type of GTH cells. Type C cells of Tixier-Vidal et al. (1975) contained very small secretory granules, which were positive to anti-LH and anti-FSH sera. We observed similar cells reactive to anti-LHβ serum, characterized by small secretory granules and a small amount of cytoplasm, in the pituitary of castrated rats (Inoue and Kurosumi 1981).

Type III cells of Childs (Moriarty and Garner 1977; Childs et al. 1980) possess morphological characteristics similar to those of ACTH cells reported by Nakane (1975) and are also positive to anti-ACTH serum. The results of our immunocytochemical studies of both GTH cells and ACTH cells at the level of light microscopy demonstrated no evidence of coexistence of ACTH and GTH (FSH and/or LH) (Inoue and Hagino 1984). Yoshimura et al. (1981) studied LH cells by comparative light and electron microscopy and drew up a complex double classification system: II/III, III, III/IV, and IV on the one hand, and types A and B on the other hand. Yoshimura's type A is a cell only containing small secretory granules like the type II cell or LH cell of our classification (Kurosumi 1974; Kurosumi and Oota 1968). Yoshimura's type B is a cell with mixed small and large granules which is similar to the type I or FSH cells of the above classification. Therefore, Yoshimura's type A and type B are the reverse of the same names in Tixier-Vidal's classification (Tixier-Vidal et al. 1975). Although Yoshimura and his associates recognized the presence of GTH cells with only small granules, they denied the existence of our type II GTH cell which has the same characteristics (Kurosumi 1974). They argued that Kurosumi-Oota LH cells are similar to the small granule PRL cells (type II PRL cells) (Nogami and Yoshimura 1980). Type II PRL cells are known to be prominent in the male pituitary but very rare in the female gland. On the other hand, type II GTH cells (formerly called LH cells) are often observed in the female pituitary. Therefore, type II PRL cells seem not to correspond to the so-called LH cells, but the latter cells are rather comparable to the type A cells of Yoshimura et al. (1981). IEM using an antibody against LHβ demonstrated the immunoreactivity of small secretory granules of both types of GTH cell. These LH-immunoreactive small granules showed the same ultrastructural characteristics as those of the so-called LH cells, but it is not yet known whether they contain FSH as well as LH.

Early reports of IEM were almost all of the peroxidase-antibody method, either the preembedding or postembedding method. Therefore, the work on GTH cells which contained both FSH and LH demonstrated that the immunoreactive secretory granules positive to one hormone were also positive to the other. But no difference in the content of GTH among large and small secretory granules in type I GTH cell was reported.

Because the antigenicity of FSHβ is easily decayed by chemical fixation, we applied rapid-freeze substitution fixation without osmium tetroxide for the substitution medium (Inoue and Kurosumi 1984). For IEM labeling we used ferritin by a modified method (Inoue et al. 1982). We used this method to examine typical GTH cells (type I GTH cells) of the adult male rat anterior pituitary, which contain both large and small granules. Ferritin particles showing the immunoreactivity to LHβ were mainly concentrated on the small granules (ferritin particles per unit area: 4250 ± 183) and rarely found on large granules (529 ± 58) (Fig. 8c), but the distribution pattern of FSHβ was the reverse, i.e., many on large granules (1153 ± 19) but few on small granules (206 ± 24) (Fig. 8d). As far as male rats are concerned, FSH is mainly localized in large granules and LH is concentrated in small granules.

With regard to the size of typical GTH cells, two groups – large cells and small cells – were found by light microscopic immunocytochemistry. The intensity of the immunocytochemical reaction for LHβ is strong in large GTH cells, while that for FSHβ is marked in small GTH cells. On the basis of this result, we called small GTH cells FSH-rich cells and large GTH cells LH-rich cells. Electron microscopy demonstrated that large GTH cells (LH-rich cells) contain numerous small granules, while small GTH cells (FSH-rich cells) contain only a few small granules as compared with the number of large granules, although both contain large and small granules. Therefore, no discrepancy was seen between the result of IEM of rapid-freeze substitution specimens and that of light microscopic immunocytochemistry (Inoue and Kurosumi 1984). However, the female GTH cells contain a very small number of large secretory granules, and localization of FSH in female GTH cells is still unknown.

3.5 Thyrotropin-Producing Cell

The identification of thyrotropes (TSH cells) as observed with the electron microscope was first established by Farquhar and Rinehart (1954b) through the use of an experimental morphological technique, specifically the observation of the anterior pituitary after thyroidectomy. Examination of the early response to thyroidectomy had enabled them to recognize the unstimulated TSH cell in the normal animal. The TSH cells thus identified have the following ultrastructural characteristics: they are angular in outline, not located on the capillary wall, and contain very small secretory granules (maximal diameter 140 nm).

The cells altered in structure after thyroidectomy have been called "thyroidectomy cells." The common idea concerning the origin of thyroidectomy cells is that they are the TSH cells of the normal pituitary, which undergo hypertrophy and vesiculation (ultrastructural dilatation of the rough ER). According to Farquhar and Rinehart (1954b), the vesicles (dilated ER) are greatly variable from one vesicle to another, and also from cell to cell. Another feature characteristic of thyroidectomy cells is the relative paucity of secretory granules. Although the number of secretory granules is markedly decreased, the size of granules is not altered as compared with that of the normal TSH cells or early thyroidectomy cells. At 10 days after thyroidectomy or later, round dense granules appeared in the dilated cisternae of the ER, which were called vesicles by the first authors (Fig. 1f). These granules may correspond to the so-called intracisternal granules found by Palade (1956) in the rough ER of guinea pig pancreatic exocrine cells. Although the rough ER of the castration cells was also strongly dilated, no intracisternal granules were found (Farquhar and Rinehart 1954a, b).

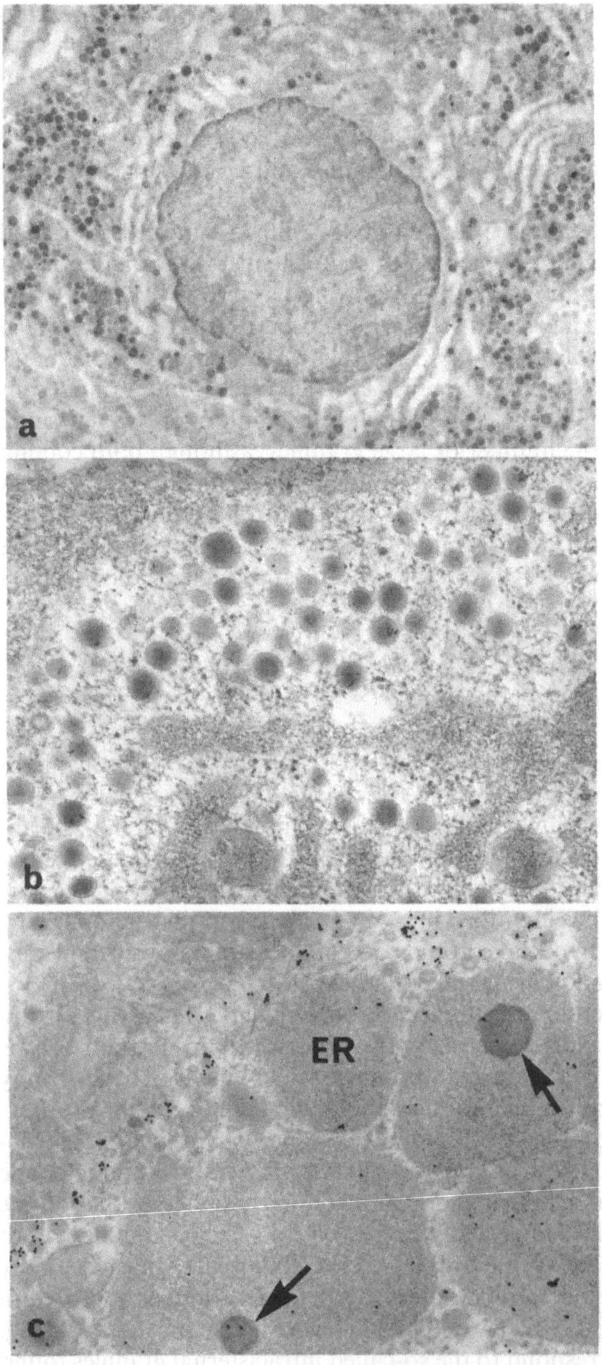
a
b
ER
c

The ultrastructures of TSH cells and thyroidectomy cells first observed by Farquhar and Rinehart (1954 b) were recognized by Kurosumi and his coworkers (Kurosumi 1968; Kurosumi and Baba 1969), who additionally reported a peculiar membrane structure filled with dark substance adjacent to the Golgi apparatus of the thyroidectomy cell (Fig. 4 c). Because the dark substance of this structure showed a positive reaction to acid phosphatase activity (Fig. 4 d), we thought that the structure might be related to the lysosome and probably corresponded to the GERL of Novikoff (1964), as the thyroidectomy cells contained many secondary lysosomes (Kurosumi and Baba 1969; Kurosumi 1972).

Although most electron microscopists engaged in research into the functional morphology of the anterior pituitary cells considered that the thyroidectomy cells were nothing but hypertrophied and hyperfunctioning TSH cells, some authors believed that they might be derived from cell types other than TSH cells, or at least arose from types besides TSH cells (Dingemans 1969; Yoshimura et al. 1973, 1977; Kiguchi 1978). IEM examination of thyroidectomy cells using anti-human TSHβ serum showed the positive immunoreaction of TSH on the secretory granules as well as the liquid substance and intracisternal granules contained in the dilated cisternae of the rough ER of the thyroidectomy cells (unpublished observations) (Fig. 9 c). These observations prove that most thyroidectomy cells are hyperfunctioning TSH cells.

Yoshimura et al. (1982) showed TSH cells in the anterior pituitaries of rats at various ages by means of a superimposition technique with light microscopic immunocytochemistry, and described TSH cells of adult rats as stellate or elongate cells with a dense arrangement of parallel arrays of rough ER and numerous secretory granules 150–250 nm in diameter. There are various steps in development of TSH cells in young animals. These immature cells contain smaller granules varying from 50 to 150 nm in diameter, depending on age. We have made a few observations of TSH cells of adult rats by the peroxidase-antibody or colloidal gold-antibody method for electron microscopy and obtained similar findings on the ultrastructure of TSH cells (Fig. 9 a, b), but immature TSH cells were not observed.

Several reports were published concerning the effects of TRH on TSH cells. For example, Shiino (1979) reported a remarkable dilatation of the ER of organ-cultured TSH cells after incubation with a TRH-containing medium. Okino et al. (1979) demonstrated the effects of TRH administered in vivo, and showed a marked increase in exocytosis of secretory granules from the TSH cells together with massive multigranular extrusion.

Fig. 9. a Immunoelectron micrograph of postembedding peroxidase-antibody method with use of anti-rat TSHβ serum. Immunoreactive TSH cell contains a large number of small secretory granules intensified in electron density as compared with other cell types. × 7000. **b** Colloidal gold IEM with anti-human TSHβ serum demonstrated the TSH cell in the rat anterior pituitary, although the reaction was weak. × 26000. **c** An IEM demonstration of TSHβ using colloidal gold method on a thyroidectomy cell. Gold particles intensely labeled secretory granules arranged along the cell periphery. A few gold particles are also shown on the intracisternal granules (*arrows*) as well as a medium-dense substance filling the dilated *ER*. × 17000

References

Aggeler J, Werb Z (1982) Initial events during phagocytosis by macrophages viewed from outside and inside the cell: mambrane-particle interactions and clathrin. J Cell Biol 94:613–623

Barnes BG (1962) Electron microscope studies on the secretory cytology of the mouse anterior pituitary. Endocrinology 71:618–628

Barnes BG (1963) The fine structure of the mouse adenohypophysis in various physiological states. In: Benoit J, Da Lage C (eds) Cytologie de l'adénohypophyse. Centre National de la Recherche Scientifique, Paris, pp 91–109

Beauvillain JC, Mazzuca M, Dubois MP (1977) The prolactin and growth-hormone producing cells of the guinea-pig pituitary. Electron microscopic study using immunocytochemical means. Cell Tiss Res 184:343–358

Bowie EP, Williams G, Shiino M, Rennels EG (1973) The corticotroph of the rat adenohypophysis: a comparative study. Am J Anat 138:499–520

Cantin M, Ballak M, Lu CL, Seidah NG, Chrétien M (1983) Corticotropin (ACTH) and the N-terminal fragment of pro-opiomelanocortin are located in the same granules in cells of rat pituitary gland. J Histochem Cytochem 31:479–487

Childs GV, Ellison DG, Garner LL (1980) An immunocytochemist's view of gonadotropin storage in the adult male rat: cytochemical and morphological heterogeneity in serially sectioned gonadotropes. Am J Anat 158:397–409

Claude A (1970) Growth and differentiation of cytoplasmic membranes in the course of lipoprotein granule synthesis in the hapatic cell. I. Elaboration of elements of the Golgi complex. J Cell Biol 47:745–766

Costoff A (1973) Ultrastructure of rat adenohypophysis, correlation with function. Academic, New York

Dacheux F (1980) Ultrastructural immunocytochemical localization of prolactin and growth hormone in the porcine pituitary. Cell Tiss Res 207:277–286

Dacheux F (1981) Ultrastructural localization of corticotropin, β-lipotropin, and α- and β-endorphin in the porcine anterior pituitary. Cell Tiss Res 215:87–101

Dacheux F, Dubois MP (1976) Ultrastructural localization of prolactin, growth hormone and luteinizing hormone by immunocytochemical techniques in the bovine pituitary. Cell Tiss Res 174:245–260

De Duve C (1963) Discussion. In: De Reuck AVS, Cameron MP (eds) Ciba Foundation symposium: lysosomes. Churchill, London, p 126

Dingemans KP (1969) On the origin of thyroidectomy cells. J Ultrast Res 26:480–500

Douglas WW (1968) The first Gaddum memorial lecture. Stimulus-secretion coupling: the concept and clues from chromaffin and other cells. Br J Pharmacol 34:451–474

Ehrenreich JH, Bergeron JJM, Siekevitz P, Palade GE (1973) Golgi fractions prepared from rat liver homogenates. I. Isolation procedure and morphological characterization. J Cell Biol 59:45–72

Farquhar MG (1957) "Corticotrophs" of the rat adenohypophysis as revealed by electron microscopy. Anat Rec 127:291(Abs)

Farquhar MG (1961) Fine structure and function in capillaries of the anterior pituitary gland. Angiology 12:270–292

Farquhar MG (1977) Secretion and crinophagy in prolactin cells. In: Dellmann HD, Johnson JA, Klachko DM (eds) Comparative endocrinology of prolactin. Plenum, New York, pp 37–94

Farquhar MG (1978) Recovery of surface membrane in anterior pituitary cells. Variations in traffic detected with anionic and cationic ferritin. J Cell Biol 77:R35–42

Farquhar MG, Rinehart JF (1954a) Electron microscopic studies of the anterior pituitary gland of castrate rats. Endocrinology 54:516–541

Farquhar MG, Rinehart JF (1954b) Cytologic alterations in the anterior pituitary gland following thyroidectomy: an electron microscope study. Endocrinology 55:857–876

Farquhar MG, Wellings SR (1957) Electron microscopic evidence suggesting secretory granule formation within the Golgi apparatus. J Biophys Biochem Cytol 3:319–322

Farquhar MG, Skutelsky EH, Hopkins CR (1975) Structure and function of the anterior pituitary and dispersed pituitary cells. In vitro studies. In: Tixier-Vidal A, Farquhar MG (eds) The anterior pituitary. Academic, New York, pp 83–135

Friend DS, Farquhar MG (1967) Functions of coated vesicles during protein absorption in the rat vas deferens. J Cell Biol 35:357–376

Fujita H, Okamoto H (1979) Fine structural localization of thiamine pyrophosphatase and acid phosphatase activities in the mouse pancreatic acinar cell. Histochemistry 64:287–295

Fujita H, Kurihara H, Miyagawa J (1983) Ultrastructural aspects of the effect of calcium ionophore A23187 on incubated anterior pituitary cells of rats. Cell Tiss Res 229:129–136

Fumagalli G, Zanini A (1985) In cow anterior pituitary, growth hormone and prolactin can be packed in separate granules of the same cell. J Cell Biol 100:2019–2024

Goldfischer S (1982) The internal reticular apparatus of Camillo Golgi: A complex heterogeneous organelle, enriched in acid, neutral, and alkaline phosphatases, and involved in glycoxylation, secretion, membrane flow, lysosome formation, and intracellular digestion. J Histochem Cytochem 30:717–733

Guy J, Leclerc R, Pelletier G (1980) Localization of a 16,000-dalton fragment of the common precursor of adrenocorticotropin and β-lipotropin in the rat and human pituitary gland. J Cell Biol 86:825–830

Haguenau F (1958) The ergastoplasm: its history, ultrastructure, and biochemistry. Int Rev Cytol 7:425–483

Harigaya T, Kohmoto K, Hoshino K (1983) Immunohistochemical identification of prolactin-producing cells in the mouse adenohypophysis. Acta Histochem Cytochem 16:51–58

Hedinger CE, Farquhar MG (1957) Elektronenmikroskopische Untersuchungen von zwei Typen acidophiler Hypophysenvorderlappenzellen bei der Ratte. Schweiz Z Allg Path Bak 20:766–768

Heidenhain M (1911) In: Plasma und Zelle. G. Fischer, Jena [cited from Abbott J, Holtzer H (1966) J Cell Biol 28:473–487]

Ichikawa A (1959) Electron microscope study on secretion of the rat adenohypophysis. Acta Anat Nippon 34:460–482 (Japanese text with English abstract)

Inoue K, Hagino N (1984) Comparative immunocytochemical demonstration of ACTH-, LH- and FSH-containing cells in the pituitary of neonatal, immature and adult rats. Cell Tiss Res 235:71–75

Inoue K, Kurosumi K (1977) Cytochemical and three-dimensional studies on Golgi apparatus and GERL of rat anterior pituitary cells by transmission electron microscopy. Cell Struct Func 2:171–186

Inoue K, Kurosumi K (1980) Electron microscopic demonstration of carbohydrate components in the cells of thyroid follicles and anterior pituitary by Concanavalin A-iron dextran technique. Acta Histochem Cytochem 13:173–180

Inoue K, Kurosumi K (1981) Mode of proliferation of gonadotrophic cells of the anterior pituitary after castration – immunocytochemical and autoradiographic studies. Arch Histol Jpn 44:71–85

Inoue K, Kurosumi K (1984) Ultrastructural immunocytochemical localization of LH and FSH in pituitary of the untreated male rat. Cell Tiss Res 235:77–83

Inoue K, Kurosumi K, Deng Z-P (1982) An improvement of the device for rapid freezing with use of liquid propane and an application of immunocytochemistry to resin section of rapid-frozen, substitution fixed anterior pituitary gland. J Electron Microsc 31:93–97

Inoue K, Tanaka S, Kurosumi K (1985) Mitotic activity of gonadotropes in the anterior pituitary of the castrated male rat. Cell Tiss Res 240:271–276

Ishikawa H, Nogami H, Kamio M, Suzuki T (1983) Single secretory granules contain both GH and prolactin in pituitary mixed type of adenoma. Virchow Arch Pathol Anat 399:221–226

Ishimura K, Egawa K, Fujita H (1980) Freeze-fracture images of exocytosis and endocytosis in anterior pituitary cells of rabbit and mice. Cell Tiss Res 206:233–241

Kanno T (1980) Process of reception, conduction and extrusion in paraneurons. Biomed Res 1: Suppl 10–16
Kawabata I (1966) Electron microscopy of the rat hypothalamic neurosecretory system. III. The supraoptic nucleus after vital staining with trypan blue. Arch Histol Jpn 26:215–240
Kiguchi Y (1978) The process of development of thyroidectomy cells from the so-called thyrotrophs. Endocrinol Jpn 25:75–86
Koyama T (1985) An immunoelectron microscopic study of the rat anterior pituitary prolactin cells by the colloidal gold-antibody method. J Clin Electron Micr 18: 261–275 (Japanese text with English abstract)
Koyama T, Kurosumi K (1986) An immunoelectron microscopic study of prolactin cells in the anterior pituitaries of the normal, castrated and steroid-treated rats. In: Hoshino K (ed) Kyoto prolactin conference monograph. vol 1. pp 97–105
Kurihara H, Fujita H (1984) Cytochemical studies on the localization and functional properties of calcium in anterior pituitary cells. Histochemistry 81:9–13
Kurosumi K (1961) Electron microscopic analysis of the secretion mechanism. Int Rev Cytol 11:1–124
Kurosumi K (1965) Electron microscopic studies on the morphology of secretion. J Electron Micr 14:12–26 (Japanese text with English abstract)
Kurosumi K (1968) Functional classification of cell types of the anterior pituitary gland accomplished by electron microscopy. Arch Histol Jpn 29:329–362
Kurosumi K (1971) Mitosis of the rat anterior pituitary cells: an electron microscope study. Arch Histol Jpn 33:145–160
Kurosumi K (1972) Cytochemistry and functional morphology of the Golgi apparatus. Acta Histochem Cytochem 5:242–245
Kurosumi K (1974) Adenohypophysis. In: Kurosumi K, Fujita H (eds) An atlas of electron micrographs. Functional morphology of endocrine glands. Igaku-Shoin, Tokyo, pp 1–100
Kurosumi K (1979) Formation and release of secretory granules during mitosis in the anterior pituitary gland. Arch Histol Jpn 42:481–486
Kurosumi K, Baba N (1969) Experimental and histochemical studies on the rat pituitary thyrotrophs by electron microscopy. Gunma Symp Endocrinol 6:197–210
Kurosumi K, Inoue K (1980) Surface pits of typical gonadotrophs and castration cells of the rat anterior pituitary suggestive of exocytosis and micropinocytosis. Arch Histol Jpn 43:373–241
Kurosumi K, Inoue K (1983) Cytochemistry and ultrastructure of Golgi apparatus and GERL of certain secretory cells. J Histochem Cytochem 31:233–234
Kurosumi K, Kobayashi Y (1966) Corticotrophs in the anterior pituitary glands of normal and adrenalectomized rats as revealed by electron microscopy. Endocrinology 78:745–758
Kurosumi K, Oota Y (1966) Corticotrophs in the anterior pituitary glands of gonadectomized and thyroidectomized rats as revealed by electron microscopy. Endocrinology 79:808–814
Kurosumi K, Oota Y (1968) Electron microscopy of two types of gonadotrophs in the anterior pituitary glands of persistent estrous and diestrous rats. Z Zellforsch 85:34–46
Kurosumi K, Kobayashi Y, Baba N (1968) The fine structure of mammary glands of lactating rats, with special reference to the apocrine secretion. Exp Cell Res 50:177–192
Kurosumi K, Kawarai K, Yukitake Y, Inoue K (1976) Electron microscopic morphometry of the rat castration cells. Gunma Symp Endocrinol 13:221–236
Kurosumi K, Inoue K, Taniguchi Y (1983) Evidence for two types of ACTH-secreting cells of the rat anterior pituitary: ultrastructural, morphometric and immunocytochemical studies. Biomed Res 4[Suppl]:117–134
Kurosumi K, Koyama T, Tokuyasu K (1984) Immunoelectron microscopy of the prolactin-cells and growth hormone-cells of the rat anterior pituitary glands. Acta Histochem Cytochem 17:710(Abs)
Kurosumi K, Taniguchi Y, Inoue K (1984) Electron microscopic immunocytochemistry by the protein A-gold method of two types of rat anterior pituitary ACTH cells. Arch Histol Jpn 47:279–291

Lacy PE, Howell SL, Young DA, Fink CJ (1968) New hypothesis of insulin secretion. Nature 219:1177–1179
Moriarty GC (1976) Immunocytochemistry of the pituitary glycoprotein hormones. J Histochem Cytochem 24:846–863
Moriarty GC, Garner LL (1977) Immunocytochemical studies of cells in the rat adenohypophysis containing both ACTH and FSH. Nature 265:356–358
Moriarty GC, Halmi NS (1972) Electron microscopic study of the adrenocorticotropin-producing cell with the use of unlabeled antibody and the soluble peroxidase-antiperoxidase complex. J Histochem Cytochem 20:590–603
Nakane PK (1970) Classification of anterior pituitary cell types with immunoenzyme histochemistry. J Histochem Cytochem 18:9–20
Nakane PK (1975) Identification of anterior pituitary cells by immunoelectron microscopy. In: Tixier-Vidal A, Farquhar MG (eds) The anterior pituitary. Academic, New York, pp 45–61
Nakayama I, Nickerson PA, Skelton FR (1969) An ultrastructural study of the adrenocorticotrophic hormone-secreting cell in the rat adenohypophysis during adrenal cortical regeneration. Lab Invest 21:169–178
Neutra M, Leblond CP (1966) Synthesis of the carbohydrate of mucus in the Golgi complex as shown by electron microscope radioautography of goblet cells from rats injected with glucose H^3. J Cell Biol 30:119–136
Nogami H (1984) Fine-structural heterogeneity and morphologic changes in rat pituitary prolactin cells after estrogen and testosterone treatment. Cell Tiss Res 237:195–202
Nogami H, Yoshimura F (1980) Prolactin immunoreactivity of acidophils of the small granule type. Cell Tiss Res 211:1–4
Novikoff AB (1964) GERL, its form and function in neurons of rat spinal ganglia. Biol Bull 127:358(Abs)
Novikoff AB, Mori M, Quintana N, Yam A (1977) Studies of the secretory process in the mammalian exocrine pancreas. I. The condensing vacuoles. J Cell Biol 75:148–165
Okino H, Matsui S, Shioda S, Nakai Y, Kurosumi K (1979) Ultrastructural and morphometric studies on the rat pituitary thyrotrophs and thyroid follicular cells following administration of thyrotropin releasing hormone. Arch Histol Jpn 42:489–505
Osamura RY, Komatsu N, Izumi S, Yoshimura S, Watanabe K (1982) Ultrastructural localization of prolactin in the rat anterior pituitary glands by preembedding peroxidase-labeled antibody method: observations in normal, castrated, or estrogen-stimulated specimen. J Histochem Cytochem 30:919–925
Palade GE (1956) Intracisternal granules in the exocrine cells of the pancreas. J Biophys Biochem Cytol 2:417–422
Parry DM, McMillen IC, Wilcox DL (1978) Immunocytochemical localisation of prolactin and growth hormone in the ovine pituitary. Cell Tiss Res 194:327–336
Pelletier G, Novikoff AB (1972) Localization of phosphatase activities in the rat anterior pituitary gland. J Histochem Cytochem 20:1–12
Pelletier G, Puviani R (1973) Detection of glycoproteins and autoradiographic localization of [^{3}H] fucose in the thyroidectomy cells of rat anterior pituitary gland. J Cell Biol 56:600–605
Pelletier G, Racadot J (1971) Identification des cellules hypophysaires secrétant l'ACTH chez le rat. Z Zellforsch 116:228–239
Pelletier G, Leclerc R, Labrie F, Cote J, Chrétien M, Lis M (1977) Immunohistochemical localization of β-lipotropic hormone in the pituitary gland. Endocrinology 100:770–776
Phifer RF, Midgley AR, Spicer SS (1973) Immunohistologic and histologic evidence that follicle-stimulating hormone and luteinizing hormone are present in the same cell type in the human pars distalis. J Clin Endocrinol Metab 36:125–141
Rennels EG (1962) An electron microscope study of pituitary autograft cells in the rat. Endocrinology 71:713–722
Rennels EG (1964) Electron microscopic alterations in the rat hypophysis after scalding. Am J Anat 114:71–91
Rennels EG, Shiino M (1968) Ultrastructural manifestation of pituitary release of ACTH in the rat. Arch Anat 51:575–590

Rinehart JF, Farquhar MG (1953) Electron microscopic studies of the anterior pituitary gland. J Histochem Cytochem 1:93–113
Shiino M (1979) Morphological changes of pituitary gonadotrophs and thyrotrophs following treatment with LH-RH or TRH in vitro. Cell Tiss Res 202:399–406
Shiino M, Yamauchi K (1985) Secretory granules of prolactin cells of neonatally thyroidectomized female rats. Cell Tiss Res 242:179–183
Shirasawa N, Kihara H, Yamaguchi S, Yoshimura F (1983) Pituitary folliculo-stellate cells immunostained with S-100 protein antiserum in postnatal, castrated and thyroidectomized rats. Cell Tiss Res 231:235–249
Siperstein ER, Allison VF (1965) Fine structure of the cells responsible for secretion of adrenocorticotrophin in the adrenalectomized rat. Endocrinology 76:70–79
Siperstein ER, Miller KJ (1970) Further cytophysiologic evidence for the identity of the cells that produce adrenocorticotrophic hormone. Endocrinology 86:451–486
Smith RE, Farquhar MG (1966) Lysosome function in the regulation of the secretory process in cells of the anterior pituitary gland. J Cell Biol 31:319–347
Tanaka S, Kurosumi K (1986) Differential subcellular localization of ACTH and α-MSH in corticotropes of the rat anterior pituitary. Cell Tiss Res 243:229–238
Tixier-Vidal A, Tougard C, Kerdelhue B, Jutisz M (1975) Light and electron microscopic studies on immunocytochemical localization of gonadotropic hormones in the rat pituitary gland with antisera against ovine FSH, LH, LHα, and LHβ. Ann NY Acad Sci 254:433–460
Tougard C, Kerdelhue B, Tixier-Vidal A, Jutisz M (1973) Light and electron microscope localization of binding sites of antibodies against ovine luteinizing hormone and its two subunits in rat adenohypophysis using peroxidase-labeled antibody technique. J Cell Biol 58:503–521
Tougard C, Picart R, Tixier-Vidal A (1980) Immunocytochemical localization of glycoprotein hormones in the rat anterior pituitary. A light and electron microscope study using antisera against rat subunits: a comparison between preembedding and postembedding methods. J Histochem Cytochem 28:101–114
Yamada K, Yamashita K (1967) An electron microscopic study on the possible site of production of ACTH in the anterior pituitary of mice. Z Zellforsch 80:29–43
Yoshimura F, Nogami H (1981) Fine structural criteria for identifying rat corticotrophs. Cell Tiss Res 219:221–228
Yoshimura F, Harumiya K, Soji T, Yokoyama M, Kumagai T (1973) Possible reversion of pituitary "thyroidectomy cells" into their original acidophils in rats. Endocrinol Jpn 20:249–262
Yoshimura F, Soji T, Takasaki Y, Kiguchi Y (1974) Pituitary acidophils with small or medium-sized granules alone in normal and adrenalectomized rats with special reference to possible ACTH secretion. Endocrinol Jpn 21:297–316
Yoshimura F, Ishihara N, Sato S (1977) Cytological changes of Siperstein's corticotrophs following thyroidectomy. Endocrinol Jpn 24:551–563
Yoshimura F, Nogami H, Shirasawa N, Yashiro T (1981) A whole range of fine structural criteria for immunohistochemically identified LH cells in rats. Cell Tiss Res 217:1–10
Yoshimura F, Nogami H, Yashiro T (1982) Fine structural criteria for pituitary thyrotrophs in immature and mature rats. Anat Rec 204:255–263
Zeigel RF, Dalton AJ (1962) Speculations based on the morphology of the Golgi systems in several types of protein-secreting cells. J Cell Biol 15:45–54

Catecholamine-Peptide Interactions in the Hypothalamus *

Y. NAKAI, S. SHIODA, H. OCHIAI, and K. KOZASA

Contents

1 Introduction

Besides the neurohypophyseal hormones oxytocin (OT) and vasopressin (VP), which belong to the classic neurosecretory systems, several other kinds of biologically active peptide have been detected in the mammalian hypothalamus. Of the latter peptides, the following are known to be present in the hypothalamus:

1. Hypophyseotropic hormones, or so-called hypothalamic hormones – thyrotropin-releasing hormone (TRH); luteinizing hormone-releasing hormone (LHRH); corticotropin-releasing hormone (CRH, CRF); somatostatin (somatotropin release-inhibiting hormone, SRIF); and growth hormone-releasing factor (GRF)
2. Anterior pituitary hormones – corticotropin (ACTH) and others
3. Opioid peptides – endorphins, enkephalins, and dynorphins

Department of Anatomy, Showa University School of Medicine, Hatanodai, Shinagawa-ku, Tokyo 142, Japan

* The work on which this study was based was supported by grants (Nos. 57570025 and 59480092) from the Japanese Ministry of Education, Science, and Culture

Current Topics in Neuroendocrinology, Vol. 7

4. Gastrointestinal peptides – neurotensin (NT), substance P, and the vasoactive intestinal polypeptides, cholecystokinin, bombesin, and glucagon
5. Peptides in a central angiotensin system.

Most of the above peptides are known to be closely related to regulating pituitary hormone secretion. The distribution and localization of the above hypothalamic peptidergic neurons have been demonstrated mostly by immunohistochemistry.

It has been demonstrated by biochemical and pharmacologic studies that monoamines play important roles in regulating the secretion of most of the neurohypophyseal and hypophyseotropic hormones and other neuropeptides in the hypothalamus. There are three mechanisms of monoamine–neuropeptide interaction in the hypothalamus: through synaptic contacts between catecholaminergic and peptidergic neurons; by interaction between monoamines and neuropeptides coexisting within the same neuron; and by release from the nerve terminals to peptide-containing neurons.

Recently, several electron microscopic studies have examined the monoaminergic synaptic inputs to the hypothalamic peptidergic neurons using the combined technique of tritiated monoamine autoradiography with neuropeptide immunocytochemistry in the same tissue section. Axon terminals labeled with [^{3}H] norepinephrine making synapses with vasopressin-containing fibers have been demonstrated in the paraventricular nucleus (PVN) of the rat hypothalamus (Silverman et al. 1983a; Silverman and Oldfield 1984). With the same technique, [^{3}H]serotonin-labeled terminals have been found to form synaptic contacts with vasoactive intestinal polypeptide-containing neurons in the suprachiasmatic nucleus (SCN), and with ACTH-containing neurons in the arcuate nucleus (AN) in the rat hypothalamus (Kiss et al. 1984). Also, synaptic contacts have been demonstrated between [^{3}H]serotonin-labeled boutons and LHRH-immunoreactive dendrites in the medial preoptic area of the rat (Kiss and Halász 1985).

Here we will consider mainly the catecholaminergic innervation of neurons containing neurohypophyseal hormones, hypothalamic hormones, and opioid peptides. These were studied using a combination of autoradiography after [^{3}H]catecholamine injection or 5-hydroxydopamine (5-OHDA) labeling to identify catecholamine-containing terminals, and immunocytochemistry of neuropeptides at the electron microscopic level.

2 Methodology

To examine the localization of catecholamines in the central nervous system at the ultrastructural level, the following techniques are generally used: autoradiography after [^{3}H]catecholamine injection; fixation with potassium permanganate; immunocytochemistry using antisera to the catecholamine-synthesizing enzymes; and uptake of the pseudotransmitters such as 5-OHDA and 6-OHDA. On the other hand, the distribution and localization of the peptidergic neurons are now usually demonstrated using various methods of immunocytochemistry. Until now, the catecholaminergic synaptic input to peptidergic neurons has been observed using a combination of autoradiography and immunocytochemistry at the electron microscopic level (Basbaum et al. 1982; Nakai et al. 1982, 1985; Silverman et al. 1983 a; Glazer and Basbaum 1984; Kiss et al. 1984; Silverman and Oldfield 1984; Takeuchi 1984; Kiss and Halász 1985; Nakada and Nakai (1985). In our laboratory, a combination of 5-OHDA labeling and neuropeptide immunocytochemistry is used in addition to this combined method, since nonspecific distribution of silver grains is sometimes found in autoradiographs.

2.1 Treatment of Animal Models

All the experiments were performed using adult male Wistar rats (150–200 g). The sodium pentobarbital-anesthetized animals were pretreated with intraventricular injections of colchicine (200 μg / 25 μl saline) in order to increase the quantity of peptide hormones in the perikarya. Two days after colchicine injection, some of the animals were reanesthetized and pretreated with a monoamine oxidase (MAO) inhibitor (phenylhydrazine hydrochloride, 100 mg/kg body weight i.p. or i.c.) to prevent the intraneuronal formation of labeled metabolites and to achieve the accumulation of [^{3}H]catecholamines in the axon terminals. The MAO inhibitor was administered 2 h prior to intraventricular microinfusion with [^{3}H]dopamine ([^{3}H]DA; sp. act. 46 Ci/mmol, Amersham, England), [^{3}H]norepinephrine ([^{3}H]NE; sp. act. 10.6 Ci/mmol, Amersham, England), or 5-OHDA. The MAO inhibitor-treated and the untreated control animals were slowly injected through a microsyringe into the lateral ventricle with 100–350 μCi of [^{3}H]DA or [^{3}H]NE in aliquots of 100–300 μl of saline containing 0.02% ascorbic acid for 20–120 min, or with 4–7 mg of 5-OHDA (Sigma, Saint Louis, USA) dissolved in 25 μl of slaine containing 0.02% ascorbic acid for 15–20 min. Immediately after the infusion of [^{3}H]DA, [^{3}H]NE, or 5-OHDA, the animals were perfused transcardially with an ice-cold mixture of 4% paraformaldehyde and 0.4% glutaraldehyde in 0.05-*M* phosphate buffer (pH 7.4).

2.2 Preparation of Tissue

After perfusion fixation, the hypothalamus was excised and immersed in the same fixative for 2–3 h. Sections cut on a Vibratome (Oxford Instruments, USA) at 30–40 μm were kept overnight in ice-cold 0.1-*M* phosphate buffer containing 7.5% sucrose.

2.3 Immunocytochemical Procedure

The sections were immunostained according to Sternberger's peroxidase–anti-peroxidase (PAP) technique as described elsewhere (Nakai et al. 1983), using arginine-VP antiserum (1 : 500, UCB-Bioproducts, Belgium), OT antiserum (1 : 4000, Calbiochem-Behring Corp., USA), LHRH antiserum (1 : 100–1 : 1000, Ferring Co., West Germany), SRIF antiserum (1 : 400–1 : 1000, UCB-Bioproducts, Belgium), TRH antiserum (1 : 3000, UCB-Bioproducts S.A.), CRF antiserum (1 : 2000, UCB-Bioproducts, Belgium), ACTH (1–24) antiserum (1 : 1000, UCB-Bioproducts, Belgium), β-endorphin (β-End) antiserum (1 : 5000–1 : 8000, Milab, Denmark), and NT antiserum (1 : 2000–1 : 5000, Milab). After immunostaining, the sections were postfixed in 1%–2% osmium tetroxide for 45–60 min at 4 °C or room temperature, dehydrated, and embedded in a mixture of Epon Araldite.

2.4 Autoradiographic Procedure

Ultrathin serial sections coated with Ilford L4 emulsion by the wire loop technique were exposed for 30–60 days, then developed and stained with lead citrate. Sections were viewed at 75 and 100 kV with Hitachi HU-12A and H-800 electron microscopes. The specificity of uptake of [^{3}H]NE into the axon terminals was confirmed by the circle analysis technique as detailed elsewhere (Nakai et al. 1983; Williams 1969).

3 Catecholamine-Neurohypophyseal Hormone Interactions

It has been suggested in biochemical, physiological, and in vitro studies that DA may play a role in the secretion of VP (Bridges et al. 1976) and OT (Bridges et al. 1976; Moos and Richard 1982). Also, NE is involved in the regulatory control of VP secretion (Kühn 1974; Sladek et al. 1983) in rats. Morphological evidence for the relationships between catecholamine- and neurohypophyseal hormone-containing neurons has been studied at the light microscopic level in the rat. It has been shown by combined monoamine histofluorescence and peptide immunostaining in a single tissue section that catecholamine-containing varicosities are in close contact with neurophysin (NP)-containing perikarya. This suggests that a functional interaction between catecholamines and neuropeptides occurs in the rat PVN (Sladek and McNeill 1980).

3.1 Catecholamine-VP Interactions

The synaptic input to the VP-like immunoreactive (VP-LI) neurons has been demonstrated at the electron microscopic level in the PVN of the rat hypothalamus (Piekut 1983; Silverman et al. 1983 b). Recently, a technique using immunocytochemistry with VP antiserum combined with autoradiography after [^{3}H]NE injection indicated that [^{3}H]NE-labeled axon terminals are rarely found to innervate VP-immunoreactive dendrites in the rat PVN (Silverman et al. 1983 a; Silverman and Oldfield 1984). Also in our laboratory, using the same combined method, [^{3}H]NE-labeled axon terminals were only rarely seen to make synaptic contacts with VP-LI nerve cell bodies and fibers in the PVN and supraoptic nucleus (SON) of the rat (Fig. 1). Examination of the specificity of silver grain localization in the

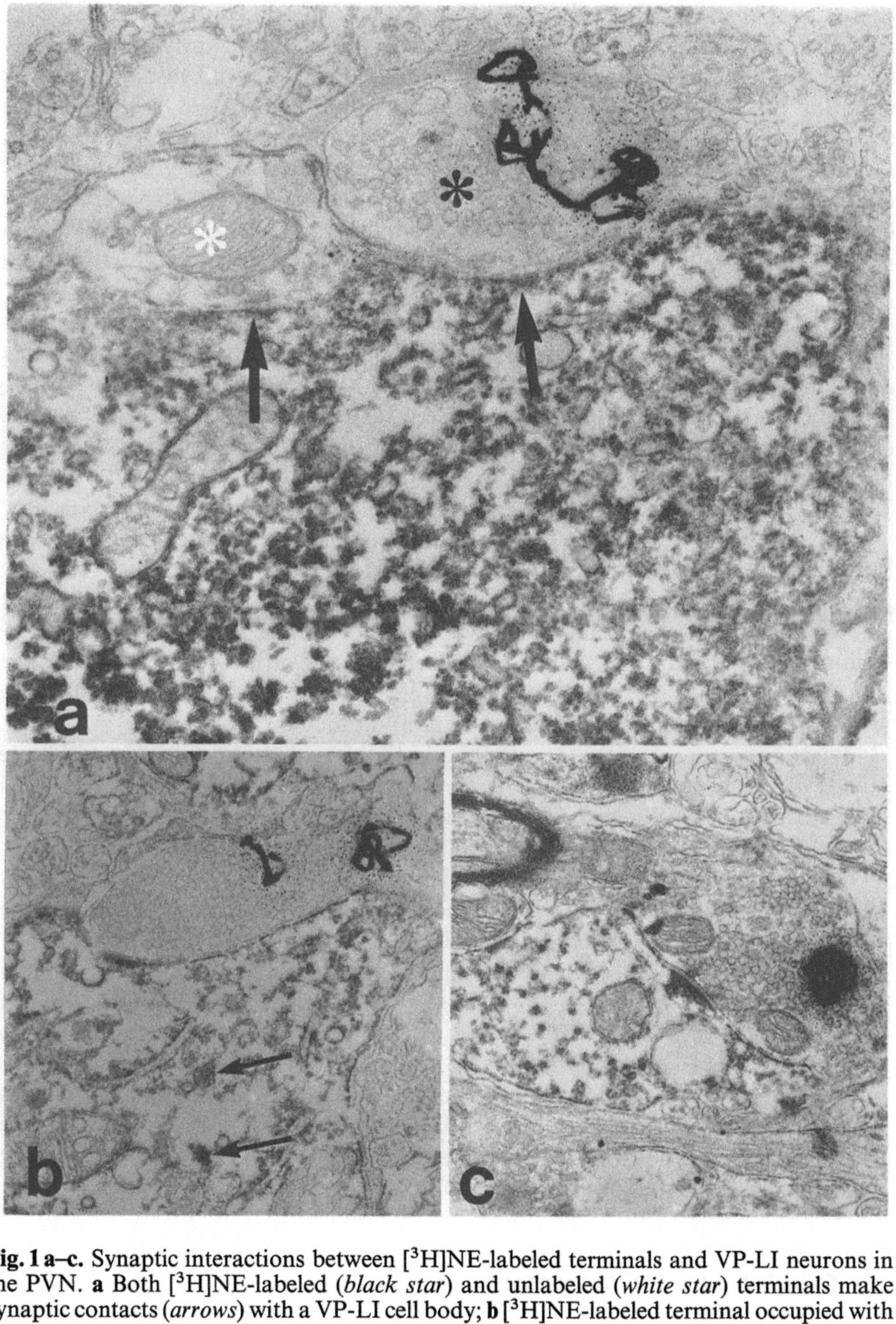

Fig. 1 a–c. Synaptic interactions between [^{3}H]NE-labeled terminals and VP-LI neurons in the PVN. **a** Both [^{3}H]NE-labeled (*black star*) and unlabeled (*white star*) terminals make synaptic contacts (*arrows*) with a VP-LI cell body; **b** [^{3}H]NE-labeled terminal occupied with numerous small, spherical, clear vesicles making synapse with VP-LI cell body containing immunoreactive neurosecretory granules (*arrows*); **c** [^{3}H]NE-labeled terminal making synapse with VP-LI nerve fiber. **a** × 35000; **b** and **c** × 26000 nerve terminals by circle analysis (Williams 1969; Nakai et al. 1983) revealed that the percentage of nerve terminals with silver grains is significantly higher than the other components, such as glial tissue or capillary lumen and that the grain distribution of [^{3}H]NE is highly nonrandom (Nakada and Nakai, 1985). Furthermore, in order to determine whether catecholaminergic terminals actually have synaptic input to VP-containing neurons, we conducted an examination using 5-OHDA labeling combined with VP immunocytochemistry and discovered that 5-OHDA-labeled terminals make synaptic contacts with VP-LI nerve cell bodies and fibers in the PVN and SON of the rat hypothalamus (Nakada and Nakai, 1985).

Most of the synapses between [^{3}H]NE- or 5-OHDA-labeled terminals and VP-LI neurons were asymmetrical types characterized by an accumulation of dense material on the cytoplasmic face beneath the postsynaptic membrane (Fig. 1 c), but the symmetrical type of synapse was also found occasionally. In many cases, however, it was very difficult to distinguish the dense postsynaptic material from the dense reaction product.

Presynaptic axon terminals labeled with [^{3}H]NE contained numerous small, spherical, clear vesicles (40–60 nm in diameter) and a few large granular vesicles (80–100 nm in diameter) in both the PVN and the SON. Besides the [^{3}H]NE-labeled presynaptic axon terminals, unlabeled presynaptic axon terminals of the same type were also found in relative abundance (Fig. 1 a). Furthermore, we occasionally observed other unlabeled and immunonegative presynaptic axon terminals containing only small, spherical, clear vesicles or numerous small, flattened, clear vesicles (30–40 and 50–60 nm in diameter).

It has been reported that the PVN and SON of the rat are innervated by relatively dense catecholaminergic terminals (Fuxe 1965; Versteeg et al. 1976; Hökfelt et al. 1974; Swanson et al. 1981), and that noradrenergic nerve fibers innervating the PVN and SON originate from the locus ceruleus (A6), ventrolateral medulla (A1), and dorsal vagal complex (A2; Sawchenko and Swanson 1981, 1982). Swanson et al. (1981) reported that noradrenergic fibers appeared to innervate the parts of the PVN and SON which contained predominantly vasopressinergic neurons in the rat. It has been reported that [^{3}H]NE is not retained in dopaminergic (Descarries et al. 1980) or serotoninergic terminals (Rothman et al. 1976), and that metabolites of NE are not retained through the autoradiographic procedure (Descarries and Dupin 1974). On the basis of the above studies and our present findings, [^{3}H]NE-labeled terminals making synaptic contacts with VP neurons may be considered to be noradrenergic ones, and it is strongly suggested that noradrenergic neurons may regulate VP secretion via synapses with other unknown neurons in the rat PVN and SON.

3.2 Catecholamine-OT Interactions

Immunonegative axon terminals making synaptic contacts with OT-like immunoreactive (OT-LI) nerve cell bodies and fibers have been demonstrated in the PVN and SON of the rat hypothalamus by means of electron microscopic immunocytochemistry (Hashimoto 1982). The following presynaptic terminals have been found:

1. Those containing only small, spherical, clear vesicles (40–50 nm in diameter)
2. Those containing a few relatively large granular vesicles (80–100 nm in diameter) with numerous small, clear vesicles
3. Those mainly occupied with numerous small, flattened, clear vesicles (F type, 30–40 and 55–60 nm in diameter)

As to catecholamine–OT interactions, [^{3}H]NE-labeled axon terminals making synaptic contacts with OT-LI nerve cell bodies and fibers containing OT-LI neurosecretory granules (160–200 nm in diameter) have been found in the rat PVN using the combined technique (Fig. 2). Hashimoto (1982), using the same combined technique as described in Sect. 3.1, demonstrated [^{3}H]NE- or [^{3}H]DA-labeled axon terminals intimately contacting OT-LI neurons in the rat PVN.

Presynaptic elements labeled with [^{3}H]DA and [^{3}H]NE and forming synapses on both VP- and OT-containing neurons belong mostly to the second type described above. The third type of presynaptic element was not labeled with [^{3}H]DA, [^{3}H]NE, or 5-OHDA. The third type of presynaptic elements containing flattened, clear vesicles may be considered to belong to the inhibitory synapse postulated by Uchizono (Uchizono 1965, 1967).

Swanson et al. (1981), using immunohistochemistry with antisera against catecholamine-synthesizing enzymes, reported that significant numbers of dopaminergic (tyrosine hydroxylase-stained) neurons are present in the rat PVN. Also, Olschowka et al. (1981) reported that at the ultrastructural level, dopamine-β-hydroxylase (DBH)-immunoreactive terminals form synapses with the dendritic processes in the PVN. This study and our present findings suggest that catecholamine-containing neurons may innervate OT neurons to control OT secretion via synapses with other unidentified neurons.

4 Catecholamine-LHRH Interactions

Physiological and other studies have shown that DA (Bennet et al. 1975; Negro-Vilar et al. 1979) and NE (Sawyer and Clifton 1980; Leung 1985) are involved in the regulation of LHRH secretion. Sawyer and Radford (1978) have suggested that NE has more of the physiologicopharmacologic characteristics of a natural central nervous activator of LHRH than epinephrine has. Jennes et al. (1982, 1983) and Sladek and Hoffman (1983), using immunohistochemical double staining on the same section with the antisera against LHRH and the enzyme-synthesizing catecholamines, demonstrated morphologically that in the rat DBH-immunoreactive fibers are in close contact with LHRH-like immunoreactive (LHRH-LI) cell bodies in the area of the organum vasculosum of the lamina terminals (OVLT), and that tyrosine hydroxylase (TH)-positive axons are apposed to LHRH-containing neurons in the diagonal band of Broca (DBB) and the preoptic region. Hökfelt et al. (1978) suggested that most DBH-positive nerve terminals may represent noradrenergic rather than adrenergic terminals. Jennes et al. (1982) reported that the DBH-positive fibers in close apposition to LHRH perikarya in the septopreoptic complex of the rat may be noradrenergic.

At the electron microscopic level, Ajika (1979, 1980) showed by the double immunostaining technique that two kinds of nerve terminal stained consecutively

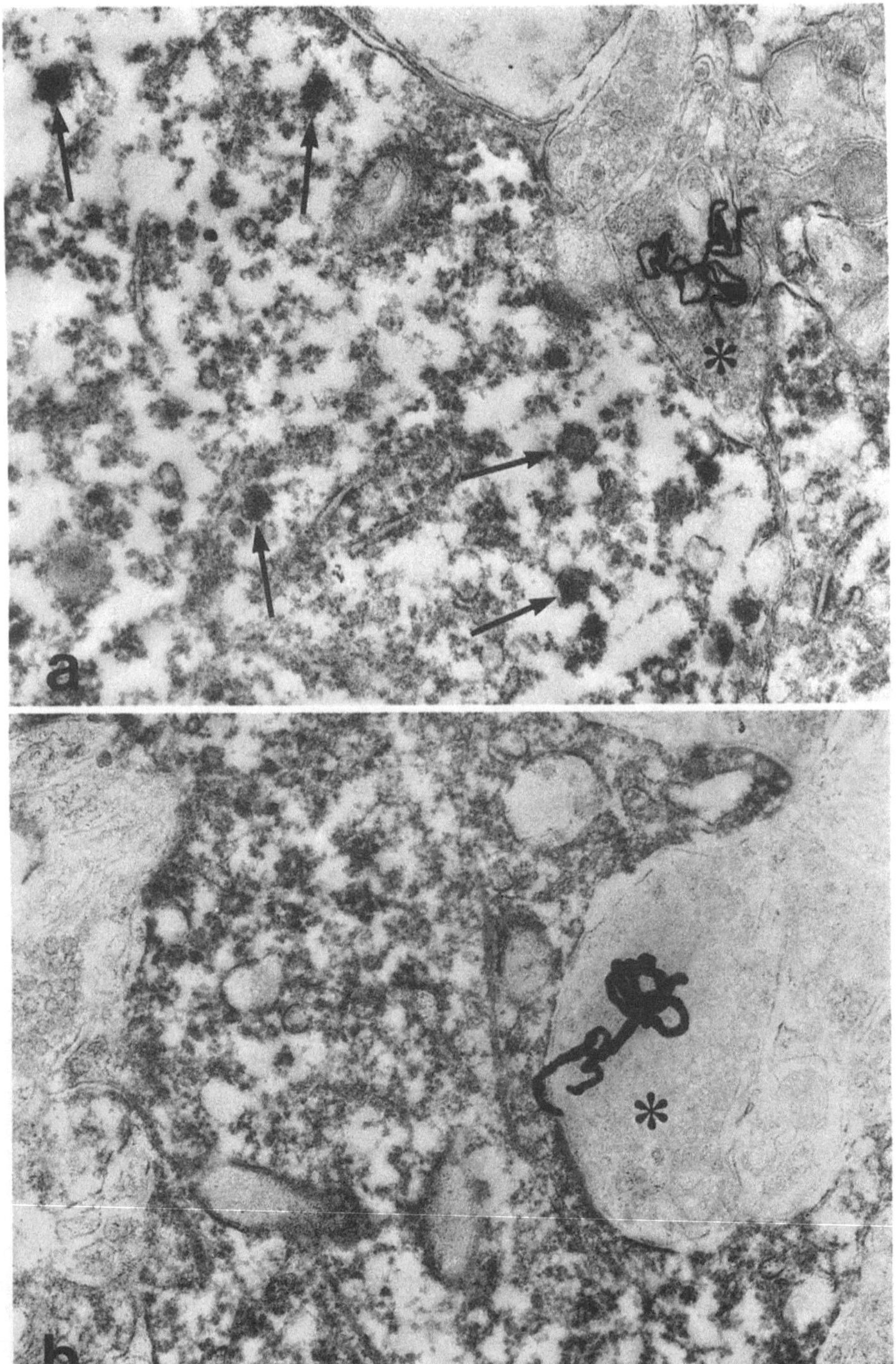
a
b

with LHRH and TH antiserum were in direct contact with each other in the median eminence of the rat, and suggested that DA released from the aminergic terminals may affect LHRH secretion by means of axoaxonic influence in the median eminence.

Although ultrastructural immunocytochemical studies of the LHRH neurons in various vertebrates have been reported by many authors, there have been only a few investigations of the afferent synaptic inputs to those neurons (Hisano et al. 1981; Jennes et al. 1985; Silverman and Witkin 1985). In the combination of LHRH immunocytochemistry with [^{3}H]NE autoradiography, axon terminals labeled with [^{3}H]NE were sometimes observed to make synaptic contacts with LHRH-LI cell bodies and processes in the preoptic area (POA; Fig. 3a), in the diagonal band of Broca (DBB), and around the OVLT (Fig. 3b). In the POA, [^{3}H]DA- or 5-OHDA-labeled axon terminals were sometimes found to make synaptic contacts with LHRH-LI perikarya (Nakai et al. 1985). Most of the above synapses were asymmetrical, but the symmetrical type was also found occasionally.

Besides the above catecholaminergic innervation of LHRH neurons, serotoninergic axons forming synaptic contacts with LHRH-LI neurons have recently been demonstrated in the medial preoptic area of the rat, using a combination of LHRH immunocytochemistry and [^{3}H]5-hydroxytryptamine autoradiography at the electron microscopic level (Kiss and Halász 1985). At the light microscopic

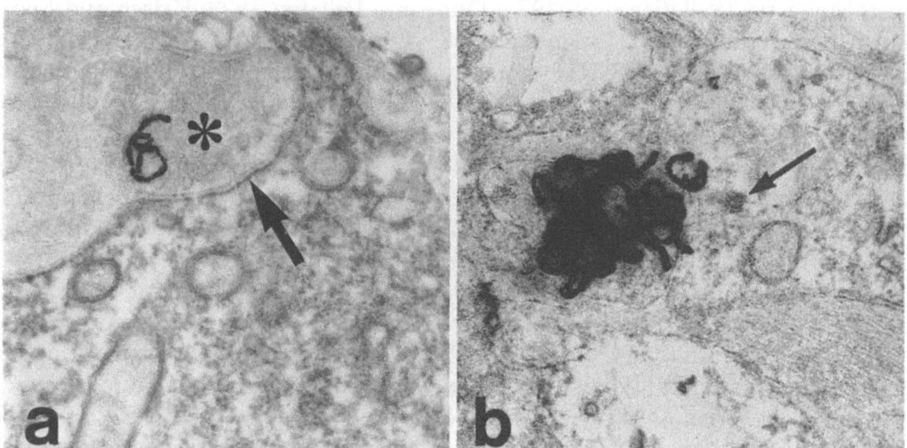

Fig. 3. a [^{3}H]NE-labeled terminal (*black star*) making synaptic contact (*arrow*) with LHRH-LI perikaryon in the POA. **b** Axon terminal densely labeled with [^{3}H]NE makes synapselike contact with LHRH-LI nerve fiber containing immunoreactive granule (*arrow*) in the OVLT. **a** × 22000; **b** × 30000

Fig. 2a, b. Synaptic interactions between [^{3}H]NE-labeled terminals and OT-LI neurons in the PVN. **a** [^{3}H]NE-labeled terminal (*black star*) making synaptic contact with OT-LI perikaryon containing immunoreactive neurosecretory granules (*arrows*). **b** [^{3}H]NE-labeled terminal (*black star*) making synaptic contact with OT-LI nerve fiber. **a** × 34000; **b** × 26000 level, serotoninergic or gamma-aminobutyric acid-(GABA)ergic terminals have been observed to be closely apposed to the LHRH-containing neurons in the septopreoptic region of the rat. In these studies immunohistochemical double staining with antiserum to serotonin or glutamic acid decarboxylase was used (Jennes et al. 1982, 1983). Recently, Leranth et al. (1985) demonstrated by a double-label electron microscopic immunostaining procedure that glutamic acid decarboxylase-immunoreactive terminals make synaptic contacts with LHRH-LI cell bodies and processes in the medial preoptic area of the rat. Although the presynaptic axon terminals occupied by numerous small, flattened vesicles, the so-called F type of vesicle, were rarely found to make synaptic contacts with LHRH neurons, it is not known whether they are equivalent to GABAergic terminals.

These morphological findings strongly suggest that catecholaminergic, indoleaminergic, or GABAergic neurons may be involved via synapses in the regulation of LHRH secretion.

5 Catecholamine-SRIF Interactions

SRIF-containing nerve cell bodies and fibers have been shown by light microscopic immunohistochemistry to be widely distributed in the rat hypothalamus. Also, their ultrastructures have been observed by electron microscopic immunohistochemistry (Pelletier et al. 1977; Dubé and Pelletier 1979; Krisch and Leonhardt 1980; Krisch 1980; Ibata et al. 1983; Ohtsuka et al. 1983; Takeuchi 1984). As to the innervation of SRIF-containing neurons, SRIF-like immunoreactive (SRIF-LI) neurons receiving synaptic inputs by immunonegative unknown axon terminals have been observed in the periventricular nucleus (Ibata et al. 1983) and in the AN of the rat hypothalamus (Ohtsuka et al. 1983). In our laboratory, the catecholaminergic terminals labeled with [^{3}H]NE or 5-OHDA have been found to make synaptic contacts with SRIF-LI nerve cell bodies and fibers both in the AN and the periventricular region of the preoptic nucleus (PON) of the rat hypothalamus (Nakai et al. 1982; Takeuchi 1984). Immunonegative terminals making synaptic contacts with the perikarya and processes of SRIF-LI neurons were often found in both the PON and the AN. The majority of the synapses appeared to be the asymmetrical type. Most of the presynaptic terminals contained numerous small, spherical, clear vesicles (40–50 nm in diameter) and a few relatively large, granular vesicles (70–100 nm in diameter). Other presynaptic terminals containing only numerous small, spherical, clear vesicles or with numerous small, flattened, clear vesicles were also recognized occasionally.

Postsynaptic SRIF-LI neurons in both the PON and the AN were generally similar in size and structure, but the size of the SRIF-LI granules seemed to be somewhat different. That is, the granules, ranging from 80 to 120 nm in the PON, were relatively smaller than those in the AN, which were about 100–160 nm in diameter.

According to the combined method of [^{3}H]NE autoradiography or 5-OHDA labeling with SRIF immunocytochemistry, [^{3}H]NE- or 5-OHDA-labeled terminals were found in both the PON (Fig. 4a) and the AN (Fig. 4b, c; Fig. 5a–c) to

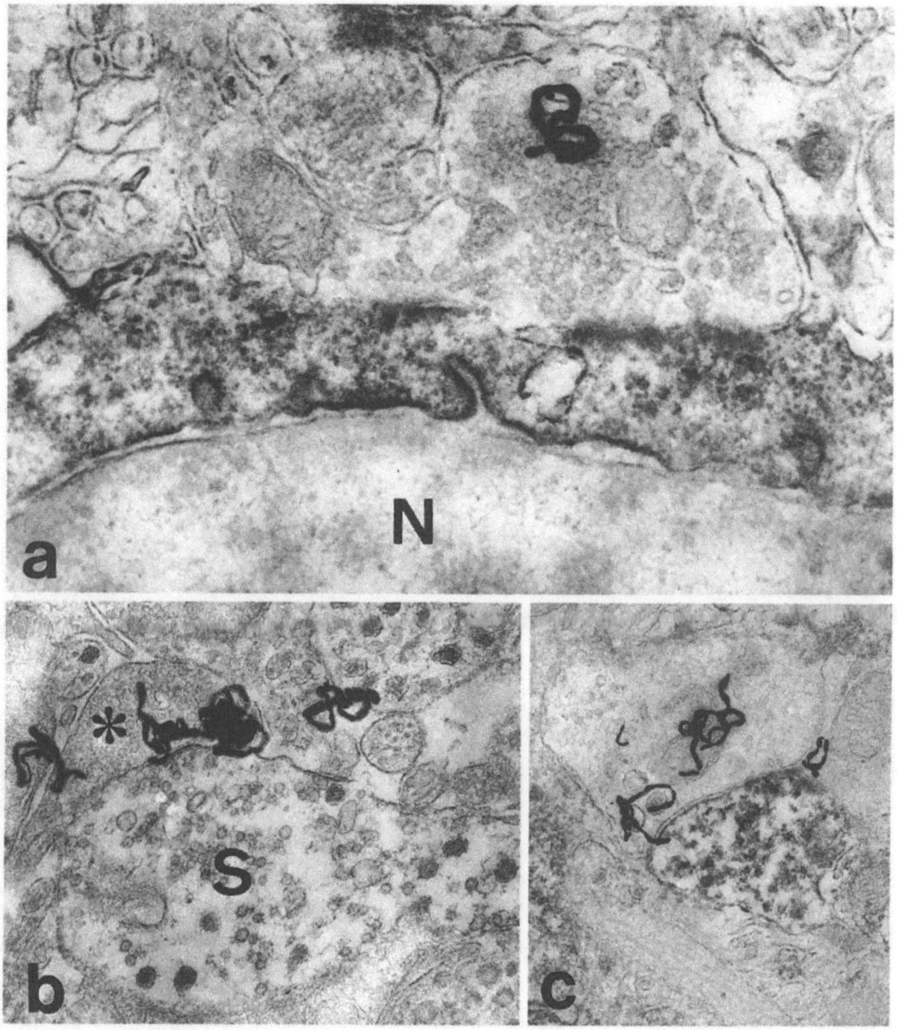

Fig. 4 a–c. Interactions between [^{3}H]NE-labeled terminals and SRIF-LI neurons. **a** [^{3}H]NE-labeled terminal making synaptic contact with SRIF-LI perikaryon in the PON. *N*, nucleus. **b** Axoaxonic synapselike structure between [^{3}H]NE-labeled terminal (*black star*) and SRIF-LI terminal (*S*) with many immunoreactive granules in the AN. **c** [^{3}H]NE-labeled terminal making synaptic contact with SRIF-LI nerve fiber in the AN. **a** × 46000; **b** × 29000; **c** × 25000

make axosomatic and axodendritic and/or axoaxonic synapses with SRIF-LI neurons, and the number of these synapses differed little between the two areas. The catecholaminergic presynaptic terminals contained numerous small, spherical, clear vesicles and a few large, granular vesicles (Figs. 4a and 5a). In the axon terminals labeled with 5-OHDA, the appearance of a dense core in small, spherical, clear vesicles and an increased electron density of the large granular vesicles were found. Most of the synapses were of the asymmetrical type (Fig. 5a, b),

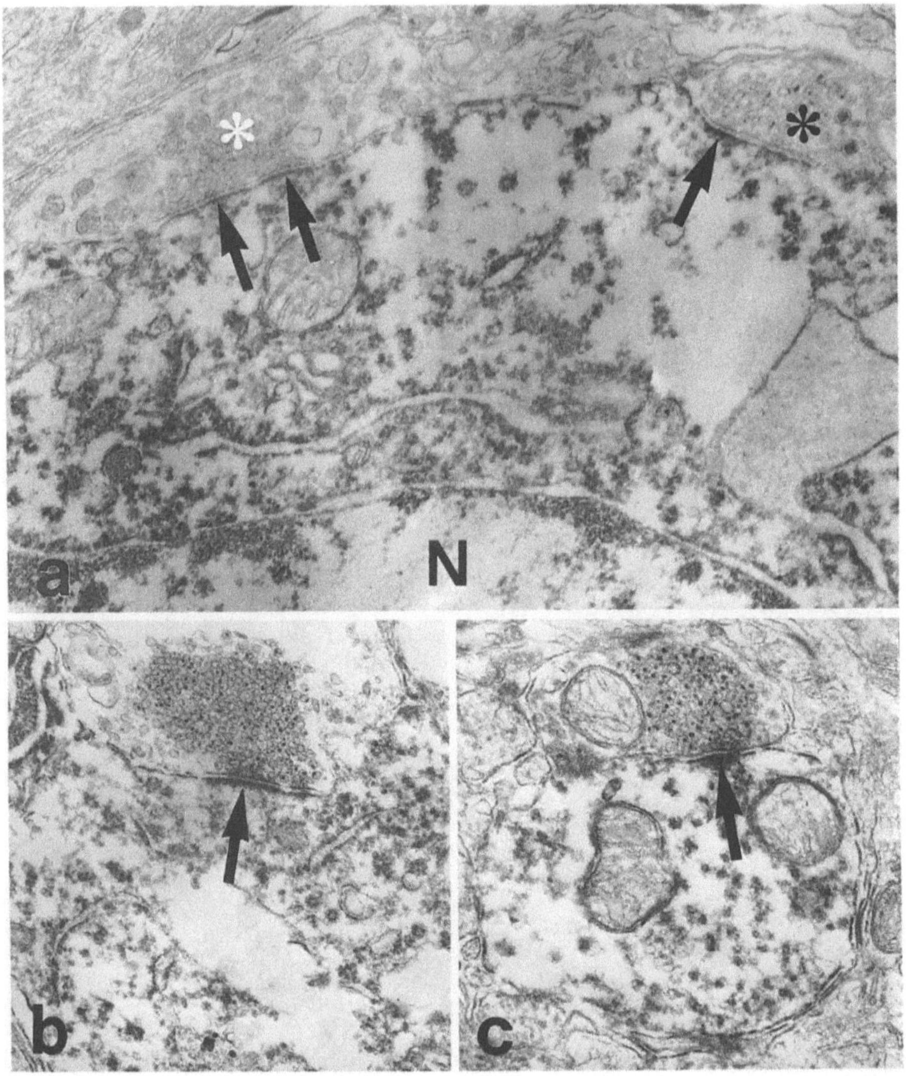

Fig. 5. a Both 5-OHDA-labeled (*black star*) and unlabeled (*white star*) terminals making synaptic contacts (*arrows*) with SRIF-LI perikaryon in the AN. *N*, nucleus. **b** and **c** 5-OHDA-labeled terminals making synaptic contacts (*arrows*) with SRIF-LI cell body (**b**) and process (**c**) in the AN. **a** × 38000; **b** × 30000; **c** × 37000

though uncertain types often appeared, owing to the accumulation of immunoreaction products beneath the postsynaptic membranes.

Chihara et al. (1979) investigated the functional interactions between monoamines and SRIF in the rat, and demonstrated that the secretion of SRIF into hypophyseal portal blood is apparently stimulated through the activation of catecholaminergic and cholinergic mechanisms, and that acetylcholine is less effective than DA or NE in stimulating SRIF secretion. The mechanism of these neurotransmitters is not clear, but it is possible that at least the secretion of SRIF neurons in the POA may be influenced directly by catecholaminergic or cholinergic neurons via synapses. Noradrenergic fibers projecting to the POA are shown to originate from the A1 and A2 neuron groups in the medulla oblongata (Dahlström and Fuxe 1964), and the group of dopaminergic cells (A14) in the rostral periventricular region are known to project their fibers to the POA (Björklund and Nobin 1973).

Unlabeled terminals also made synaptic contacts with the same SRIF-LI neurons with 5-OHDA- (Fig. 5 a) or [^{3}H]NE-labeled terminals (Nakai et al. 1985). The nature of these unlabeled terminals is unclear, but they are probably cholinergic, serotoninergic, or peptidergic presynaptic terminals. The presynaptic terminals containing large granular vesicles such as those seen in Fig. 5 a may be considered to be peptidergic.

Kawano et al. (1982) demonstrated that in the rat, SRIF neurons in the rostral periventricular area project fibers to the median eminence, but that those in the AN do not. Thus, they postulated that two kinds of SRIF neuronal system exist in the rat hypothalamus: one is involved in the production of hormonal SRIF, and the other serves to regulate neuronal activities in the restricted hypothalamic nuclei. Thus, secretion of SRIF neurons located in the periventricular POA may be regulated by catecholaminergic neurons, but the physiological significance of catecholaminergic innervation of SRIF neurons in the AN is unknown.

6 Catecholamine-TRH Interactions

TRH-like immunoreactive (TRH-LI) cell bodies are widely distributed in regions of the rat hypothalamus, including the PVN, dorsomedial nucleus, PON, periventricular region, lateral hypothalamic area, and perifornical area (Hökfelt et al. 1978; Johansson and Hökfelt 1980; Lechan and Jackson 1982). At the electron microscopic level, the immunonegative terminals forming synaptic inputs to TRH-LI neurons have been observed in the dorsomedial nucleus and in the area dorsal to the chiasma opticum of the rat hypothalamus (Johansson et al. 1980). It has been demonstrated by immunohistochemistry, using antisera to catecholamine-synthesizing enzymes, that noradrenergic fibers (Hökfelt et al. 1978) and adrenergic fibers (Hökfelt et al. 1978; Swanson et al. 1981) are distributed throughout the entire PVN. Biochemical and physiological studies have identified that TRH release may be stimulated by NE (Krulich et al. 1977; Moss et al. 1978) or inhibited by it (Dyer and Dyball 1974) or by DA (Krulich et al. 1977). As to the monoamine–TRH interactions, monoaminergic terminals labeled with [^{3}H]DA, [^{3}H]NE, or [^{3}H]serotonin in close contact with TRH-LI terminals in the vicinity of the primary portal capillaries in the rat median eminence have been demonstrated using a technique combining immunocytochemistry with autoradiography (Nakai et al. 1983; Shioda et al. 1983). In the median eminence, however, typical synapses were not found between them. Recently, using the combined method of TRH immunocytochemistry with [^{3}H]NE autoradiography or 5-OHDA labeling, we found that the axon terminals labeled with [^{3}H]NE or 5-

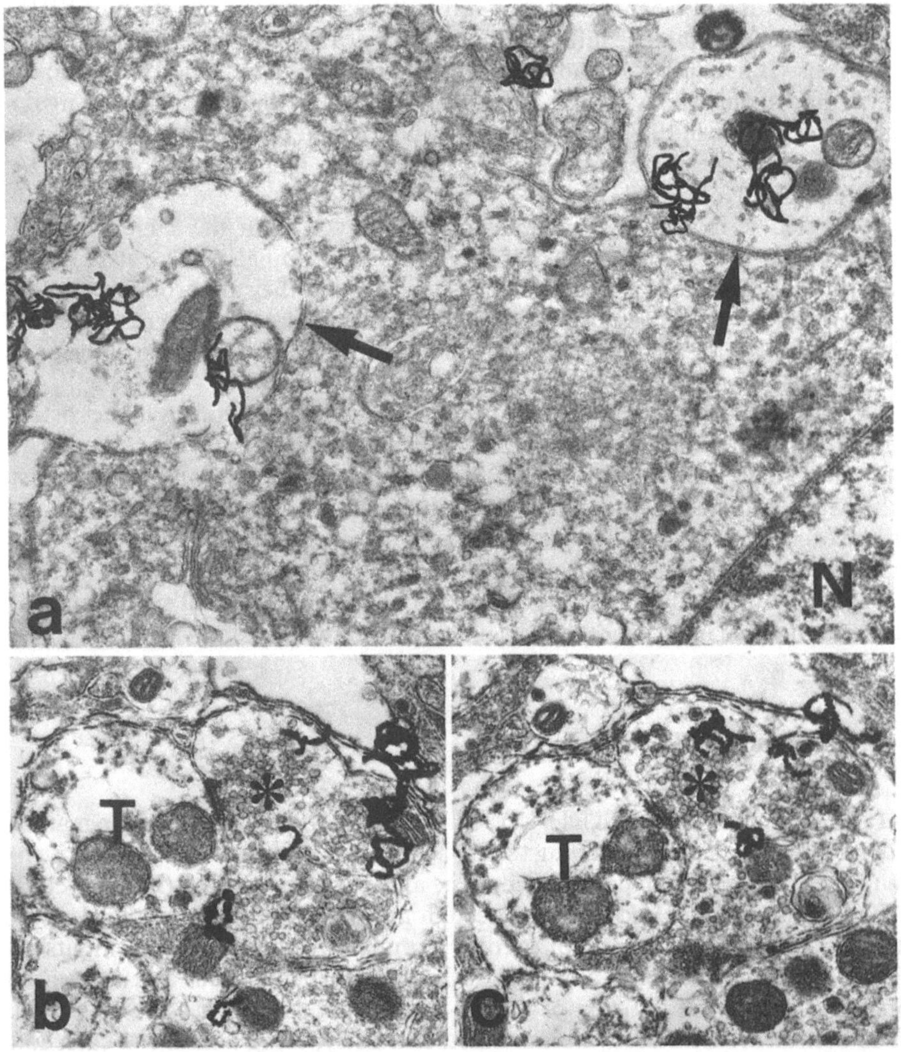

Fig. 6. a Two [^{3}H]NE-labeled terminals making synaptic contacts (*arrows*) with TRH-LI perikaryon with many immunoreactive granular vesicles in the PVN. *N*, nucleus. **b** and **c** Serial sections showing [^{3}H]NE-labeled terminal (*black star*) making synaptic contacts with TRH-LI nerve fibers (*T*) in the PVN. **a** × 20000; **b** and **c** × 23000

OHDA make synapses with TRH-LI perikarya and cell processes mainly in the parvocellular part of the PVN of the rat hypothalamus (Figs. 6 and 7). The postsynaptic TRH neurons contained dense immunoreactive granular vesicles (about 90–140 nm in diameter). The presynaptic terminals labeled with [^{3}H]NE or 5-OHDA mostly contained a large number of small, spherical, clear vesicles (30–50 nm in diameter) and a few dense granular vesicles ranging 70–90 nm in diameter (Fig. 6c, 7b). Most of these catecholaminergic synapses appeared to be of the asymmetrical type, though symmetrical types were also found occasionally.

Fig. 7 a, b. Axon terminals labeled with 5-OHDA (*black star*) make synapses (*arrows*) with **a** TRH-LI nerve cell body and **b** fiber in the PVN. A large number of TRH-LI granules are seen in the cytoplasm of the cell body and the fiber. **a** × 30 000; **b** × 28 000

This suggests that catecholaminergic neurons regulate TRH secretion by TRH-containing neurons via synapses, and strongly supports the findings of the above biochemical studies (Dyer and Dyball 1974; Krulich et al. 1977; Moss et al. 1978).

7 Catecholamine-CRF Interactions

In the rat hypothalamus, CRF-containing cell bodies have been immunohistochemically demonstrated to be localized in the parvocellular part of the PVN (Bugnon et al. 1982; Merchenthaler et al. 1982; Olschowka et al. 1982; Antoni et al. 1983; Paull and Gibbs 1983; Kawata et al. 1983; Swanson et al. 1983). Moreover, in the rat, CRF-like immunoreactive (CRF-LI) fibers arising from the PVN project into the external layer of the median eminence as revealed by the deafferentation technique (Antoni et al. 1983; Merchenthaler et al. 1984). Unidentified immunonegative terminals have been found to form asymmetrical synapses on CRF-LI nerve cell bodies and fibers in the rat PVN (Liposits et al. 1983; Shioda et al. 1985). Here we present some evidence of catecholamine–CRF synaptic interactions recently observed in the rat hypothalamic PVN using the combined technique of CRF immunocytochemistry with [^{3}H]NE autoradiography. In the PVN, axon terminals labeled with [^{3}H]NE made synaptic contacts with CRF-LI nerve cell bodies (Fig. 8a) and with nerve fibers (Fig. 8b) containing CRF-LI granular vesicles ranging from 80 to 120 nm in diameter. The presynaptic terminals labeled with [^{3}H]NE contained both numerous small, spherical, clear vesicles (40–60 nm in diameter) and a few granular vesicles (60–100 nm in diameter). The presynaptic terminals labeled with [^{3}H]NE are probably either noradrenergic or adrenergic ones. These axosomatic and axodendritic and/or axoaxonic synapses appeared to be mostly of the asymmetrical type.

Pharmacologic studies have suggested the inhibitory effect of NE (Jones and Hillhouse 1977) and epinephrine (Roth et al. 1981; Mezey et al. 1984) in the regulation of CRF secretion.

These findings by our combined method and the other studies indicate the possibility that catecholaminergic neurons, probably containing NE or epinephrine, may play a role in the inhibitory regulation of secretory activity of CRF-containing neurons through direct synaptic input.

8 Catecholamine-ACTH/β-End Interactions

Opioid peptides are classified into three systems: ACTH-β-lipotropin (pro-opiomelanocortin), and the proenkephalin A and proenkephalin B (prodynorphin) systems (Marx 1983). From these three systems the three separate opioid systems, β-End, enkephalin, and dynorphin, are derived. These are present in brain as indicated by immunohistochemical studies.

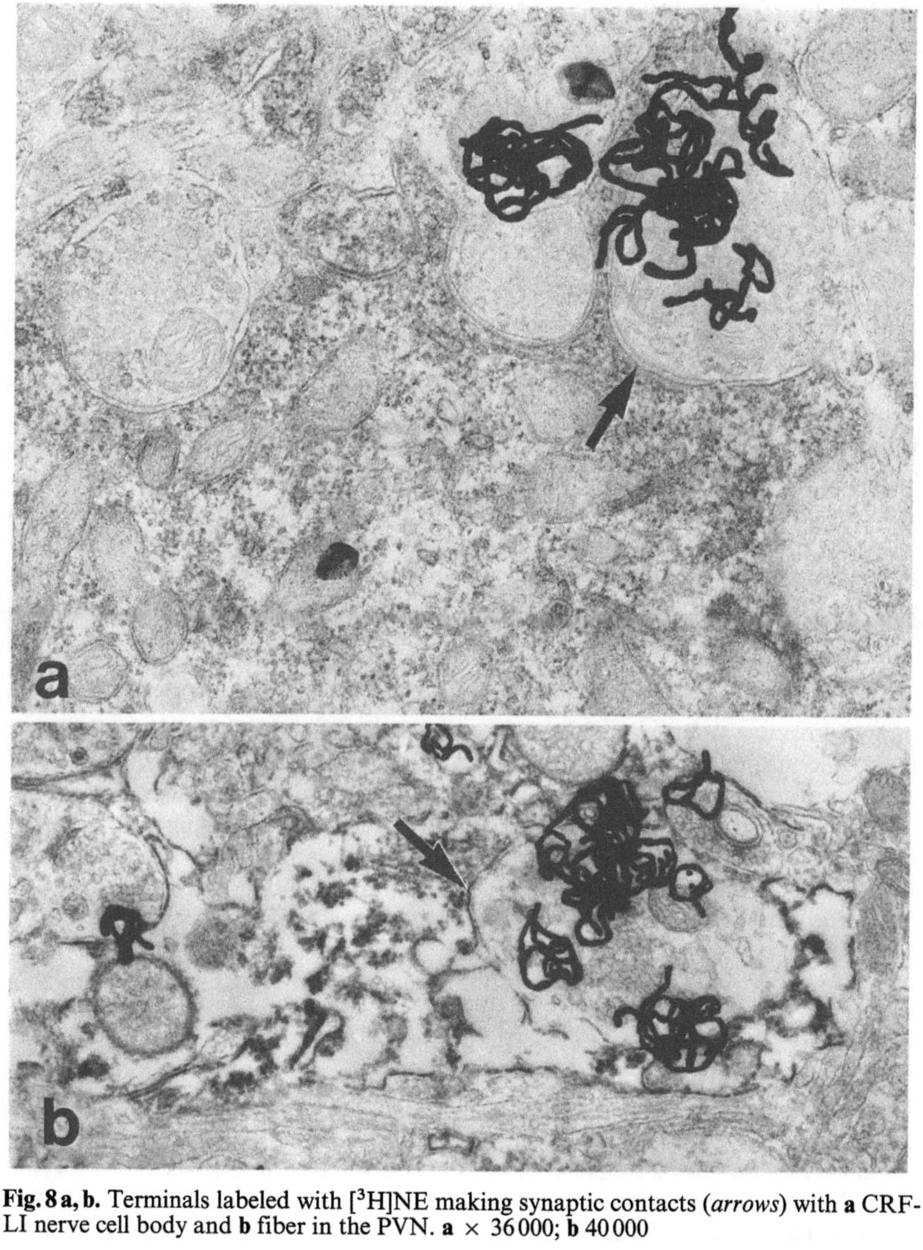

Fig. 8 a, b. Terminals labeled with [^{3}H]NE making synaptic contacts (*arrows*) with **a** CRF-LI nerve cell body and **b** fiber in the PVN. **a** × 36000; **b** 40000

Neurons containing β-End exist separately from those containing enkephalin in rat brain (Bloom et al. 1978). Two opioid systems in the brain, an enkephalin system and a β-lipotropin (β-LPH)/β-End system, have been distinguished on the basis of biochemical and anatomical evidence. Also, immunocytochemical evidence for the coexistence of ACTH with β-LPH/β-End has been demonstrated in rat brain (Watson et al. 1978). Dynorphin, which was extracted and purified from porcine pituitary (Goldstein et al. 1979), has been revealed by radioimmunoassay and immunohistochemistry to be widely distributed in the rat central nervous system. It is suggested that dynorphin occurs in neuronal systems that are immunocytochemically distinct from those containing other opioid peptides in the rat central nervous system (Khachaturian et al. 1982). We are studying the synaptic interactions between catecholaminergic neurons and ACTH/β-End-, methionine-enkephalin-, or dynorphin-containing neurons, using the combined method, but in the present paper only the observations on catecholaminergic innervation of ACTH/β-End neurons are reported.

It has been shown by light microscopic immunocytochemistry that ACTH and β-End are contained in the same neurons in the hypothalamic AN (Watson et al. 1978; Bloch et al. 1979; Bugnon et al. 1979; Nilaver et al. 1979; Sofroniew 1979). At the electron microscopic level, neurons immunoreactive with ACTH (Hisano et al. 1982) and β-End (Hisano et al. 1982) or β-End/β-LPH (Ibata et al. 1985) have been observed in the AN of the colchicine-treated rats. No difference has been recognized in the perikaryal ultrastructures of ACTH- and β-End-immunoreactive neurons in the AN. Two unknown types of immunonegative axon terminal, one containing many membrane-limited granules and the other occupied by numerous small clear vesicles, were found to make axosomatic synapses with the ACTH/β-End-immunoreactive neurons (Hisano et al. 1982). Bugnon et al. (1981) reported an unknown axon terminal making axosomatic synapses with ACTH-containing neurons in the rat AN in adjacent ultrathin sections of immunostained thick sections. Ibata et al. (1985) also reported that β-End/βLPH-like-immunoreactive neurons received axosomatic and axodendritic synaptic inputs from unidentified immunonegative neurons in the rat AN.

Concerning the synaptic interactions between monoaminergic axon terminals and ACTH- or β-End-containing neurons, Kiss et al. (1984), using a combination of ACTH immunocytochemistry with [^{3}H]serotonin autoradiography at the electron microscopic level, have recently found terminals showing selective uptake of [^{3}H]serotonin forming synaptic contacts with ACTH-like-immunoreactive (ACTH-LI) perikarya or dendrites in the AN of the rat hypothalamus. Also, in our laboratory, with the technique combining immunocytochemistry using antisera to ACTH$^{1\text{-}24}$ or β-End with [^{3}H]DA autoradiography or 5-OHDA labeling, we found catecholaminergic synaptic inputs to ACTH-LI or β-End-like-immunoreactive (β-End-LI) neurons in the AN of the rat hypothalamus (Figs. 9 and 10). Axon terminals labeled with [^{3}H]DA have been observed to make synaptic contacts with ACTH-LI cell bodies (Fig. 9) or β-End-LI nerve fibers (Fig. 10). Similar synaptic interactions between 5-OHDA-labeled terminals and ACTH-LI or β-End-LI perikarya have also been recognized. The above findings of Kiss et al. (1984) and our laboratory are considered to provide ultrastructural evidence that serotoninergic, dopaminergic, or other catecholaminergic nerve fibers may innervate the perikarya or processes of both ACTH- and β-End containing neurons to regulate ACTH or β-End secretion. In the present studies, ACTH-LI or β-End-LI granules were found in the perikarya, and the diameter of ACTH-LI and β-End-LI granules was almost the same, ranging from 80 to 100 nm.

Concerning the interactions between endogenous opiates and DA, it has been shown that in the rat administration of endogenous opiates reduces hypothalamic

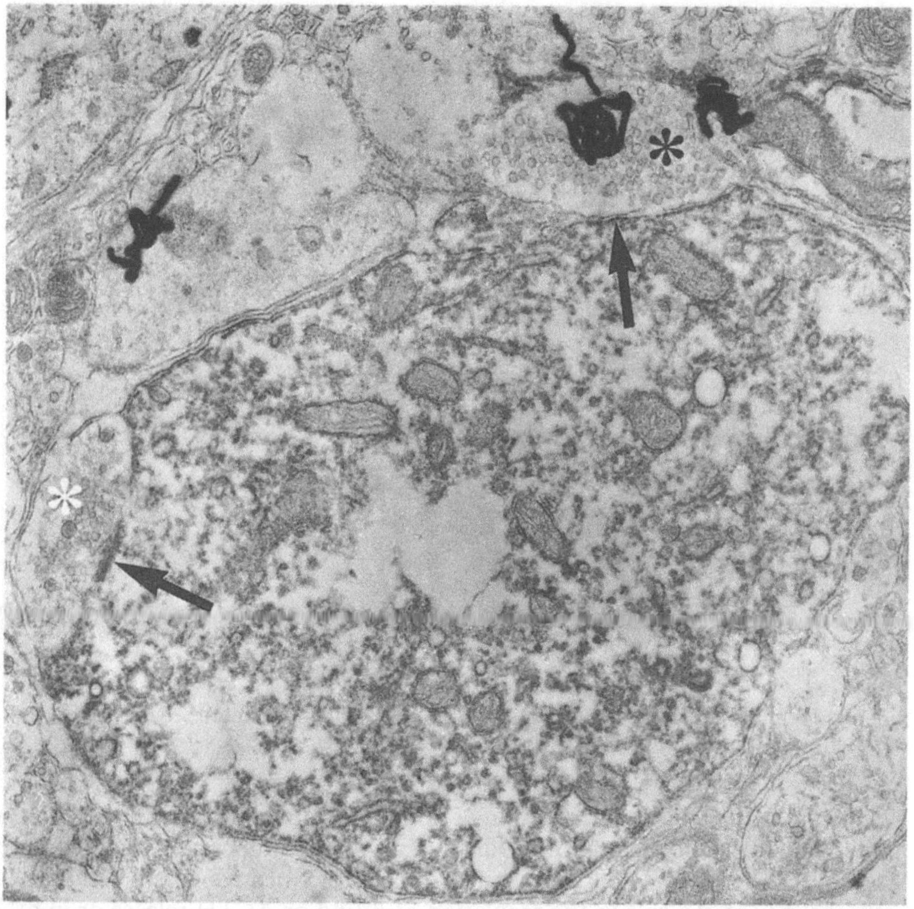

Fig. 9. Both [^{3}H]DA-labeled (*black star*) and unlabeled (*white star*) terminals making synaptic contacts (*arrows*) with an ACTH-LI cell body in the AN. × 24000

DA activity (Van Vugt and Meites 1980) and inhibits the release of DA from the tuberoinfundibular neurons into the pituitary portal vessels (Gudelsky and Porter 1979; Haskins et al. 1981).

It is well-known that in the rat intraventricular administration of β-End increases prolactin secretion from the anterior pituitary, and it is postulated that β-End may enhance prolactin secretion by inhibiting the release of DA from the tuberoinfundibular DA neurons. However, to our knowledge there is no morphological evidence of synaptic input by β-End neurons onto DA neurons in the hypothalamus. Also, the coexistence of ACTH and DA in the same neurons in the hypothalamus has not been shown, though immunoreactive ACTH and β-End are stored in the same neurons of the AN of the rat hypothalamus (Bugnon et al. 1979).

Recently, Locatelli et al. (1983) have reported that β-End concentrations in both the medial basal hypothalamus and the median eminence of the rat decrease

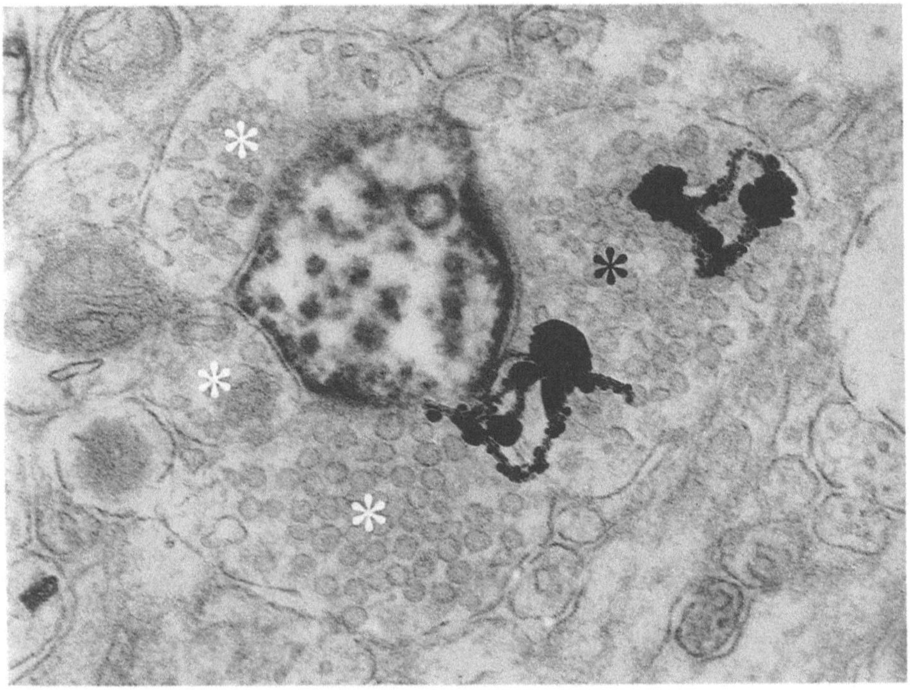

Fig. 10. Terminal labeled with [^{3}H]DA (*black star*) and three unlabeled terminals (*white stars*) make synaptic contacts with a β-End-LI nerve fiber in the AN. × 33000

significantly after the administration of dopaminergic agents, but the mechanism underlying this action is obscure. Our findings of dopaminergic synaptic input to both β-End- and ACTH-containing neurons, as described above, suggest that DA neurons may play a role in the regulation of β-End or ACTH release or in other functions of these neurons via synapses.

9 Catecholamine-NT Interactions

It has been demonstrated by light microscopic immunocytochemistry that NT-like immunoreactive (NT-LI) perikarya are present mainly in the PVN, the AN, the periventricular nucleus, the lateral hypothalamic area, and the medial preoptic area of the rat hypothalamus. At the electron microscopic level, NT-LI neurons have been found to receive synaptic inputs from unidentified immunonegative axon terminals in the AN and the PVN of the rat hypothalamus (Ibata et al. 1984a, b).

Concerning catecholamine–NT interactions, it has been suggested from the morphological finding of a marked decrease in the number of NT-LI perikarya after transection of the ascending noradrenergic pathway that the ascending noradrenergic neurons influence the NT neurons in the rat PVN (Ibata et al. 1984b;

Kawakami et al. 1984). Recently, using the technique combining NT immunocytochemistry with [^{3}H]DA autoradiography or 5-OHDA labeling, NT-LI cell bodies and processes have been found in our laboratory to receive synaptic inputs from catecholaminergic terminals labeled with [^{3}H]DA or 5-OHDA and from unlabeled terminals in the PVN and the AN of the rat hypothalamus (Doi and Nakai, 1986).

Ibata et al. (1984a) demonstrated by immunohistochemistry that in the rat NT-LI terminals extending from NT-LI neurons in the AN are located very close to the pericapillary spaces of the hypophyseal portal capillaries in the median eminence. From this study it appears that NT is released into the hypophyseal portal capillaries to control the release of some pituitary hormones. Although the exact physiological significance of the catecholaminergic innervation of NT neurons in the hypothalamus is still unknown, it is suggested that catecholamine neurons may regulate NT secretion via synapses.

10 Summary

The synaptic interactions between catecholaminergic terminals and peptide-containing neurons in the rat hypothalamus were examined at the electron microscopic level using the combined technique of neuropeptide immunocytochemistry with [^{3}H]DA or [^{3}H]NE autoradiography or 5-OHDA labeling in the same tissue section. In the present paper the catecholaminergic synaptic inputs to the peptidergic neurons which contain VP, OT, LHRH, SRIF, TRH, CRF, ACTH, β-End, and NT have been described and discussed.

Axon terminals labeled with [^{3}H]NE were found to make axosomatic and axodendritic and/or axoaxonic synapses on the VP-LI neurons in the PVN and the SON, on the LHRH-LI neurons in the POA and the DBB, on the SRIF-LI neurons in the PON and the AN, and on the TRH-LI and CRF-LI neurons in the PVN. Terminals labeled with [^{3}H]DA were found to make synaptic contacts with ACTH-LI and β-End-LI cell bodies in the AN, and with NT-LI cell bodies and processes in the PVN and the AN. Furthermore, VP-, SRIF-, TRH-, and NT-containing neurons were recognized to receive synaptic inputs by axon terminals showing uptake of 5-OHDA.

It is considered that these findings suggest catecholaminergic neurons play some important roles in the regulation of the secretory or other functions of the above neuropeptide-containing neurons. Also, it is suggested that in the rat hypothalamus catecholaminergic fibers and peptidergic neurons containing hypothalamic hormones such as LHRH, SRIF, TRH, or CRF probably interact with each other via synapses to control the secretion of anterior pituitary hormones.

Acknowledgements. The authors wish to thank Prof. D. W. Pfaff of Rockefeller University for reviewing the manuscript, and also Miss Y. Ishimura for preparation of the manuscript.

References

Ajika K (1979) Simultaneous localization of LHRH and catecholamines in rat hypothalamus. J Anat 128:331–347
Ajika K (1980) Relationship between catecholaminergic neurons and hypothalamic hormone-containing neurons in the hypothalamus. In: Martini L, Ganong WF (eds) Frontiers in neuroendocrinology, vol 6. Raven, New York, pp 1–32
Antoni FA, Palkovits M, Makara GB, Linton EA, Lowry PJ, Kiss JZ (1983) Immunoreactive corticotropin-releasing hormone in the hypothalamoinfundibular tract. Neuroendocrinology 36:415–423
Basbaum AI, Glazer EJ, Lord BAP (1982) Simultaneous ultrastructural localization of tritiated serotonin and immunoreactive peptides. J Histochem Cytochem 30:780–784
Bennett GW, Edwardson JA, Holland D, Jeffcoate SL, White N (1975) Release of immunoreactive luteinising hormone-releasing hormone and thyrotropin-releasing hormone from hypothalamic synaptosomes. Nature 257:323–324
Björklund A, Nobin A (1973) Fluorescence, histochemical and microspectrofluorometric mapping of dopamine and noradrenaline cell groups in the rat diencephalon. Brain Res 51:193–205
Bloch B, Bugnon C, Fellmann D, Lenys D, Gouget A (1979) Neurons of the rat hypothalamus reactive with antisera against endorphins, ACTH, MSH, and β-LPH. Cell Tissue Res 204:1–15
Bloom F, Battenberg E, Rossier J, Ling N, Guillemin R (1978) Neurons containing β-endorphin in rat brain exist separately from those containing enkephalin: immunocytochemical studies. Proc Natl Acad Sci USA 75:1591–1595
Bridges TE, Hillhouse EW, Jones MT (1976) The effect of dopamine on neurohypophysial hormone release in vivo and from the rat neural lobe and hypothalamus in vitro. J Physiol (Lond) 260:647–666
Bugnon C, Bloch B, Lenys D, Gouget A, Fellmann D (1979) Comparative study of the neuronal populations containing β-endorphin, corticotropin and dopamine in the arcuate nucleus of the rat hypothalamus. Neurosci Lett 14:43–48
Bugnon C, Bloch B, Lenys D (1981) Ultrastructural study of presumptive pro-opiocortin producing neurons in the rat hypothalamus. Neuroscience 6:1299–1313
Bugnon C, Fellman D, Gouget A, Cardot J (1982) Corticoliberin in rat brain: immunocytochemical identification and localization of a novel neuroglandular system. Neurosci Lett 30:25–30
Chihara K, Arimura A, Schally AV (1979) Effect of intraventricular injection of dopamine, norepinephrine, acetylcholine, and 5-hydroxytryptamine on immunoreactive somatostatin release into rat hypophyseal portal blood. Endocrinology 104:1656–1662
Dahlström A, Fuxe K (1964) Evidence for existence of monoamine-containing neurons in the central nervous system. I. Demonstration of monoamines in the cell bodies of brain stem neurons. Acta Physiol Scand 62[Suppl 232]:1–55
Descarries L, Dupin JC (1974) Retention of noradrenaline-^{3}H in brain and preferential extraction of labeled metabolites by glutaraldehyde fixation. Experientia 30:1164–1165
Descarries L, Bosler O, Berthelet F, Des Rosiers MH (1980) Dopaminergic nerve endings visualized by high resolution autoradiography in adult rat neostriatum. Nature 284:620–622
Doi H, Nakai Y (1985) Electron microscopic cytochemistry of catecholaminergic innervation of neurotensin neurons in the rat hypothalamus. J Clin Electron Microscopy 18(4):367–374
Dubé D, Pelletier G (1979) Effect of colchicine on the immunohistochemical localization of somatostatin in the rat brain: light and electron microscopic studies. J Histochem Cytochem 27:1577–1581
Dyer RG, Dyball REG (1974) Evidence for a direct effect of LRF and TRH on single unit activity in the rostral hypothalamus. Nature 252:486–488
Fuxe K (1965) Evidence for the existence of monoamine neurons in the central nervous system. IV. The distribution of monoamine nerve terminals in the central nervous system. Acta Physiol Scand 247:37–85

Glazer EJ, Basbaum AI (1984) Axons which take up [^{3}H]serotonin are presynaptic to enkephalin-immunoreactive neurons in cat dorsal horn. Brain Res 298:386–391

Goldstein A, Tachibana S, Lowney LI, Hunkapiller M, Hood L (1979) Dynorphin(1-13), an extraordinarily potent opioid peptide. Proc Natl Acad Sci USA 76:6666–6670

Gudelsky GA, Porter JC (1979) Morphine and opioid peptide-induced inhibition of the release of dopamine from tuberoinfundibular neurons. Life Sci 25:1697–1702

Hashimoto A (1982) Electron microscopic cytochemistry on the innervation of oxytocin-producing cells (in Japanese, abstract in English). J Showa Med Assoc 42:805–818

Haskins JT, Gudelsky GA, Moss RL, Porter JC (1981) Iontophoresis of morphine into the arcuate nucleus: effects on dopamine concentrations in hypophysial portal plasma and serum prolactin concentrations. Endocrinology 108:767–771

Hisano S, Kawano H, Maki Y, Daikoku S (1981) Electron microscopic study of immunoreactive LHRH perikarya with special reference to neuronal regulation. Cell Tissue Res 220:511–518

Hisano S, Kawano H, Nishiyama T, Daikoku S (1982) Immunoreactive ACTH/β-endorphin neurons in the tubero-infundibular hypothalamus of rats. Cell Tissue Res 224:303–314

Hökfelt T, Fuxe K, Goldstein M, Johansson O (1974) Immunohistochemical evidence for the existence of adrenaline neurons in the rat brain. Brain Res 66:235–251

Hökfelt T, Elde R, Fuxe K, Johansson O, Ljungdahl Å, Goldstein M, Luft R, Efendic S, Nilsson G, Terenius L, Ganten D, Jeffcoate SL, Rehfeld J, Said S, Perez de la Mora M, Possani L, Tapin R, Teran L, Palacios R (1978) Aminergic and peptidergic pathways in the nervous system with special reference to the hypothalamus. In: Reichlin S, Baldessarini RJ, Martin JB (eds) The hypothalamus. Raven, New York, pp 69–135

Ibata Y, Obata HL, Kubo S, Fukui K, Okamura H, Ishigami T, Imagawa K, Sin S (1983) Some cellular characteristics of somatostatin neurons and terminals in the periventricular nucleus of the rat hypothalamus and median eminence. Electron microscopic immunohistochemistry. Brain Res 258:291–295

Ibata Y, Kawakami F, Fukui K, Obata-Tsuto HL, Tanaka M, Kubo T, Okamura H, Morimoto N, Yanaihara C, Yanaihara N (1984a) Light and electron microscopic immunocytochemistry of neurotensin-like immunoreactive neurons in the rat hypothalamus. Brain Res 302:221–230

Ibata Y, Kawakami F, Fukui K, Okamura H, Obata-Tsuto HL, Tsuto T, Terubayashi H (1984b) Morphological survey of neurotensin-like immunoreactive neurons in the hypothalamus. Peptides 5:109–120

Ibata Y, Kawakami F, Okamura H, Obata-Tsuto HL, Morimoto N, Zimmermann EA (1985) Light and electron microscopic immunocytochemistry of β-endorphin/β-LPH-like immunoreactive neurons in the arcuate nucleus and surrounding areas of the rat hypothalamus. Brain Res 341:233–242

Jennes L, Beckman WC, Stumpf WE, Grzanna R (1982) Anatomical relationships of serotoninergic and noradrenergic projections with the GnRH system in septum and hypothalamus. Exp Brain Res 46:331–338

Jennes L, Stumpf WE, Tappaz ML (1983) Anatomical relationships of dopaminergic and GABAergic systems with the GnRH-system in the septo-hypothalamic area. Immunohistochemical studies. Exp Brain Res 50:91–99

Jennes L, Stumpf WE, Sheedy ME (1985) Ultrastructural characterization of gonadotropin-releasing hormone (GnRH)-producing neurons. J Comp Neurol 232:534–547

Johansson O, Hökfelt T (1980) Thyrotropin releasing hormone, somatostatin, and enkephalin: distribution studies using immunohistochemical techniques. J Histochem Cytochem 28:364–366

Johansson O, Hökfelt T, Jeffcoate SL, White N, Sternberger LA (1980) Ultrastructural localization of TRH-like immunoreactivity. Exp Brain Res 38:1–10

Jones MT, Hillhouse EW (1977) Neurotransmitter regulation of corticotropin-releasing factor in vitro. Ann NY Acad Sci 297:536–560

Kawakami F, Fukui K, Okamura H, Nakajima T, Yanaihara N, Ibata Y (1984) Influence of ascending noradrenergic fibers on the neurotensin-like immunoreactive neurons in the rat paraventricular nucleus. Neurosci Lett 44:149–154

Kawano H, Daikoku S, Saito S (1982) Immunohistochemical studies of intrahypothalamic somatostatin-containing neurons in rat. Brain Res 242:227–232

Kawata M, Hashimoto K, Takahara J, Sano Y (1983) Immunohistochemical identification of neurons containing corticotropin-releasing factor in the rat hypothalamus. Cell Tissue Res 230:239–246

Khachaturian H, Watson SJ, Lewis M, Coy D, Golstein A, Akil H (1982) Dynorphin immunocytochemistry in the rat central nervous system. Peptides 3:941–954

Kiss J, Halász B (1985) Demonstration of serotoninergic axons terminating on luteinizing hormone-releasing hormone neurons in the preoptic area of the rat using a combination of immunocytochemistry and high resolution autoradiography. Neuroscience 14:69–78

Kiss J, Léránth CS, Halász B (1984) Serotoninergic endings on VIP-neurons in the suprachiasmatic nucleus and on ACTH-neurons in the arcuate nucleus of the rat hypothalamus. A combination of high resolution autoradiography and electron microscopic immunocytochemistry. Neurosci Lett 44:119–124

Krisch B (1980) Differing immunoreactivities of somatostatin in the cortex and the hypothalamus of the rat. A light and electron microscopic study. Cell Tissue Res 212:457–464

Krisch B, Leonhardt H (1980) Luliberin and somatostatin fiber-terminals in the subfornical organ of the rat. Cell Tissue Res 210:33–45

Krulich L, Giachetti A, Marchlewska-Koj A, Helco E, Jameson HE (1977) On the role of the central noradrenergic and dopaminergic systems in the regulation of TSH secretion in the rat. Endocrinology 100:496–505

Kühn ER (1974) Cholinergic and adrenergic release mechanisms of vasopressin in the male rat: a study with injections of neurotransmitters and blocking agents into the third ventricle. Neuroendocrinology 16:255–264

Lechan RM, Jackson IMD (1982) Immunohistochemical localization of thyrotropin-releasing hormone in the rat hypothalamus and pituitary. Endocrinology 111:55–65

Leranth C, MacLusky NJ, Sakamoto H, Shanabrough M, Naftolin F (1985) Glutamic acid decarboxylase-containing axons synapse on LHRH neurons in the rat medial preoptic area. Neuroendocrinology 40:536–539

Leung PCK (1985) Interaction of norephinephrine and LHRH on luteinizing hormone release in vivo. In: Kobayashi H, Bern H, Urano A (eds) Neurosecretion and the biology of neuropeptides. Japan Sci Soc Press, Tokyo/Springer, Berlin Heidelberg New York Tokyo, pp 196–201

Liposits Z, Görcs T, Sétáló G, Lengvári I, Flerkó B, Vigh S, Schally A (1983) Ultrastructural characteristics of immunolabeled, corticotropin releasing factor (CRF)-synthesizing neurons in the rat brain. Cell Tissue Res 229:191–196

Locatelli V, Petraglia F, Penalva A, Panerai AE (1983) Effect of dopaminergic drugs on hypothalamic and pituitary immunoreactive β-endorphin concentrations in the rat. Life Sci 33:1711–1717

Marx JL (1983) Synthesizing the opioid peptides. Science 220:395–397

Merchenthaler I, Vigh S, Petrusz P, Schally AV (1982) Immunocytochemical localization of corticotropin-releasing factor (CRF) in the rat brain. Am J Anat 165:385–396

Merchenthaler I, Hynes MA, Vigh S, Schally AV, Petrusz P (1984) Corticotropin releasing factor (CRF): origin and course of afferent pathways to the median eminence (ME) of the rat hypothalamus. Neuroendocrinology 39:296–306

Mezey É, Kiss JZ, Skirboll LR, Goldstein M, Axelrod J (1984) Increase of corticotropin-releasing factor staining in rat paraventricular nucleus neurones by depletion of hypothalamic adrenaline. Nature 310:140–141

Moos F, Richard P (1982) Excitatory effect of dopamine on oxytocin and vasopressin reflex release in the rat. Brain Res 241:249–260

Moss RL, Dudley CA, Kelly M (1978) Hypothalamic polypeptide releasing hormones: modifiers of neuronal activity. Neuropharmacology 17:87–93

Nakada H, Nakai Y (1985) Electron microscopic examination of the catecholaminergic innervation of neurophysin- or vasopressin-containing neurons in the rat hypothalamus. Brain Res 361:247–257

Nakai Y, Ochiai H, Shioda S, Kudo J, Nakada H (1982) Simultaneous demonstration of monoamines and neuropeptides by combined autoradiography and immunocytochemistry in the same tissue section. In: Electron microscopy 1982. 10th international congress of electron microscopy 3:279–280 (abstract)
Nakai Y, Shioda S, Ochiai H, Kudo J, Hashimoto A (1983) Ultrastructural relationship between monoamine- and TRH-containing axons in the rat median eminence as revealed by combined autoradiography and immunocytochemistry in the same tissue section. Cell Tissue Res 230:1–14
Nakai Y, Ochiai H, Kitazawa S, Shioda S, Watanabe T (1985) Electron microscopic cytochemistry of the relationship between catecholaminergic axons and peptidergic neurons in the rat hypothalamus. In: Kobayashi H, Bern HA, Urano A (eds) Neurosecretion and the biology of neuropeptides. Japan Sci Soc Press, Tokyo/Springer, Berlin Heidelberg New York Tokyo, pp 121–129
Negro-Vilar A, Ojeda SR, McCann SM (1979) Catecholaminergic modulation of luteinizing hormone-releasing hormone release by median eminence terminals in vitro. Endocrinology 104:1749–1757
Nilaver G, Zimmerman EA, Defendini R, Liotta AS, Krieger AS, Brownstein MJ (1979) Adrenocorticotropin and β-lipotropin in the hypothalamus. J Cell Biol 81:50–58
Ohtsuka M, Hisano S, Daikoku S (1983) Electron microscopic study of somatostatin-containing neurons in rat arcuate nucleus with special reference to neurons in rat arcuate nucleus with special reference to neuronal regulation. Brain Res 263:191–199
Olschowka JA, Molliver ME, Grzanna R, Rice FL, Coyle JT (1981) Ultrastructural demonstration of noradrenergic synapses in the rat central nervous system by dopamine-β-hydroxylase immunocytochemistry. J Histochem Cytochem 29:271–280
Olschowka JA, O'donohue TL, Mueller GP, Jacobowitz DM (1982) The distribution of corticotropin releasing factor-like immunoreactive neurons in rat brain. Peptides 3:995–1015
Paull WK, Gibbs FP (1983) The corticotropin releasing factor (CRF) neurosecretory system in intact, adrenalectomized, and adrenalectomized-dexamethasone-treated rats. An immunocytochemical analysis. Histochemistry 78:303–316
Pelletier G, Dubé D, Puviani R (1977) Somatostatin: electron microscope immunohistochemical localization in secretory neurons of rat hypothalamus. Science 196:1469–1470
Piekut DT (1983) Ultrastructural characteristics of vasopressin-containing neurons in the paraventricular nucleus of the hypothalamus. Cell Tissue Res 234:125–134
Roth KA, Katz RJ, Sibel M, Mefford IN, Barchas JD, Carroll BJ (1981) Central epinergic inhibition of corticosterone release in rat. Life Sci 28:2389–2394
Rothman TP, Ross LL, Gershon MD (1976) Separately developing axonal uptake of 5-hydroxytryptamine and norepinephrine in the fetal ileum of the rabbit. Brain Res 115:437–456
Sawchenko PE, Swanson LW (1981) Central noradrenergic pathways for the integration of hypothalamic neuroendocrine and autonomic responses. Science 214:685–687
Sawchenko PE, Swanson LW (1982) The organization of noradrenergic pathways from the brain stem to the paraventricular and supraoptic nuclei in the rat. Brain Res Rev 4:275–325
Sawyer CH, Clifton DK (1980) Aminergic innervation of the hypothalamus. Fed Proc 39:2889–2895
Sawyer CH, Radford HM (1978) Effects of intraventricular injection of norepinephrine on brain-pituitary-ovarian function in the rabbit. Brain Res 146:83–93
Shioda S, Nakai Y, Ochiai H, Hashimoto A (1983) Monoamine-TRH interactions by means of simultaneous autoradiography and immunocytochemistry. In: Sano Y, Ibata Y, Zimmerman EA (eds) Structure and function of peptidergic and aminergic neurons. Japan Sci Soc Press, Tokyo/VNU Sci Press BV, Utrecht, pp 99–108
Shioda S, Nakai Y, Kitazawa S, Sunayama H (1985) Immunocytochemical observations of corticotropin-releasing-factor-containing neurons in the rat hypothalamus with special reference to neuronal communication. Acta Anat (Basel) 124:58–64

Silverman AJ, Oldfield BJ (1984) Synaptic input to vasopressin neurons of the paraventricular nucleus (PVN). Peptides 5:139–150

Silverman AJ, Witkin JW (1985) Synaptic interactions of luteinizing hormone-releasing hormone (LHRH) neurons in the guinea pig preoptic area. J Histochem Cytochem 33:69–72

Silverman AJ, Hou-Yu A, Oldfield BJ (1983 a) Ultrastructural identification of noradrenergic nerve terminals and vasopressin-containing neurons of the paraventricular nucleus in the same thin section. J Histochem Cytochem 31:1151–1156

Silverman AJ, Hou-Yu A, Zimmerman EA (1983 b) Ultrastructural studies of vasopressin neurons of the paraventricular nucleus of the hypothalamus using a monoclonal antibody to vasopressin: analysis of synaptic input. Neuroscience 9:141–155

Sladek CD, Armstrong WE, Sladeck JR Jr (1983) Relationship between noradrenergic and osmotic control of vasopressin release. In: Sano Y, Ibata Y, Zimmerman EA (eds) Structure and function of peptidergic and aminergic neurons. Japan Sci Soc Press, Tokyo. VNU Sci Press BV, Utrecht, pp 149–162

Sladek JR Jr, McNeill TH (1980) Simultaneous monoamine histofluorescence and neuropeptide immunocytochemistry. IV. Verification of catecholamine-neurophysin interactions through single-section analysis. Cell Tissue Res 210:181–189

Sladek JR Jr, Hoffman GE (1983) Simultaneous analysis of monoaminergic and peptidergic neurons. In: Barker JL, Mckelvy JF (eds) Current methods in cellular neurobiology, vol 1. Anatomical techniques. Wiley, New York, pp 203–217

Sofroniew MV (1979) Immunoreactive β-endorphin and ACTH in the same neurons of the hypothalamic arcuate nucleus in the rat. Am J Anat 154:283–289

Swanson LW, Sawchenko PE, Bérod A, Hartman BK, Helle KB, Vanorden DE (1981) An immunohistochemical study of the organization of catecholaminergic cells and terminal field in the paraventricular and supraoptic nuclei of the hypothalamus. J Comp Neurol 196:271–285

Swanson LW, Sawchenko PE, Rivier J, Vale WW (1983) Organization of ovine corticotropin-releasing factor immunoreactive cells and fibers in the rat brain: an immunohistochemical study. Neuroendocrinology 36:165–186

Takeuchi M (1984) Electron microscopic cytochemistry on the innervation of somatostatin-containing neurons in the periventricular region of the preoptic nucleus of the rat hypothalamus (in Japanese, abstract in English). J Showa Med Assoc 44:413–423

Uchizono K (1965) Characteristics of excitatory and inhibitory synapses in the central nervous system of the cat. Nature 207:642–643

Uchizono K (1967) Inhibitory synapses on the stretch receptor neuron of the crayfish. Nature 214:833–834

Van Vugt DA, Meites J (1980) Influence of endogeneous opiates on anterior pituitary function. Fed Proc 39:2533–2538

Versteeg DHG, Van der Gugten J, Jong WD, Palkovits M (1976) Regional concentrations of noradrenaline and dopamine in rat brain. Brain Res 113:563–574

Watson SJ, Akil H, Richard III CW, Barchas JD (1978) Evidence for two separate opiate peptide neuronal systems. Nature 275:226–228

Williams MA (1969) The assessment of electron microscopic autoradiographs. Adv Opt Electron Microscop 3:219–272

Neuroendocrine Projections to the Median Eminence

G. E. HOFFMAN [1], C. J. PHELPS [1], H. KHACHATURIAN [2], and J. R. SLADEK, JR. [1]

Contents

1 Introduction

Advertisements in the Sunday newspaper boast of how to command your pituitary gland with a simple tablet, but in fact, regulation of anterior pituitary function is the result of complex interactions of cells in the pituitary with hormones from the gland itself, from peripheral target tissues, and from neurons in the central nervous system. Within the brain, neuronal signals are integrated with bloodborne hormonal messages. These are directed to a set of neuroendocrine neurons whose axons extend to the floor of the third ventricle, the median eminence, where they terminate within the perivascular spaces of the primary capillary plexus of the pituitary portal system. The neurohormones are then carried via the portal

1 Department of Neurobiology and Anatomy, University of Rochester, School of Medicine and Dentistry, Rochester, NY 14642, USA
2 Mental Health Research Institute, University of Michigan, Ann Arbor, MI 48109, USA

Current Topics in Neuroendocrinology, Vol. 7

system to the anterior pituitary where they effect or inhibit release of the pituitary hormones.

Evidence in support of the existence of neurosecretory neurons that regulate anterior pituitary function has been accumulating since 1927 (Smith 1927), yet, it has taken nearly 60 years to delineate the major regulatory systems for anterior pituitary function. Unlike the hypothalamo-neurohypophyseal neurosecretory system, the releasing factors for the anterior pituitary were not amenable to localization by Gomori stains and Golgi techniques. Moreover, studies of degenerating fibers were either of tenuous reliability for demonstration of these pathways or only useful for local or selected circuits (Zaborszky 1982). The chemical makeup of some of the pathways, in that they lack the rich cysteine content that allows visualization with Gomori stains, and the fine caliber of the axons to the median eminence, which prevents detection with stains such as the Fink-Heimer preparation, contributed to the problem. In addition, studies of neuronal degeneration after lesions may have failed to reveal degenerating median eminence terminals because of either very rapid degeneration, as is seen for dopamine terminals (Holzwarth-McBride et al. 1976), or very slow degeneration which may not have taken place by the time of killing. The latter occurs with the preoptico-infundibular luteinizing hormone releasing hormone (LHRH) path. Loss of some LHRH immunoreactive terminals was not observed as late as 14 days after surgical deafferentation of the median eminence, but terminals were absent 20 days after surgical intervention (Scott and Knigge 1981). Although the authors interpreted this as evidence for intrinsic LHRH pathways within the medial basal hypothalamus, a more plausible explanation is either a slow process of degeneration or the escape from severance of certain LHRH pathways (Hoffman and Gibbs 1982).

Based on the few successful attempts using Golgi techniques, neurons of the arcuate nucleus, periventricular area, and retrochiasmatic area were found to send axons to the external or contact zone of the median eminence (Szentagothai 1964). The impact of these findings was heightened by the ability of these same regions to maintain function of pituitary transplants (Halasz et al. 1962, 1965). In fact, for many years the dominance of the medial basal hypothalamus as the source of neuronal afferents to the median eminence overshadowed existent, although fragmentary, evidence that other brain areas might contribute directly to median eminence function. What was missed were some of the important afferent populations to the median eminence. The introduction of improved tract-tracing methods, using retrograde transport of horseradish peroxidase and anterograde transport of labeled amino acids, into the technical repertoire of the neuroscientist expanded the view of the hypophysiotropic area to include brain regions both inside and outside the medial basal hypothalamus. These are summarized in Table 1. Yet the application of tract-tracing methods brought into question a number of projections thought to terminate in the median eminence as suggested either by chemical measurement of the putative transmitter or by electron microscopic examination of the median eminence following brain lesions. These include adrenergic, noradrenergic, and serotonergic projections from the various monoamine cell groups in the brain stem (see Zaborszky 1982 for review).

Table 1. Projections to the contact zone of the median eminence based on anterograde (A) and retrograde (R) transport

Region	Method	Reference
Forebrain		
Olfactory tubercle	A	Halasz et al. 1979
Medical preoptic area	A	Daikoku et al. 1976
	R	Lechan et al. 1980
		Wiegand and Price 1980
Hypothalamus		
Periventricular stratum	A	Saper et al. 1978
	R	Lechan et al. 1980
		Wiegand and Price 1980
Paraventricular nucleus	A	Conrad and Pfaff 1976
	R	Lechan et al. 1980
		Wiegand and Price 1980
		Swanson et al. 1980
Retrochiasmatic area	R	Swanson et al. 1980
Arcuate nucleus	A	Krieger et al. 1979
	R	Lechan et al. 1980
		Wiegand and Price 1980
Ventromedial nucleus	A	Krieger et al. 1979

Many laboratories attempted to define the brain regions that governed release of specific pituitary hormones using a variety of methods including stereotaxically placed lesions and knife cuts, pituitary transplantation, electrical stimulation, pharmacological manipulation, and exploration of changes in electrical activity of neurons or groups of neurons under a variety of stimuli affecting pituitary hormone release. The text edited by Haymaker et al. (1969), as well as a number of other comprehensive treatises on the hypothalamus, contain excellent reviews of these earlier studies, and hence the results will not be repeated here. However, the major problem noted with these studies was that most could not discriminate between the neurohormone system and neural systems that are afferent to the releasing hormone network.

We are now beginning to understand the organization of the releasing hormone systems. Figures 1–8 illustrate the location of neuronal cell bodies for the major hypophysiotropic neuroendocrine systems that regulate thyroid stimulating hormone (TSH), luteinizing hormone/follicle stimulating hormone (LH/FSH), prolactin (PRL), growth hormone (GH), and adrenocorticotropic hormone (ACTH) in the rat. There was an assumption inherent in the early studies that the hypothalamic neuroendocrine systems would fit into discrete nuclei and that projections to the median eminence would be along previously defined paths. In part, this assumption was fostered by the neuronal pattern of the magnocellular system. In general, the magnocellular neurosecretory neurons that contain vasopressin and oxytocin are confined to the supraoptic and paraventricular hypothalamic nuclei (with a few scattered cells within the accessory nuclei found between the two nuclei), and the axons from the magnocellular neurons course in

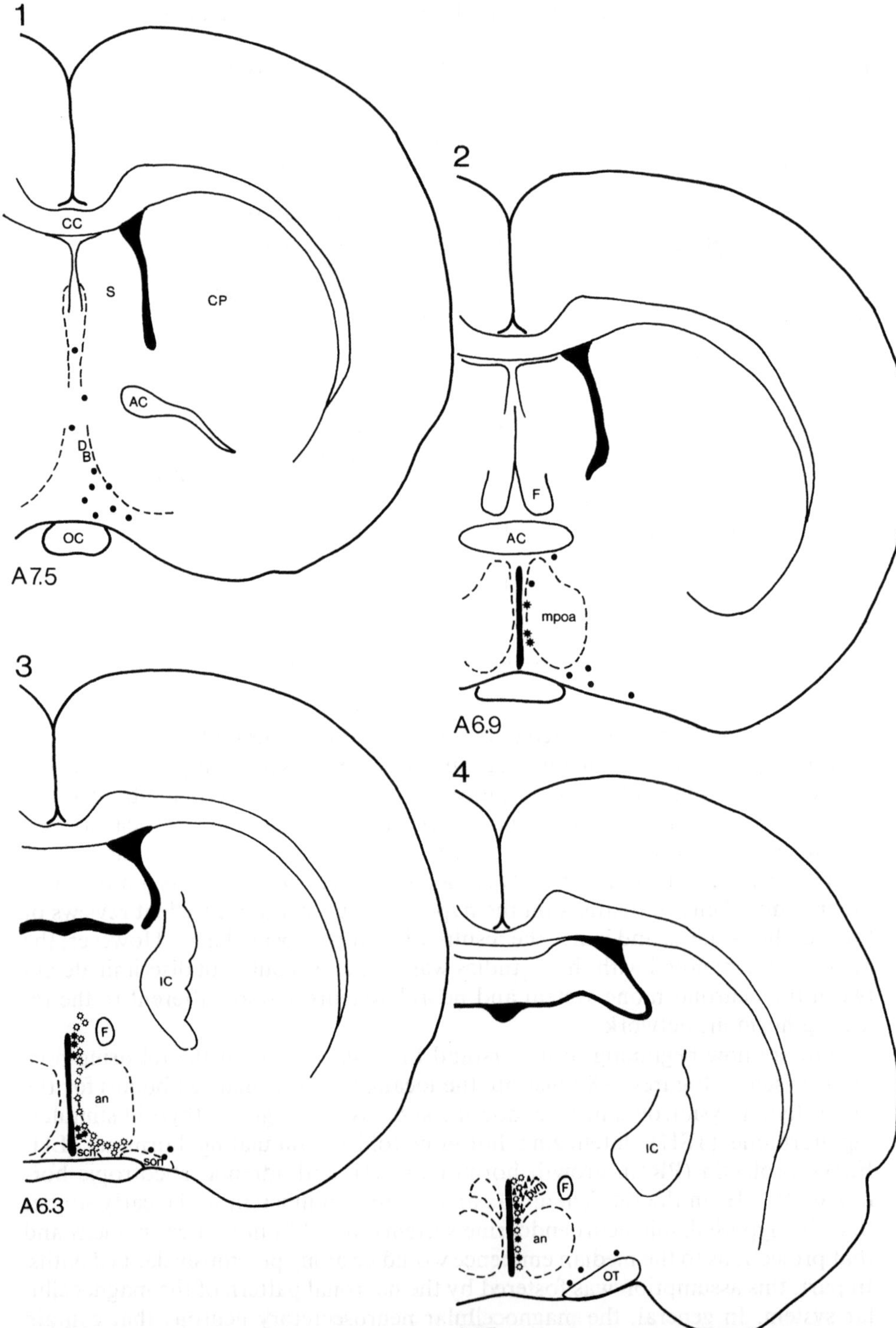

Figs. 1–8. Location of neuronal cell bodies from neuroendocrine systems which regulate the anterior pituitary. For the sake of comparison, all are drawn on a rostral-to-caudal coronal series in the rat. The stereotaxic levels refer to planes anterior (*A*) to an interaural line, as in the atlas of König and Klippel (1963). ● = LHRH, ✡ = TRH; ✸ = SRIF, ★ = GHRH; ↗ = DA; ▲ = CRH; *a*, arcuate nucleus; *AC*, anterior commissure; *an*, anterior hypothalamic

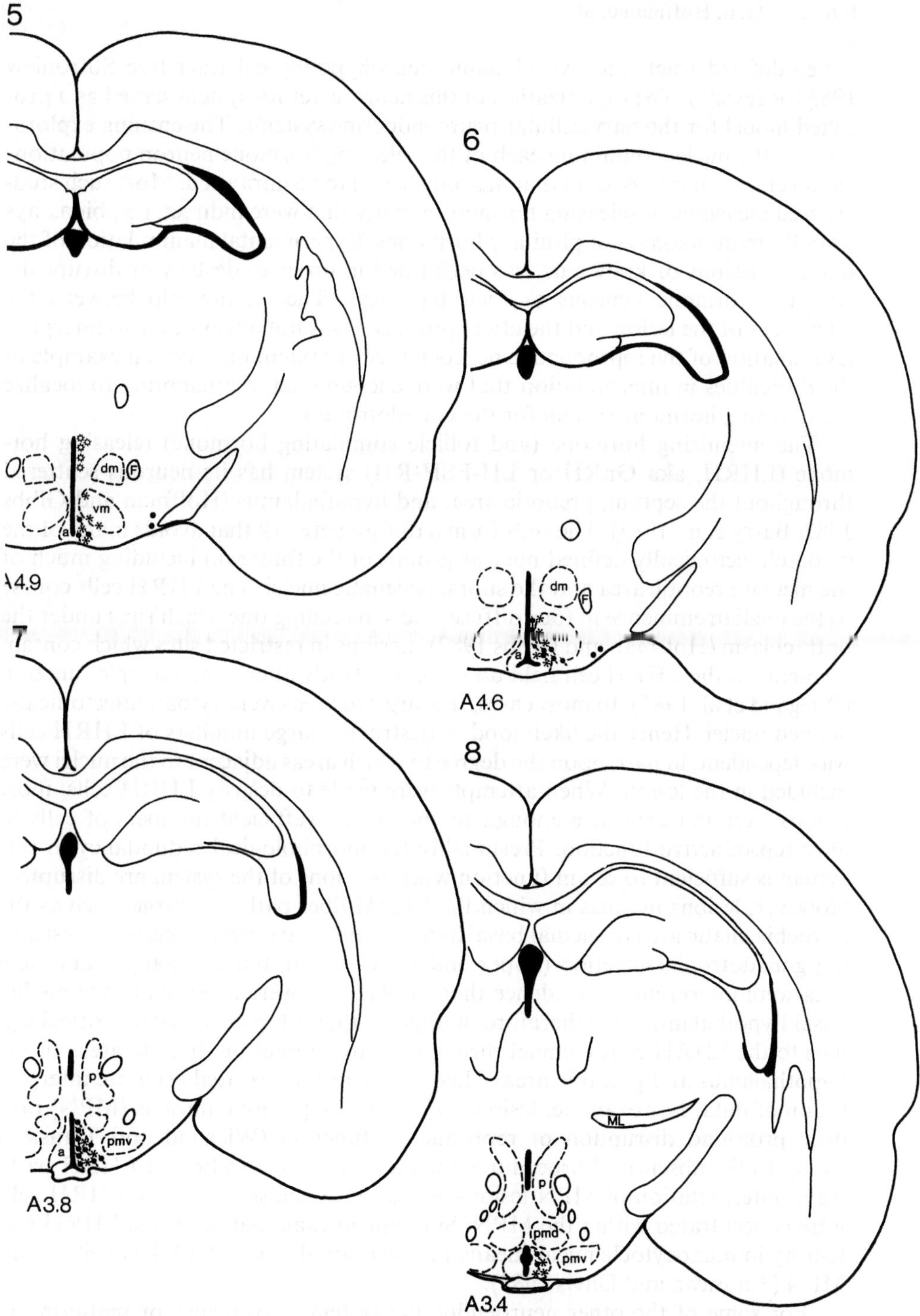

nucleus; *DB*, diagonal band of Broca; *CC*, corpus callosum; *CP*, caudate-putamen; *dm*, dorsomedial nucleus; *F*, fornix; *IC*, internal capsule; *ML*, medial lemniscus; *mpoa*, medial preoptic area; *OC*, optic chiasm; *OT*, optic tract; *p*, posterior hypothalamic nucleus; *pmd*, dorsal premammillary nucleus; *pmv*, ventral premammillary nucleus; *pvn*, paraventricular nucleus; *S*, septum; *scn*, suprachiasmatic nucleus; *son*, supraoptic nucleus; *vm*, ventromedial nucleus a well-defined tract, the hypothalamo-neurohypophyseal tract (see Sofroniew 1985 for review). The organization of this neurosecretory system served as a projected model for the parvicellular neuroendocrine systems. The ensuing exploration for the nuclei containing each of the releasing hormone neuron populations left an elegant tangle of data that has only begun to be unraveled. Most such studies used measures of releasing hormone activity that were indirect; i.e., bioassays or radioimmunoassays of pituitary hormones. Experimental manipulation of the brain by lesions or knife cuts was performed in order to destroy or disrupt the releasing hormone neurons or their pathways. The relationship between the placement of the lesion and the effect produced was not always easy to interpret. Examination of the reproductive neuroendocrine system provides an example of the difficulties in interpretation that were encountered in attempting to localize the releasing hormone system for the gonadotropins.

The luteinizing hormone (and follicle stimulating hormone) releasing hormone (LHRH, aka GnRH or LH-FSH-RH) system has its neurons scattered throughout the septum, preoptic area, and hypothalamus (Hoffman and Gibbs 1982; Barry et al. 1985). The cells form a diffuse network that avoids many of the cytoarchitectonically defined nuclear groups of the forebrain including much of the medial preoptic area and the suprachiasmatic nuclei. The LHRH cells course to the median eminence in four diffuse tracts, including one which runs under the optic chiasm (Hoffman and Gibbs 1982). Lesions in restricted sites which contain segments of the LHRH cell field do not consistently alter gonadotropic function (Wiegand et al. 1985). In most cases, the targeted areas were cytoarchitectonically defined nuclei. Hence the likelihood of destroying large numbers of LHRH cells was dependent, in part, upon the degree to which areas adjacent to the nuclei were included in the lesion. When attempts were made to destroy LHRH cells, most lesions were not extensive enough to encompass sufficient numbers of cells to alter reproductive function. Presumably the morphological redundancy in this system is sufficient to retain function when portions of the system are disrupted. However, lesions in areas in which the LHRH fiber paths converge, such as the retrochiasmatic area or medial basal hypothalamus, are more effective in disrupting gonadotropin secretion (Soper and Weick 1980), but for many years these data were interpreted as evidence that LHRH cells were localized to the medial basal hypothalamus. Furthermore, it appears that afferents carrying critical signals to the LHRH cells channel through, or are present in, discrete areas of the hypothalamus and preoptic area. This feature has obfuscated accurate interpretation of data. For example, lesions in the medial preoptic nucleus (MPN) produce profound disruption of reproductive function (Wiegand and Terasawa 1978). In the absence of procedures for the direct visualization of LHRH pathways, interpretation of MPN lesions led to the conclusion that the LHRH cells were concentrated within the MPN. Subsequent examination of the LHRH system by immunocytochemical means revealed an absence of LHRH cells in the MPN (Terasawa and Davis 1983).

For some of the other neuroendocrine systems, divergence or scattering of axonal projections from the cells of origin to the median eminence, in addition to the existence of polysynaptic pathways, led to the assumption that a particular neuroendocrine pathway was organized into a diffuse neuronal network when, in fact, the releasing hormone was localized to a relatively discrete brain region. This feature is well illustrated by corticotropin-releasing hormone (CRH). CRH neurons that project to the median eminence lie within the paraventricular nuclei (PVN) of the hypothalamus (Makara et al. 1981; Liposits et al. 1983, 1985). Their axons fan laterally and course toward the supraoptic nuclei along with the fibers of the magnocellular hypothalamo-neurohypophyseal tract. Lesions of the hypothalamus implicated a variety of regions in the control of ACTH release. In one series of experiments (Brodish 1964) the hypothalamus was divided into four zones extending from the optic chiasm to the mammillary bodies. Plasma corticosterone levels in response to ether stress were measured after bilateral destruction of each zone. Significant impairment of release was observed for lesions in each zone. Lesions encompassing more than one zone produced greater impairment of the response. The degree of impairment was graded and depended upon the extent of the damage. The author concluded that the CRH system must be a diffusely organized network extending across much of the hypothalamus.

Similar disparities have plagued the other releasing hormone systems. It was not until chemical characterization of the releasing hormones (Burgus et al. 1969; Schally et al. 1969; Amos et al. 1971; Matsuo et al. 1971; Brazeau et al. 1974; Vale et al. 1981; Guillemin et al. 1982; Rivier et al. 1982; Spiess et al. 1983; Nikolics et al. 1985) enabled the generation of specific antibodies for immunocytochemical localization that some closure on the question of where the releasing hormone pathways were located could be achieved. Over the past 10 years our knowledge of the organization of neuroendocrine systems has expanded tremendously. The remainder of this chapter will deal with available morphologic information on the specific neuroendocrine regulatory systems.

2 Thyrotropin Releasing Hormone

The first of the hypothalamic releasing hormones to be isolated and purified was thyrotropin releasing hormone, TRH (Burgus et al. 1969; Schally et al. 1969). Although antibodies capable of sensitive measurement of brain TRH were available since the early 1970s, the number of studies that have explored the anatomical localization of TRH have been relatively few. TRH, a tripeptide, is easily extracted from brain during tissue processing. Although this feature has enabled studies detailing the differential content of TRH in various brain regions (Winokur and Utiger 1974; Brownstein et al. 1974; Palkovits and Brownstein 1985), immunocytochemical studies that accurately reflected the organization of TRH were difficult to achieve.

2.1 Cell Location

Hökfelt and his coworkers described TRH-containing nerve fibers in the external zone of the median eminence and the presence of a few weakly stained cell bodies in the dorsomedial nucleus and perifornical region of the hypothalamus (Hökfelt et al. 1975b). Yet it was clear, based on indirect evidence, that these cells did not project to the median eminence. Until recently, the source of the median eminence projection evaded direct identification. It was only after more rapid and effective means of fixing tissue for immunocytochemistry became available that visualization of the low concentrations of TRH found in many neuronal structures was possible. Using acrolein as a fixative, TRH cell bodies were identified in numerous sites within the forebrain and diencephalon (Lechan and Jackson 1982; Ishikawa et al. 1984). These included the diagonal band of Broca, medial preoptic area, periventricular regions of the preoptic area and anterior hypothalamus, parvicellular PVN, lateral hypothalamus, dorsomedial nuclei, posterior arcuate nucleus, and lateral premammillary nucleus. The cells of the PVN and periventricular hypothalamus are implicated as the source of the projection to the median eminence based on the relationship of these cells to the pattern of retrograde labeling of the hypothalamus from median eminence injections (Lechan et al. 1980; Wiegand and Price 1980). Furthermore, lesions of the periventricular zone of the anterior hypothalamus in the vicinity of the parvicellular PVN significantly reduced TRH content in the median eminence (Jackson and Reichlin 1977), eliminated all immunoreactive axons containing TRH in the median eminence (Elde et al. 1976), and reduced TSH responses to hypothalamic stimulation (Martin and Reichlin 1972). The TRH neuroendocrine system is illustrated in Figs. 3–5.

3 Luteinizing Hormone Releasing Hormone

Great controversy has surrounded the issue of the location of cells that contain the releasing hormone for the gonadotropins. Since this decapeptide was among the first of the hypothalamic releasing hormone systems to be localized by immunocytochemistry, it suffered from some of the pitfalls of the early immunocytochemical technology: lack of adequate antigen retention in the tissue when "routine" fixation and embedding in paraffin were used, poor selection of controls, and confusion in interpretation of data (see Barry et al. 1985 for review). Once technical advances were achieved, a consistent picture of the organization of this key neuroendocrine system began to emerge. An example of LHRH cells and their projection to the median eminence is shown in Figs. 9 and 10. As alluded to above, many of the preconceived notions of how neuroendocrine systems should be organized were shattered by this system. The LHRH cells did not fall into classical cytoarchitectonic nuclear boundaries. They were relatively few in number. There seem to be significant species differences in the location of the LHRH cells. Even the paths of projection to the median eminence were not always conventional ones.

3.1 Cell Location

LHRH cells have been found in a broad range of species (see Barry et al. 1985 for review). The vast majority of species show marked similarities in the organi-

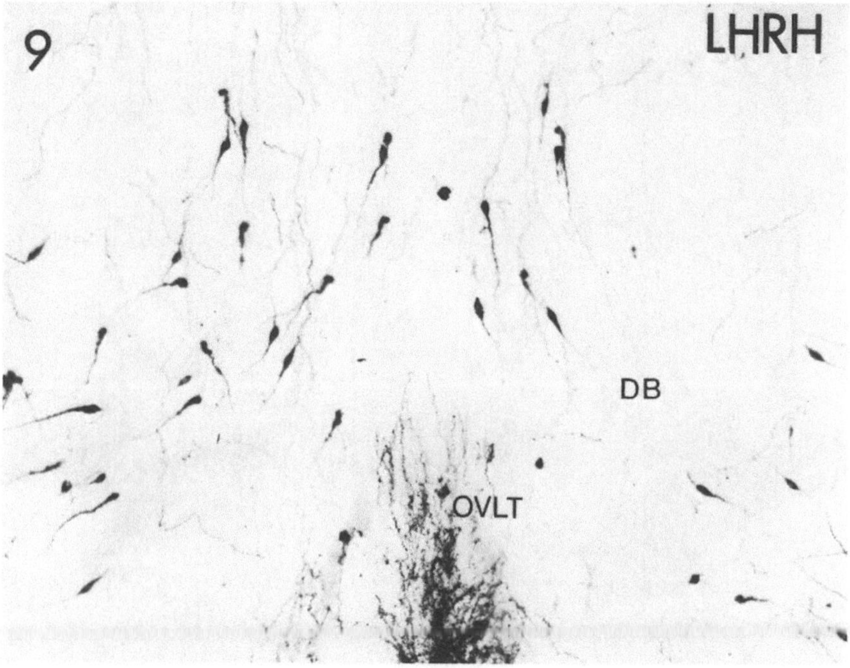

Fig. 9. The location of LHRH neurons at a level similar to that illustrated in Fig. 1 in a 30 μm thick, immunocytochemically stained coronal section of rat brain (from Watson et al. 1986). *OVLT,* organum vasculosum of the lamina terminalis; *DB,* diagonal band of Broca; × 25

zation of the LHRH neuroendocrine system. As exemplified by the LHRH cell distribution seen in the rat (Figs. 1–6), the cells form a Y-shaped field oriented with the base along the midline of the rostral forebrain in a rostral-caudal plane. Its arms first extend upwards through the diagonal band of Broca to the medial septum, and then turn ventrally into the preoptic area as they fan out laterally. They pass through the junction of the medial and lateral preoptic areas, the region just lateral to the suprachiasmatic nuclei, and then extend into the ventrolateral anterior hypothalamus where they lie just medial to the supraoptic nuclei. Only a few cells are found in the medial basal hypothalamus. This general pattern is preserved throughout vertebrate classes, including birds, reptiles, fish, and most mammals. However, the precise parcellation of cells that are found in the various portions of the field vary in certain species. For example, in primates (Marshall and Goldsmith 1980; Silverman et al. 1982), there is a greater number of cells (perhaps as many as one-third of the total) in the ventrolateral hypothalamus in the tuberal region, while in most rodents fewer than 1% of the cells are located caudal to the anterior hypothalamus (Jennes and Stumpf 1980). In the dog the distribution of cells in the various portions of the cell field appears to be quite even (Hoffman et al. 1978). In the ferret nearly all the cells are found in the tuberal region, with many extending into the median eminence itself, and only a few cells lie more anteriorly (King and Anthony 1984). Perhaps the only exception to the scheme

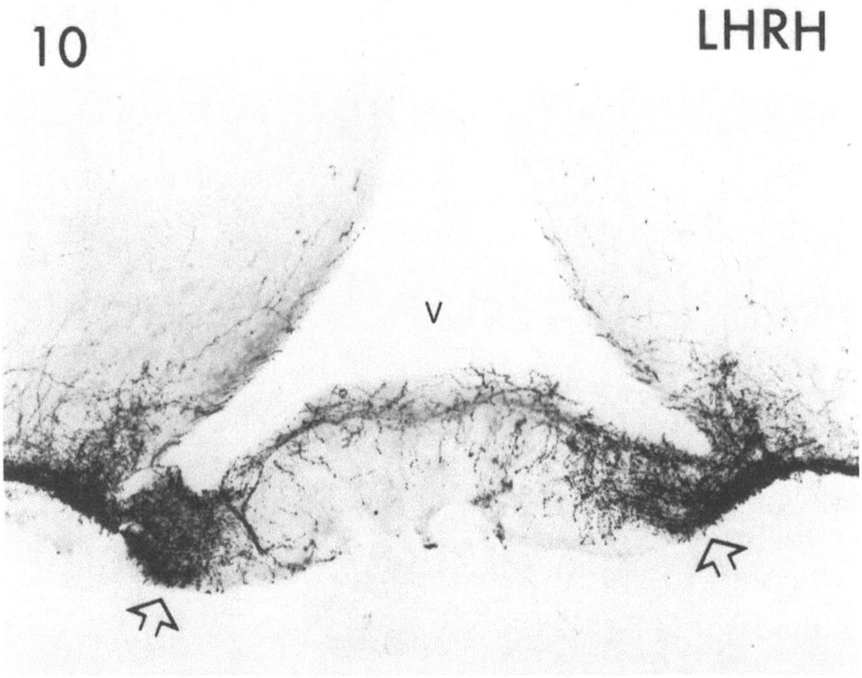

Fig. 10. The LHRH axons that terminate in the median eminence are densely concentrated in the lateral median eminence (*arrows*), with fewer fibers in the more medial zone (from Watson et al. 1986). *V*, third ventricle; × 25

provided above is the insectivore *Myotis lucifugus lucifugus*. In this bat, the cells are found in the medial basal hypothalamus with a large portion of the cells associated with the arcuate nucleus (King et al. 1984).

3.2 Cell Morphology

The cells are generally small, simple, bipolar, fusiform neurons, but recent studies indicate some interesting somal and dendritic modifications. In the rat the cells are of two basic morphologic types, one smooth in contour and the other with spines on the soma and dendrites (Wray and Hoffman 1986). The spiny cells comprise approximately 70% of the total in the adult. Spiny LHRH cells possess more synaptic contacts than do smooth cells, suggesting that the spiny neurons are more intimately involved in integration of multiple neuronal signals (Jennes et al. 1985). Studies combining immunocytochemical localization of LHRH with retrograde tracings reveal that both smooth and spiny cells project to the median eminence, and the distribution of these cells is fairly uniform throughout the LHRH field (Silverman and Renaud 1985).

In the primate another cell surface modification has been noted. Certain of the LHRH cells appear to send fine bridges to other LHRH neurons in the vicinity (Marshall and Goldsmith 1980). Although these "contacts" have not been examined at the ultrastructural level, it is likely that they offer a means of synchronizing firing of the LHRH cells. Dendritic fusion and contacts on LHRH neurons by LHRH axonal processes provide other possible modifications aimed at synchronous firing (Hoffman 1983). Deviation from the simple bipolar neuronal morphology has been noted in a few species, notably the cat, which possesses some cells with rather elaborate dendritic branches, and the rabbit, which in addition to smooth and spiny cells has some LHRH neurons with flattened, fanlike dendritic extensions (Hoffman, unpublished observation).

3.3 Cell Number

One surprising feature in comparison to other releasing hormone systems is that there are relatively few LHRH neurons in the mammalian brain. Even with the most sensitive techniques, and with colchicine treatment to increase the amount of peptide within the cell body, LHRH cells number only 1300 in the rat (Wray and Hoffman 1986), 750 in the mouse (Hoffman and Finch 1986), 800 in the hamster (Jennes and Stumpf 1980), and 2400 in the baboon (Marshall and Goldsmith 1980).

3.4 Projections to the Median Eminence

In the rat, the projections of the LHRH system to the median eminence have been extensively studied (Ibata et al. 1979; Bennett-Clarke and Joseph 1982; Hoffman and Gibbs 1982; King et al. 1982). The LHRH axons are most concentrated in the lateral median eminence. In general, there are four paths to the contact zone of the median eminence: along the walls and floor of the third ventricle, along the ventrolateral hypothalamus close to the fibers of the medial forebrain bundle, and between the two optic nerves and the ventral surface of the optic chiasm (Fig. 11). In the rat the ventrolateral path carries the majority of fibers (Wiegand et al. 1985). However, even the relatively sparse subchiasmatic projection is sufficient to maintain gonadotropin secretion after the other paths are severed (Hoffman and Gibbs 1982).

3.5 Release of Multiple Peptides: Regulation of Other Pituitary Hormones

Thus far, the decapeptide LHRH stimulates only the gonadotropes of the anterior pituitary. LHRH is synthesized as a prohormone, which is cleaved, and more than one peptide product is probably released upon stimulation. One of these, a 56 amino acid chain, has potent prolactin-inhibiting activity (Nikolics et al. 1985); this peptide is termed GAP for gonadotropin-releasing hormone-associated peptide. Since prolactin has antigonadotropic activity (Beck and Wuttke 1977), simultaneous release of both LHRH and GAP would prevent dampening of

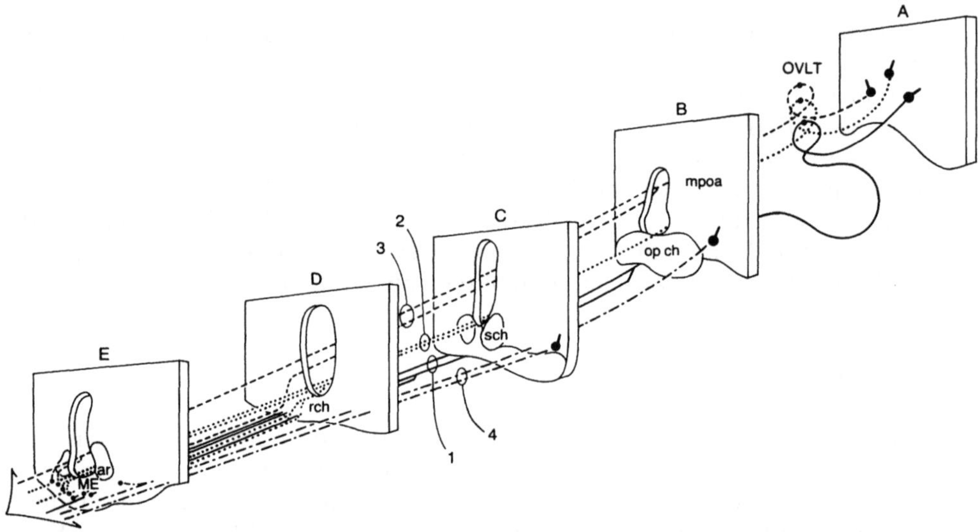

Fig. 11. Two groups of LHRH neurons give rise to four distinct tracts to the median eminence, one of which (1) courses ventral to the optic chiasm. Fibers from each tract begin to merge at the retrochiasmatic area (*rch*) prior to entry to the rostral median eminence (from Hoffman and Gibbs 1982)

LHRH-induced gonadotropic activity by prolactin. However, the variety of physiological endocrine states precludes such a simple neuroendocrine regulatory scheme (Fink 1985).

4 Prolactin Inhibiting Factor

Among all the endocrine neurohormones, only one (dopamine) has been established physiologically that is not peptide in nature. Although the physiological status of dopamine, a catecholamine neurotransmitter, as the primary factor in the control of prolactin (PRL) secretion is well documented and accepted, the terminology for a nonpeptidergic neurohormone remains uncomfortable. Dopamine (DA) is usually referred to as such, rather than as PIF, PIH, or PRIH, although all these terms do appear in the literature.

The foremost characteristic in the unusual status of DA is that PRL secretion is predominantly inhibited by its action. This was initially established when it was discovered that PRL secretion was enhanced by transplantation of the pituitary to ectopic sites (Everett 1954; Nikitovitch-Winer and Everett 1958, 1959) or by physical disruption of hypothalamic-pituitary contiguity (Nikitovitch-Winer 1965). Basal hypothalamic lesions proved to release PRL secretion (Bishop et al. 1972). PRL-secreting cells are, further, unusually autonomous in their capacity to continue basal secretory rates, both in vivo (Turpen and Dunn 1976) and in vitro (Phelps and Hymer 1986) after removal of hypothalamic influence. The inhibitory role of DA has ascended to routine clinical therapy; for example, PRL-secreting pituitary tumors are treated with the DA agonist bromocriptine (Thorner et al. 1980). For a review of the established role of DA in PRL secretion, see MacLeod (1976) and Neill (1980).

4.1 Cell Location and Projections to the Median Eminence

Using sensitive formaldehyde induction of histofluorescence (Falck et al. 1962) the cytoarchitectonic distribution of CNS catecholamine (CA) transmitter, including DA, was determined before certain DA-containing cell groups were known to influence PRL secretion. In 1964, Fuxe described intense CA fluoresence in the median eminence external palisade zone, which was localized to distinct axon terminals abutting capillary loops of the pituitary portal plexus; he described a similar pattern in six mammalian species. The cells of origin for these fluorescent terminals were first discerned in an elegant study by Lichtensteiger and Langemann (1966) wherein exogenous systemic CA was sequestered and retrogradely transported in brain areas devoid of the blood-brain barrier in rats. Subsequent histofluorescence showed the entire extent of axons emanating from cell bodies in the hypothalamic arcuate nucleus and extending to the median eminence (Figs. 5–8). The development of microspectofluorometry (Bjorklund et al. 1968) allowed differentiation of DA- and norepinephrine-containing cell bodies and fibers in the diencephalon (Bjorklund et al. 1970; Bjorklund and Nobin 1973). Biochemical measurements confirmed the differential distribution of the two CA (Kavanagh and Weisz 1973). Four diencephalic DA cell groups exist in the hypothalamus, designated A11-14 in the system devised by Fuxe (see Fuxe and Hökfelt 1969). The perikarya in the arcuate nucleus, A12, were shown to be the major afferent path to the external median eminence by studies employing basal hypothalamic deafferentation (Jonsson et al. 1972; Turpen and Sladek 1978), selective knife cuts (Bjorklund et al. 1970), and electrolytic lesions (Smith and Fink 1972). Although small contributions of A14 anterior periventricular axons to the external median eminence were not ruled out by any of the above studies, a reevaluation of nonarcuate DA afferents by Bjorklund et al. (1973) appeared to eliminate this possibility. They found that the arcuate-periventricular (i.e., A12) projections could be divided into two groups. One, in the rostral arcuate nucleus, projects widely to the entire median eminence, whereas the second appears to project dorsoventrally to connect each rostrocaudal level of the arcuate nucleus with corresponding portions of the underlying median eminence. This tuberoinfundibular system has been localized more recently using immunocytochemical methods (Hoffman et al. 1982). Antiserum to tyrosine hydroxylase, the rate-limiting enzyme in DA synthesis, illustrates median eminence-afferent arcuate perikarya, as shown in Fig. 12.

Histofluorescence in the internal zone was consistently attributed to norepinephrine from extrahypothalamic, lower brainstem neurons (A1), constituting a reticuloinfundibular pathway associated with the medial forebrain bundle (Bjorklund et al. 1973). Microdissection ("punch") and biochemical measurement of DA after lesions of the substantia nigra have been interpreted to suggest that some ME-afferent fibers might arise from the A8–10 groups of the mesen-

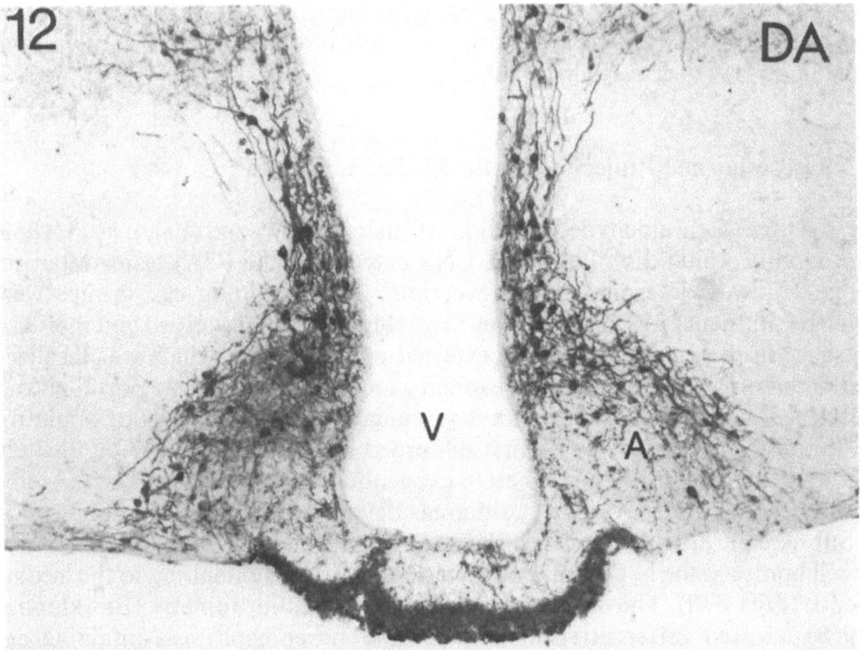

Fig. 12. Dopamine neurons of the arcuate nucleus (*A*) and their projections to the median eminence in the mouse brain are revealed by immunocytochemical localization of the synthetic enzyme, tyrosine hydroxylase. Section thickness is 30 μm; *V*, third ventricle; × 25

cephalon (Kizer et al. 1976). DA is heterogeneously distributed in the median eminence, the medial regions having a higher DA content than lateral areas (Selmanoff 1981); the topographic projections of A12 neurons to medial and lateral zones have not been established. The external zone is characterized by dense patterns of DA fluorescence which follow the contours of the portal vasculature. The DA network is particularly dense laterally and caudally. Moreover, considerable overlap exists between these patterns of DA terminal fibers in the external zone of the median eminence and other releasing hormones, particularly LHRH and somatostatin. The reader is referred to several studies which describe these chemically specific patterns in morphological detail (Sladek et al. 1978; Hökfelt et al. 1978; McNeill and Sladek 1978; Ajika 1979; Hoffman and Sladek 1980; McNeill et al. 1980; Ibata et al. 1981).

Corresponding endocrinological studies performed simultaneously have implicated DA in PRL regulation. DA treatment depressed PRL secretion in vivo (Kamberi et al. 1971; Lu and Meites 1971) and in vitro (MacLeod and Lehmeyer 1974). Hypothalamic deafferentation disturbed neither DA concentration within the islands (Weiner et al. 1972) nor peripheral PRL levels (Blake et al. 1972; Turpen and Dunn 1976). More recent studies adressed the effect of PRL on tuberoinfundibular DA neurons, which are stimulated (Selmanoff 1981; MacLeod 1976), perhaps to toxic levels (Sarkar et al. 1981), in hyperprolactinemia, confirming the reciprocal relationship between DA and PRL. That all pituitary hormones should have peptidergic CNS regulators has long been an hypothesis of neuroendocrinologists, and it seems to have been confirmed with the recent characterization of GAP (discussed above under LHRH). It is interesting, in retrospect, that the preoptic area was implicated in PRL inhibition in early studies: severing preoptic-hypothalamic connections induced lactation (Grosz and Rothballer 1961) and electrical stimulation of that area lowered PRL levels (Clemens et al. 1971). It was reported that PRL inhibiting activity was difficult to separate from LH releasing factor (Schally et al. 1964; Dhariwal et al. 1968).

The search for a physiological PRL-releasing hormone continues. Among the substances implicated, vasoactive intestinal peptide (VIP) has been shown to directly stimulate both PRL levels in vivo (Vijanian et al. 1979) and PRL mRNA in (PRL-producing) GH_3 cells in vitro (Carrillo et al. 1985). VIP has been immunocytochemically localized in the median eminence (Mezey and Kiss 1985); the cellular origin of this VIP remains to be elucidated, although suprachiasmatic and paraventricular nuclei have been implicated (Mezey and Kiss 1985).

5 Growth Hormone Release Inhibiting Hormone

In 1974 Brazeau and coworkers characterized a tetradecapeptide (somatostatin) that was an active inhibitor of growth hormone. Some time later this peptide was also found to inhibit TSH and, under certain conditions, PRL (Vale et al. 1974). Somatostatin-containing cells are found in a variety of peripheral organs (Hökfelt et al. 1975 a) where they are thought to play a role in the regulation of a multitude of secretory events (Bloom SR et al. 1974; Koerker et al. 1974; Barros et al. 1975; Gomez-Pan et al. 1975).

5.1 Cell Location and Projections to the Median Eminence

Unlike TRH and LHRH, somatostatin neurons were relatively easy to localize with immunocytochemical technique. The peptide is well retained in most common tissue preparations (Hoffman et al. 1978) and is present in sufficient concentrations in the perikaryon to be visualized without colchicine treatment. Somatostatin neurons are found broadly distributed throughout the brain (Finley et al. 1978; Forssmann 1978; Krisch 1978; Hoffman and Hayes 1979; Krisch and Leonhardt 1979; Bennett-Clarke et al. 1980; Richoux and Dubois 1980; Takatsuke et al. 1981). In contrast, the somatostatin neuroendocrine system is restricted to a narrow band of cells in the periventricular area of the anterior hypothalamus (Elde et al. 1976; Epelbaum et al. 1977; Critchlow et al. 1978; Crowley and Terry 1980) which sends axons principally to the external zone of the median eminence (Figs. 2–4 and 13). There, dense fibers are found throughout the entire median eminence (Fig. 14). In general, the organization of the neuroendocrine regulatory somatostatin system does not vary among species.

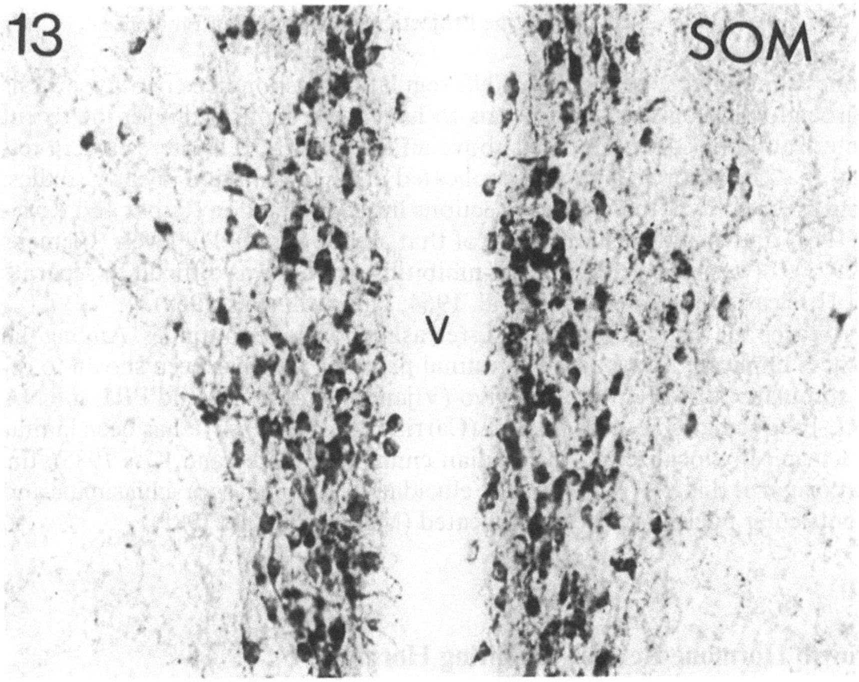

Fig. 13. Numerous somatostatin (*SOM*) cells are found in the periventricular area of the anterior hypothalamus of the rat (from Don Carlos et al. 1986), × 25

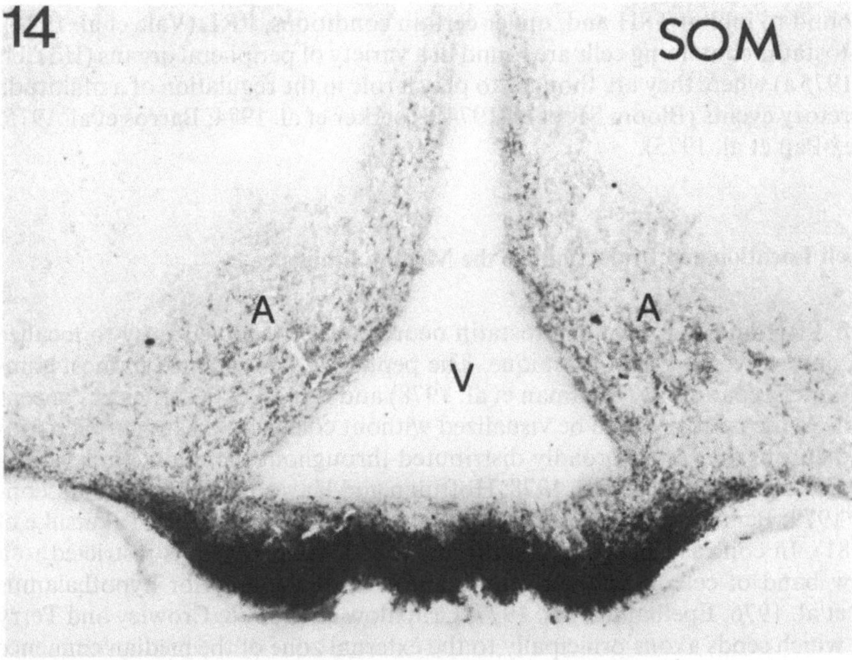

Fig. 14. Somatostatin distribution in the median eminence. This projection is depicted in rat brain, and encompasses the entire median eminence contact zone (from Don Carlos et al. 1986). *V*, third ventricle; *A*, arcuate nucleus; × 25

6 Corticotropin Releasing Hormone

One of the first factors present in extracts of the median eminence that had clear biological activity was the corticotropin releasing factor (CRF) (Guillemin and Rosenberg 1955; Saffran and Schally 1955). The search for its structure was long, finally culminating in the characterization of a 41-amino acid peptide, corticotropic releasing hormone, CRH, derived from ovine brain (Vale et al. 1981). Although there had been some suggestion that a variety of substances endogenous to the CNS, including vasopressin, oxytocin, norepinephrine, epinephrine, and angiotensin II, had CRF activity (see Vale et al. 1983 for review) and might each be the elusive CRF, it is now clear that the peptide CRH provides the major stimulus for ACTH release by the anterior pituitary. Among the releasing factors CRH is unique in that it was the first factor identified that was co-localized with other peptides capable of influencing ACTH release.

6.1 Cell Location

As had been mentioned previously, studies employing lesions to indicate the location of CRF neurons suggested that CRF had a diffuse organization within the hypothalamus. Rarely was the paraventricular nucleus considered a likely site of CRF projections to the median eminence. To some extent, possible consideration of the PVN of the hypothalamus for the location of both vasopressin and CRF was discounted since it was difficult to separate the effects of vasopressin from another CRF, and the PVN was considered the realm of vasopressin and oxytocin (Palkovits 1977). After the availability of antibodies against CRH, immunocytochemical evidence convincingly demonstrated that the PVN housed the CRH neurons.

CRH immunoreactive neurons have been identified in the PVN, parvicellular division (Figs. 4, 15), or its analogue in sheep, monkey, cat, dog, pig, dormouse, pigeon, duck, sparrow, blackbird, red-eared turtle, frog, toad, newt, trout, goldfish, tench, guinea pig, and rat (Bloom FE et al. 1982; Bugnon et al. 1982; Kawata et al. 1982; Merchenthaler et al. 1982; Olschowka et al. 1982; Paull et al. 1982; Bugnon et al. 1984). In the colchicine-treated rat, there are approximately 2000 CRH neurons in the PVN (Swanson and Kuypers 1980). The presence of degenerating CRH axons in the median eminence after PVN lesions (Liposits et al. 1983) supports the concept that the PVN gives rise to the median eminence projection (Figs. 16, 17) and hence is the source of the ACTH neuroendocrine regulator. Similar lesion placements result in diminished secretion of ACTH to a variety of stimuli (Makara et al. 1981). Further implication of CRH parvicellular neurons of the PVN in ACTH regulation via a projection to the median eminence derives from evidence of increased neurosecretory granules, hypertrophied rough endoplasmic reticulum with dilated endoplasmic cisternae, and well developed Golgi complexes in CRH neurons of the PVN following adrenalectomy and dexamethasone treatment (Liposits et al. 1985). At the light microscopic level, adrenalectomy also produces profound increases in the CRH contained within PVN neurons as well as within terminals in the median eminence (Swanson et al. 1983).

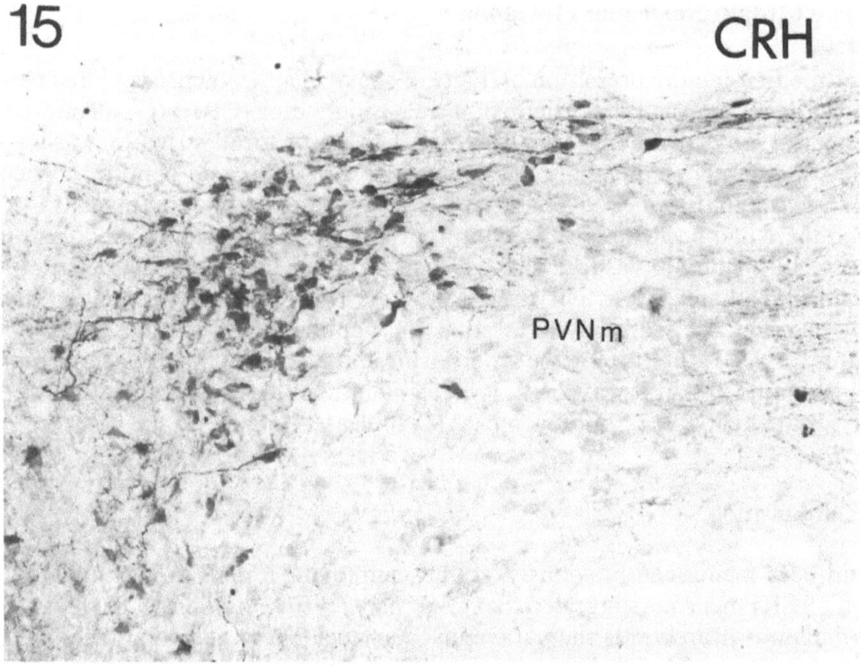

Fig. 15. CRH is localized within parvicellular neurons of the rat PVN. The larger neurons of the magnocellular division of the PVN (*PVNm*) are unstained. × 25

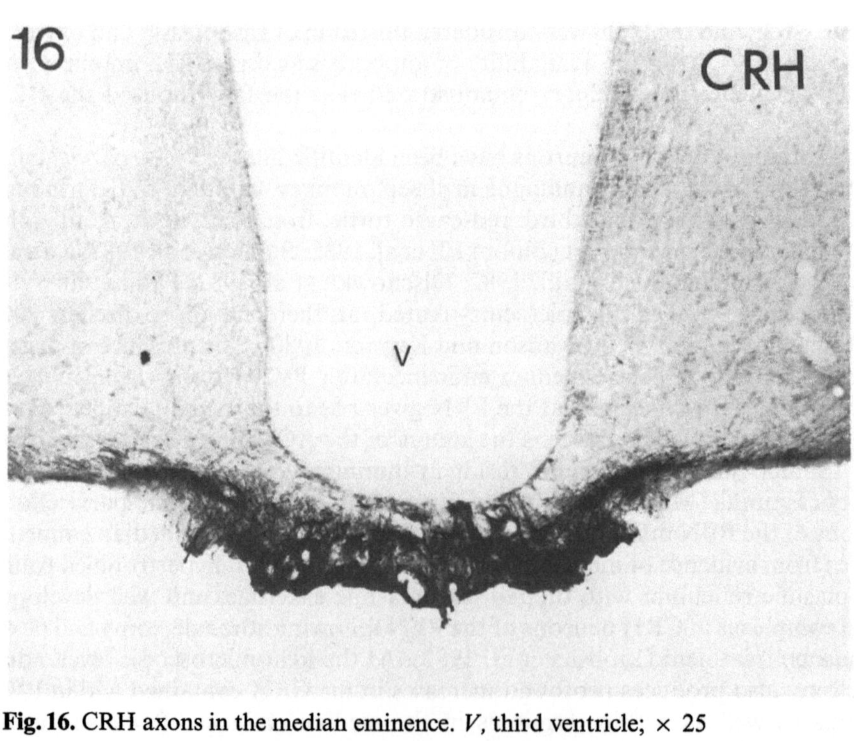

Fig. 16. CRH axons in the median eminence. *V*, third ventricle; × 25

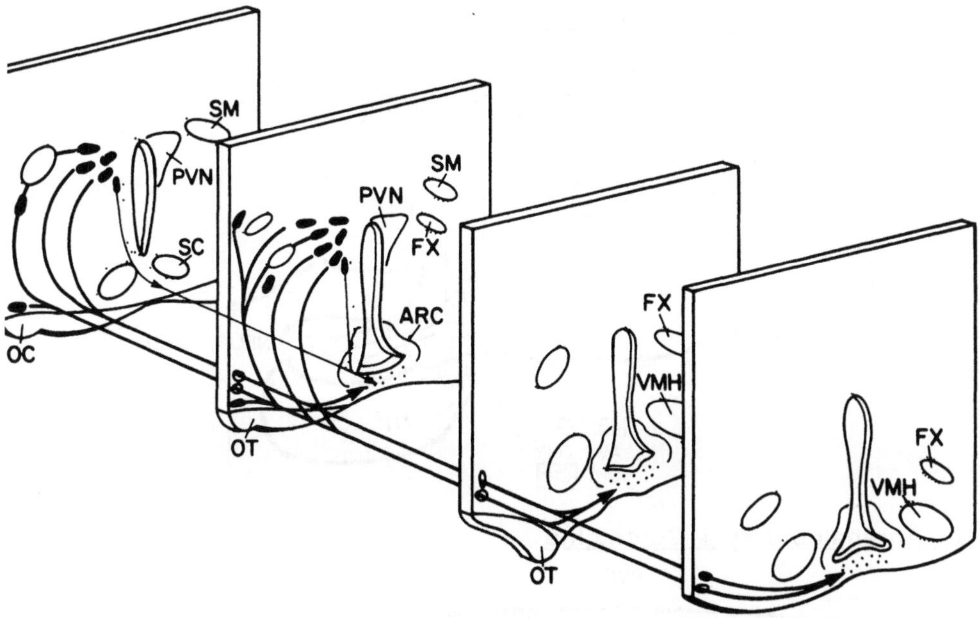

Fig. 17. The "paraventriculo-infundibular" system from the parvicellular PVN to the external zone of the median eminence (from Merchenthaler et al. 1983). *ARC*, arcuate nucleus; *FX*, fornix; *OC*, optic chiasm; *OT*, optic tract; *PVN*, paraventricular nucleus; *SC*, suprachiasmatic nucleus; *SM*, stria medullaris; *VMH*, ventromedial nucleus

6.2 Release of Multiple Peptides: Coregulation of ACTH Release with Vasopressin and Other Peptides

Although it was recognized that vasopressin and CRH both influence ACTH release, only recently have data appeared that elucidate the probable means by which these multiple peptides regulate adrenal endocrine function: both are released from the same neurons. This discovery has significantly altered the concept of neuroendocrine regulation. In addition, the description of this unique system has cleared much of the controversy surrounding the issue of how vasopressin and CRH were differentially distributed within the PVN, but yet acted in concert to affect ACTH release.

Certain cells of the PVN and terminals in the contact zone of the median eminence were found to contain both CRH and vasopressin (Roth et al. 1982; Tramu et al. 1983; Kiss et al. 1984; Sawchenko et al. 1984a, b). In terminals the two peptides are found in the same synaptic vesicles (Whitnall et al. 1985). Vasopressin has the ability to potentiate the actions of CRH on pituitary corticotropes (Gillies et al. 1982; Rivier and Vale 1984; Rivier et al. 1984), yet is not always expressed in the CRH neurons. In the normal rat, only 1%–2% of the CRH cells in the PVN and only a few terminals in the contact zone of the median eminence contain vasopressin (Fig. 18a). The number of CRH-vasopressin cells increases dramatically following adrenalectomy as do terminals in the median eminence (Kiss et al. 1984; Sawchenko et al. 1984a), as shown in Fig. 18b. Since CRH and vasopressin come

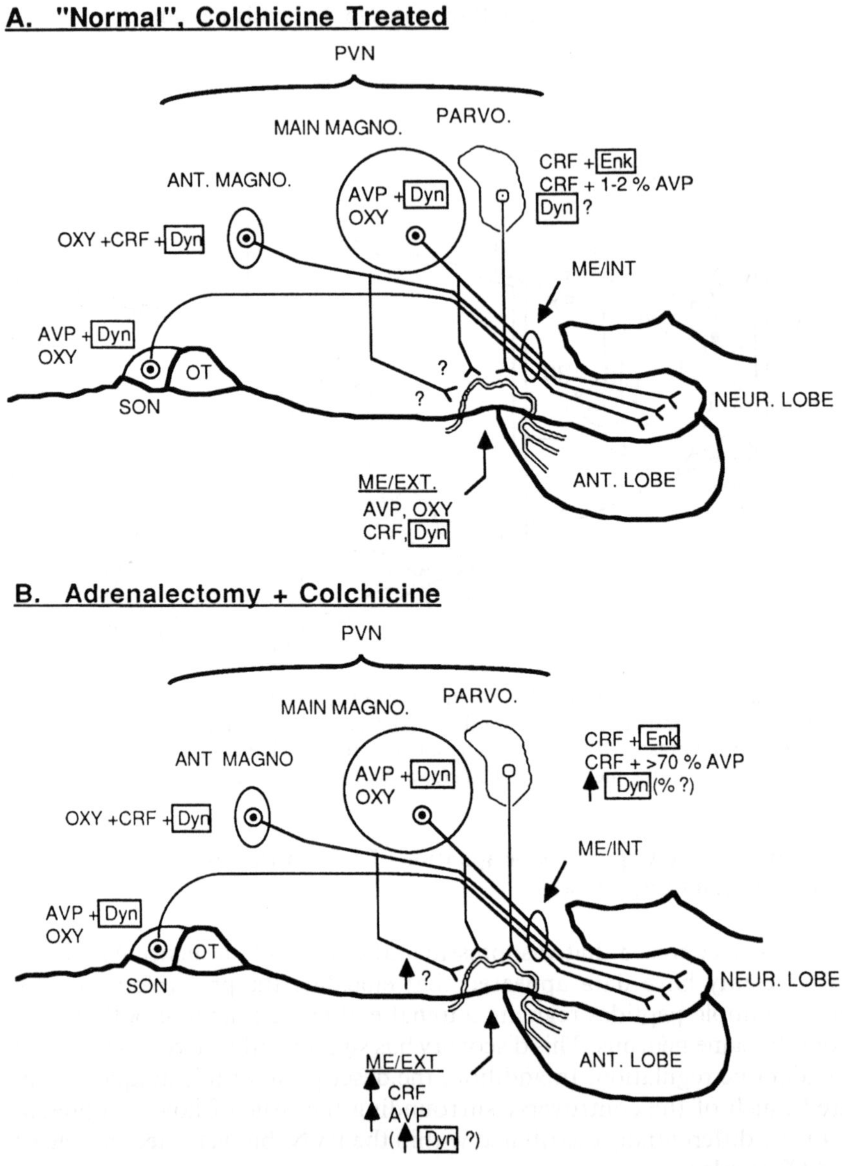

Fig. 18 a, b. The distribution of opioid and related peptides (in boxes) in the rat hypothalamus under normal (**a**) and adrenalectomy (**b**) conditions. **a** In the normal animal, vasopressin (*AVP*) and oxytocin (*OXY*) cells project to the neural lobe of the pituitary via the internal layer of the median eminence (*ME/INT*), with some projections to the external layer of median eminence (*ME/EXT*). The majority of CRF cells in the paraventricular nucleus (*PVN*) also project to the ME/EXT. Normally, AVP and dynorphin (*Dyn*) are co-stored in some magnocellular neurons, *Enk* and *CRF* are co-localized in some parvicellular neurons, while 1%–2% of CRF cells also contain AVP. Furthermore, in the anterior PVN, some magnocellular cells co-store OXY, Dyn and CRF. The exact source of magnocellular projections to the ME/EXT or the coexistence of Dyn in some CRF-AVP cells is not known. **b** In the adrenalectomized animal, there is a dramatic rise in the number of parvicellular PVN cells which co-store CRF and AVP, and some evidence suggests a rise in Dyn levels in these cells as well. Also, in the ME/EXT, there is an increase in CRF, AVP and possibly Dyn fiber immunoreactivity. The exact neuronal origin of these fibers has not yet been determined from different precursor proteins, this system is capable of providing multiple sites for releasing factor regulation. The interaction between vasopressin and CRH in the control of corticotropin release is further strengthened by (a) the demonstration of vasopressin messenger RNA in CRH neurons after adrenalectomy (Wolfson et al. 1985), and (b) the observation that adrenalectomy causes a significant rise in vasopressin levels in the median eminence (Stillman et al. 1977; Swanson and Kuypers 1980). The secretion of vasopressin and CRH are differentially regulated during various stressful conditions (Gibbs 1984), suggesting that such finely tuned regulation occurs in vivo.

Other neuropeptides are found to coexist within CRH neurons of the hypothalamus. Both the peptide, PHI, and enkephalin reside in one subset of CRH neurons (Hökfelt et al. 1983). PHI, like vasopressin, can potentiate the actions of CRH on corticotropin secretion (Tilders et al. 1984) and has a spectrum of activity similar to VIP. The opioids are present at all levels of the hypothalamic-pituitary-adrenal axis. At the level of the hypothalamus, opioid involvement in stress physiology is poorly understood; however, several observations lend support for their interaction with CRH and vasopressin in the regulation of anterior lobe corticotropes. The association of the opiate peptide enkephalin with CRH neurons (Hökfelt et al. 1983) offers further expansion of the possible activity of these cells to include regulation of somatotropes in addition to effects on lactotropes. Opiatergic activity in concert with CRH release is further supported by the identification of a subpopulation of CRH cells that also contain the opiate peptide dynorphin (Roth et al. 1983). Dynorphin immunoreactive terminals have been observed in the external layer of the median eminence as well (Khachaturian, unpublished). Here, dynorphin terminals are seen in apposition to the portal capillaries (Fig. 19). The exact origin of these dynorphin terminals is currently not known. Since some magnocellular vasopressin neurons do project to the external layer, it is conceivable that they also contain dynorphin peptides. Dynorphin and CRH have been co-localized in a subset of anterior PVN neurons in normal rats (Schafer et al. 1985) as seen in Fig. 20. Again, the exact projections of these latter neurons have not been elucidated. In addition to the opiate peptides and vasopressin, a small number of CRH neurons are found to contain neurotensin (Sawchenko et al. 1984b). Neurotensin can stimulate release of TSH (Tixier-Vidal and Gourdji 1981), further expanding the interrelationships of CRH neurons with other neuroendocrine regulatory systems.

7 Growth Hormone Releasing Factor

As was typical for earlier studies on CNS localization of hypophysiotropic factors, lesions and electrical stimulation of the medial basal hypothalamus (see Martin et al. 1978 for review) indicated that growth hormone releasing factor (GRF)-producing perikarya existed in the ventromedial and arcuate nuclei. GRF activity in hypothalamic extracts was reported as early as 1964 (Deuben and Meites 1964). Dunn and Arimura (1974) proposed that medial hypothalamic ablation interfered with both stimulatory and inhibitory influences on GH secre-

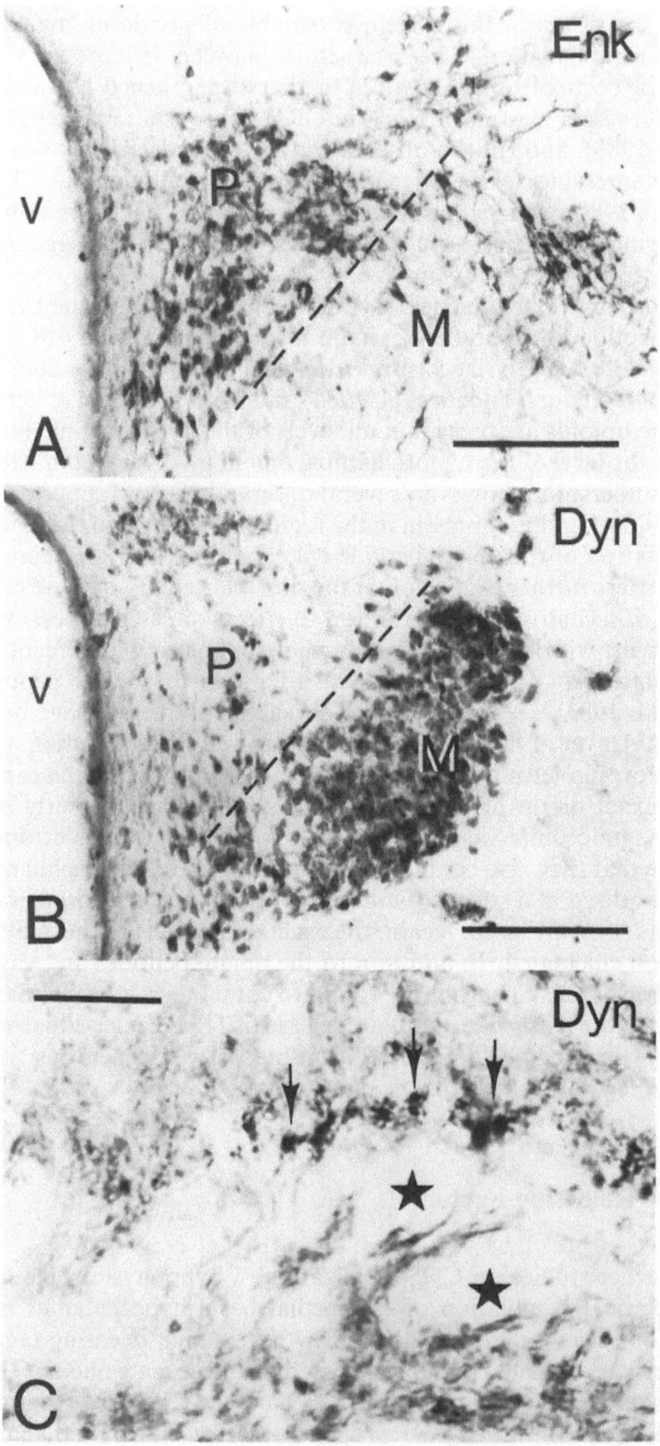
Enk
V
P
M
A
Dyn
V
P
M
B
Dyn
C

tion; current immunocytochemical data indicate that such lesions would destroy median eminence-afferent somatostatin pathways and, indeed, GRF-containing perikarya.

The designation "GRF" is used to indicate the tentative status of this peptide as a physiological hormone. Some authors currently use the abbreviation GHRH; GRF/GHRH is also referred to as somatocrinin. Elucidation of the GRF peptide structure did not result from examination of hypothalamic extracts but from analysis of a pancreatic islet tumor removed in 1982 from an acromegalic patient (Thorner et al. 1982). Three such "hpGRFs" were characterized, sequenced, and synthesized within the same year (Esch et al. 1982; Guillemin et al. 1982; Rivier et al. 1982; Spiess et al. 1982). The major original pancreatic peptides are 40- and 44-residue molecules with biological GH releasing activity for rat pituitary in vitro (Cronin et al. 1983) and in vivo in humans (Thorner et al. 1983); several synthetic peptides with GH releasing activity have also been reported (Badger et al. 1984). Subsequent to the characterization of hpGRF, a 43-amino acid GRF from rat hypothalamic extracts (rGRF) was sequenced (Spiess et al. 1983); rGRF and hpGRF differ in 30% of residues. Rat anterior pituitary cells bind hpGRF, rGRF, and synthetic analogues in direct proportion to the biological potency of each peptide (Seifert et al. 1985).

Rabbit antisera to hpGRF40, hpGRF44, and rGRF43 have been developed in several laboratories. Antigenicity of the molecules appears to lie in the C-terminal ending (40–44 sequence) although some cross-reactivity exists between the significantly different rat and human sequences. Reactivity of at least one antiserum to hpGRF44 was not diminished by preincubation with hpGRF37 or hpGRF40 (Bresson et al. 1984). Immunocytochemical localization of GRF somata and fibers has been reported in adult human and squirrel monkey (Bloch et al. 1983) and rat (Merchenthaler et al. 1984 b) using antibodies to hpGRF40, and in garden dormouse (Fellmann et al. 1983), cat (Bugnon et al. 1983), rat (Jacobowitz et al. 1983), human fetuses, infants, and adults (Bresson et al. 1984) using antibody to hpGRF44. Bloch and colleagues (1983) reported that anti-hpGRF44 was not immunoreactive in rat brain. Localization in rat brain using antiserum to rGRF43 (Bloch et al. 1983; Merchenthaler et al. 1984 a) has been reported more recently.

A consistent finding among immunocytochemical studies in all species, using antisera to different GRF molecules, is the presence of GRF immunoreactivity in cell bodies of the hypothalamic arcuate nucleus (Figs. 5–8, 21, 23), although the necessity of colchicine pretreatment for visualization is reported as variable; i.e., colchicine is precluded in the study of human autopsy material (Bloch et al.

Fig. 19 a–c. Relationship of enkephalin and dynorphin neurons in the PVN. **a, b** are adjacent 100 μm sections through the paraventricular nucleus. Note that the majority of Leu-enkephalin (*Enk*) positive cells are concentrated in the parvicellular (*P*) division while the majority of dynorphin B (*Dyn*) positive cells reside in the magnocellular (*M*) division (the dashed line is an approximate separation of the two divisions). However, considerable overlap is evident in both regions. In particular, the parvicellular Enk cells, some of which are known to contain CRF, might also co-store Dyn peptides. **c** Dyn B immunoreactive varicosities (*arrows*) can be seen in the external layer of the median eminence in the vicinity of portal vessels (*stars*). *v*, third ventricle; *scale* **a, b**, 200 μm; *scale* **c**, 25 μm

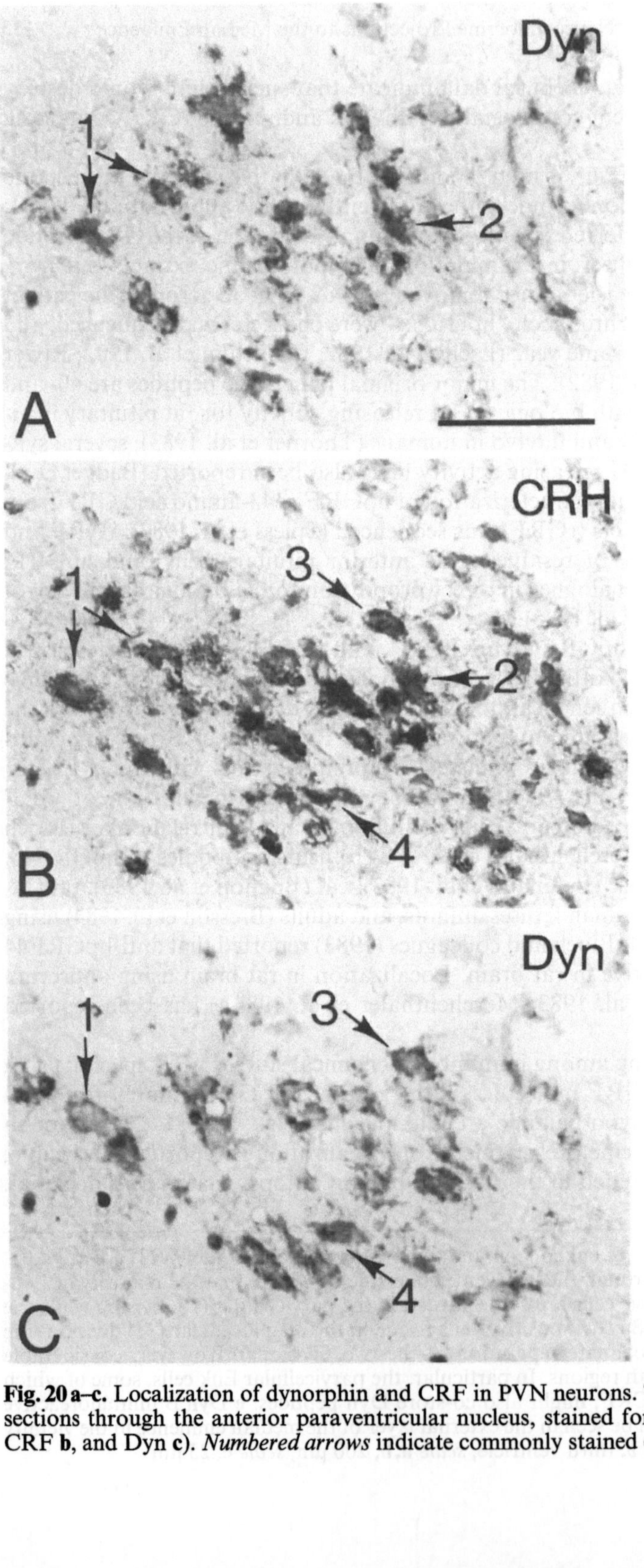

Fig. 20 a–c. Localization of dynorphin and CRF in PVN neurons. Three consecutive 5 μm sections through the anterior paraventricular nucleus, stained for dynorphin B (Dyn, **a**, CRF **b**, and Dyn **c**). *Numbered arrows* indicate commonly stained cells. *Scale*, 50 μm

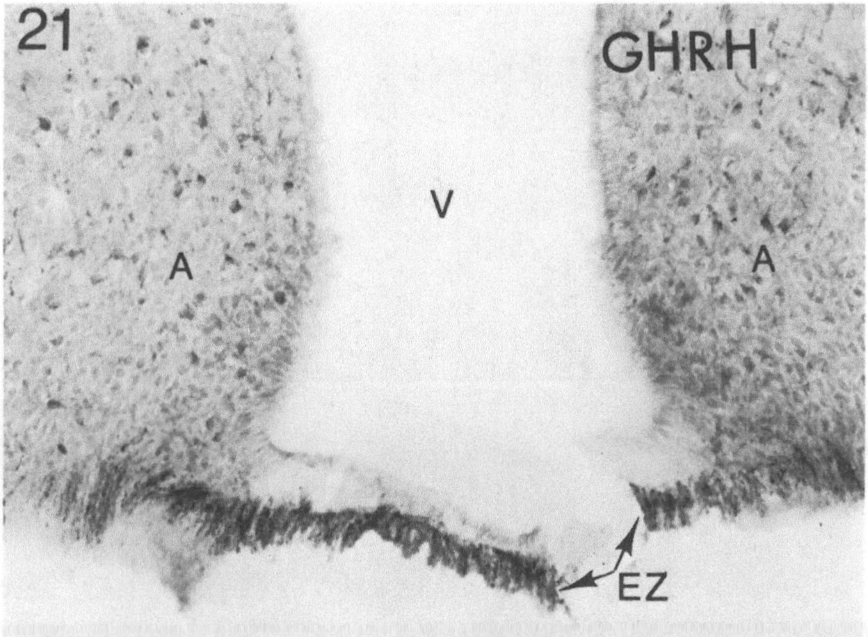

Fig. 21. Numerous cells containing GHRH are found in the arcuate nucleus (*A*) of the colchicine pretreated mouse brain. These cells appear to send axons to the external zone (*EZ*) of the median eminence (from Phelps and Hoffman 1986). *V*, third ventricle, × 50

1983; Bresson et al. 1984) but appears to be necessary in the rat (Figs. 22 and 23). Cells in the arcuate nucleus appear to constitute the major median eminence-afferent pathway in the rat, based on the absence of morphine-stimulated GH responses (Millard et al. 1982) and selective disappearance of median eminence GRF immunoreactivity (Bloch et al. 1984) in rats treated with monosodium glutamate, which causes selective loss of arcuate neurons.

The distribution of GRF within the median eminence has been reported to be confined to the perivascular external zone. GRF terminals have been shown in most studies to extend throughout the zona externa, but Bresson and associates (1984) reported that GRF was found predominantly in the lateral median eminence, similar to, but morphologically distinguishable from, somatostatin terminals.

GRF-immunoreactive perikarya in most species studied appear to be confined to the hypothalamus, although, in the primate brain, extrahypothalamic areas may not have been examined. Within the hypothalamus reports of cell body distribution are not consistent, even within a species. For example, in the rat Jacobowitz and colleagues (1983), using antiserum to hpGRF44, found GRF-containing perikarya in the arcuate nucleus, medial forebrain bundle, zona incerta, dorsal dorsomedial nucleus (DM), and dorsal PVN. Using an hpGRF40 antiserum, Merchenthaler and coworkers (1984b) described major perikarya collections in the medial perifornical regions, as well as the arcuate nucleus, and scat-

22

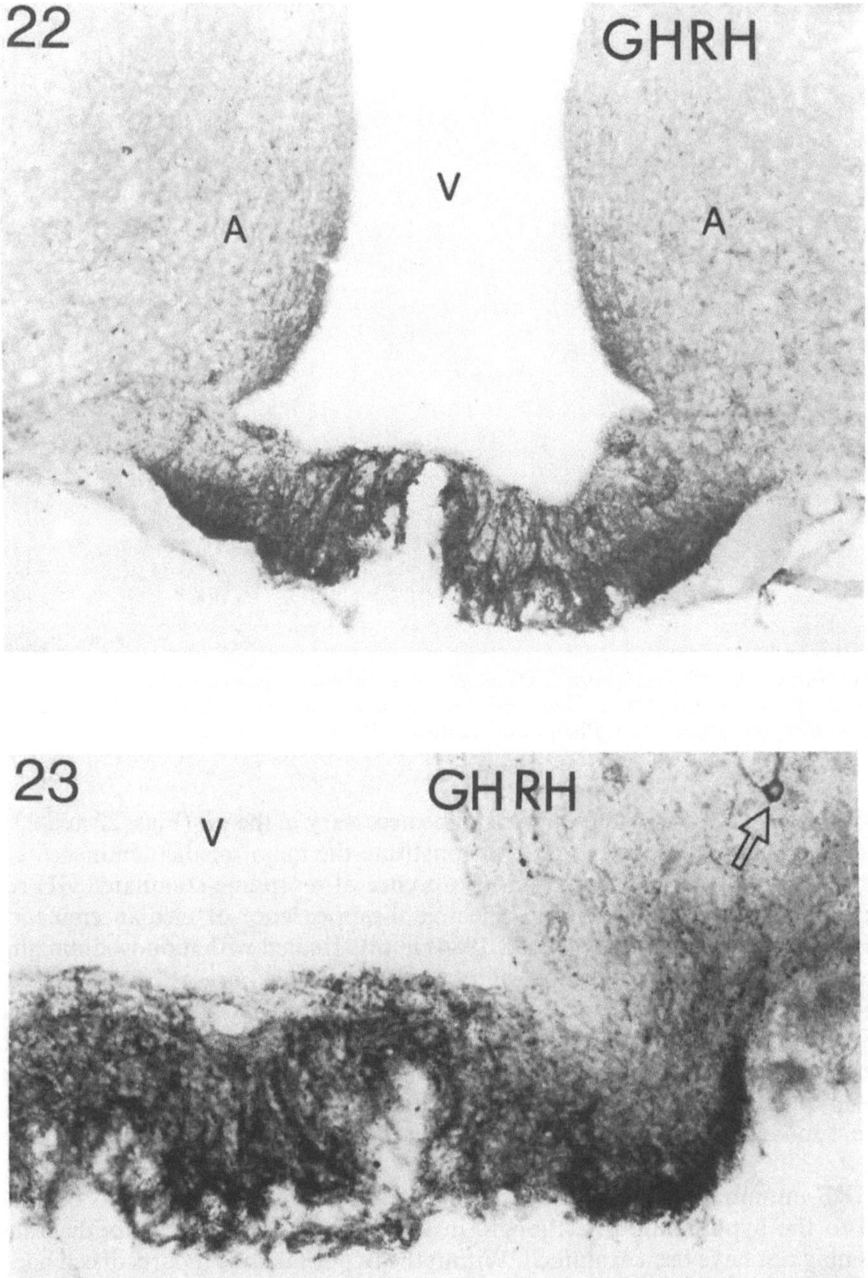

23

Figs. 22, 23. The localization of immunoreactive GHRH in the medial basal hypothalamus of a normal (Fig. 22) and colchicine-pretreated (Fig. 23) rat. In both instances, the median eminence has a notable innervation by GHRH axons. Immunoreactive cells (*arrow*) are easily visualized in the arcuate nucleus (*A*) with colchicine pretreatment. *V*, third ventricle; × 25 (Fig. 22), × 50 (Fig. 23) tered cells in PVN and DM. Using antibody generated against rGRF43, Merchenthaler et al. (1984a) described additional cells in the ventral premammillary nucleus, basal preoptic and suprachiasmatic regions, periventricular areas, and stria terminalis. Jacobowitz et al. (1983) described cells in the dorsal and ventral perimeter of the ventromedial nucleus (VM); Merchenthaler et al. (1984b) described cells in the medial and lateral VM. Our preliminary data using anti-rGRF43 in mouse brain suggest that GRF-containing perikarya are located in more diverse CNS areas, including cerebral cortex, hippocampus, and medial and lateral preoptic areas (Phelps and Hoffman 1986). Since mouse GRF has not been characterized, the extent of true cross-reactivity is unknown; yet, it is likely that extrahypothalamic GRF, like somatostatin, does not project to the median eminence.

8 Other Potential Afferents to the Median Eminence

In addition to the substances discussed above, there are a number of other putative releasing or inhibitory factors which are found in axonal projections to the median eminence. These include serotonin (Palkovits et al. 1977), norepinephrine (Palkovits et al. 1980b), epinephrine (Palkovits et al. 1980a), angiotensin II (Hökfelt et al. 1978), gastrin (Hökfelt et al. 1978), components of the proopiomelanocortin system, α-melanocyte-stimulating hormone, β-endorphin, ACTH, β-lipotropin (Watson SJ et al. 1978; Zimmerman et al. 1978), GABA (Jennes et al. 1983), and the list is continuously expanding. Pharmacologic studies suggest that each of these can influence pituitary function (Tixier-Vidal and Gourdji 1981). However, since many of these same factors are likely to present afferents to the releasing factor neurons discussed above, the precise role of median eminence projections remains unclear.

References

Ajika K (1979) Simultaneous localization of LHRH and catecholamines in rat hypothalamus. J Anat 128:331–347

Amos M, Burgus R, Blackwell R, Vale W, Fellows R, Guillemin R (1971) Purification, amino acid composition and N-terminus of the hypothalamic luteinizing hormone-releasing factor (LRF) of bovine origin. Biochem Biophys Res Commun 44:205–210

Badger TM, Millard WJ, McCormick GF, Bowers CY, Martin JB (1984) The effects of growth hormone (GH)-releasing peptides on GH secretion in perfused pituitary cells of adult male rats. Endocrinology 115:1432–1438

Barros D'sa AKJ, Bloom SR, Baron JH (1975) Direct inhibition of gastric acid by growth-hormone release-inhibiting hormone in dogs. Lancet 1:886–887

Barry J, Hoffman GE, Wray S (1985) LHRH-containing systems. In: Björklund A, Hökfelt T (eds) Handbook of chemical neuroanatomy, vol 4: GABA and neuropeptides in the CNS. Elsevier, Amsterdam

Beck W, Wuttke W (1977) Desensitization of the dopaminergic inhibition of pituitary LH release by prolactin in ovariectomized rats. J Endocrinol 74:67–74

Bennett-Clarke C, Joseph SJ (1982) Immunocytochemical distribution of LHRH neurons and processes in the rat: hypothalamic and extrahypothalamic locations. Cell Tissue Res 221:493–504
Bennett-Clarke C, Romagnano MA, Joseph SA (1980) Distribution of somatostatin in the rat brain: telencephalon and diencephalon. Brain Res 188:473–486
Bishop W, Fawcett CP, Krulich L, McCann SM (1972) Acute and chronic effects of hypothalamic lesions on the release of LH, FSH, and prolactin in intact and castrated rats. Endocrinology 91:643–656
Björklund A, Nobin A (1973) Fluorescence histochemical and microspectrofluorimetric mapping of dopamine and noradrenaline cell groups in the rat diencephalon. Brain Res 51:193–205
Björklund A, Enemar A, Falck B (1968) Monoamines in the hypothalamo-hypophyseal system of the mouse with special reference to the ontogenic aspect. Z Zellforsch 89:590–607
Björklund A, Falck B, Hromek F, Owman C, West KA (1970) Identification and terminal distribution of the tubero-hypophyseal monoamine fibre system in the rat by means of stereotaxic and microspectrofluorimetric techniques. Brain Res 17:1–23
Björklund A, Moore RY, Nobin A, Stenevi U (1973) The organization of tubero-hypophyseal and reticulo-infundibular catecholamine neuron systems in the rat brain. Brain Res 51:171–191
Blake CA, Weiner RI, Sawyer CH (1972) Pituitary prolactin secretion in female rats made persistently estrous or diestrous by hypothalamic deafferentation. Endocrinology 90:862–866
Bloch B, Brazeau P, Ling N, Boehlen P, Esch F, Wehrenberg WB, Benoit R, Bloom F, Guillemin R (1983) Immunohistochemical detection of growth hormone releasing factor in brain. Nature 301:607–608
Bloch B, Ling N, Benoit R, Wehrenberg WB, Guillemin R (1984) Specific depletion of immunoreactive growth hormone-releasing factor by monosodium glutamate in rat median eminence. Nature 307:272–273
Bloom FE, Battenberg ELF, Rivier J, Vale W (1982) Corticotropin releasing factor (CRF): immunoreactive neurons and fibers in rat hypothalamus. Reg Pept 4:43–48
Bloom SR, Mortimer CH, Thorner MD, Besser GM, Hall R, Gomez-Pan A, Roy VM, Russel RCG, Coy DH, Kastin AJ, Schally AV (1974) Inhibition of gastrin and gastric acid secretion by growth-hormone release-inhibiting hormone. Lancet 2:1106–1109
Brazeau P, Vale W, Burgus R, Ling N, Butcher M, Rivier J, Guillemin R (1974) Hypothalamic polypeptide that inhibits the secretion of immunoreactive pituitary growth hormone. Science 179:77–79
Bresson JL, Clavequin MC, Fellmann D, Bugnon C (1984) Ontogeny of the neuroglandular system revealed with hpGRF44 antibodies in human hypothalamus. Neuroendocrinology 39:68–73
Brodish A (1964) Role of the hypothalamus in the regulation of ACTH release. In: Bajusz E, Jasmin G (eds) Major problems in neuroendocrinology. Karger, Basel
Brownstein MJ, Palkovits M, Saavedra JM, Bassari RM, Utiger RD (1974) Thyrotropin-releasing hormone in specific nuclei of rat brain. Science 185:267–269
Bugnon C, Fellman D, Gouget A, Cardot J (1982) Ontogeny of the corticoliberin neuroglandular system in rat brain. Nature 298:159–161
Bugnon C, Gouget A, Fellmann D, Clavequin MC (1983) Immunocytochemical demonstration of a novel peptidergic neuron system in cat brain with an anti-growth hormone-releasing factor serum. Neurosci Lett 38:131–137
Bugnon C, Fellmann D, Gouget A, Bresson JL, Clavequin MC, Hadjiyiassemis M, Cardot J (1984) Corticoliberin neurons: Cytophysiology, phylogeny and ontogeny. J Steroid Biochem 20:183–195
Burgus R, Dunn TF, Desiderio D, Guillemin R (1969) Structure moleculaire du facteur hypothalamique hypophysiotrope TRF d'origine ovine: mise evidence par spectrometrie de masse de la sequence. CR Acad Sci (D) 269:1870–1893
Carrillo A, Pool TB, Sharp ZD (1985) Vasoactive intestinal peptide increases messenger ribonucleic acid content in GH_3 cells. Endocrinology 116:202–206

Clemens JA, Shaar CJ, Tandy WA, Roush ME (1971) Effects of hypothalamic stimulation on prolactin secretion in steroid-treated rats. Endocrinology 89:1317–1319

Conrad LCA, Pfaff DW (1976) Efferents from the medial basal forebrain and hypothalamus in the rat. II. An autoradiographic study of the anterior hypothalamus. J Comp Neurol 169:221–262

Critchlow V, Rice RW, Abe K, Vale W (1978) Somatostatin content of the median eminence in female rats with lesion-induced disruption of the inhibitory control of growth hormone secretion. Endocrinology 103:817–825

Cronin MJ, Rogol AD, MacLeod RM, Keefer DA, Login IS, Borges JLC, Thorner MO (1983) Biological activity of a growth hormone-releasing factor secreted by a human tumor. Am J Physiol 244:E346–E353

Crowley WR, Terry LC (1980) Biochemical mapping of somatostatinergic systems in rat brain: effects of periventricular hypothalamic and medial basal amygdaloid lesions on stomatostatin-like immunoreactivity in discrete brain nuclei. Brain Res 200:283–291

Daikoku S, Matsumura J, Shinokura Y (1976) Efferent projections of the nucleus preopticus medialis to the median eminence in rats. Neuroendocrinology 21:130–138

Deuben RR, Meites J (1964) Stimulation of pituitary growth hormone release by a hypothalamic extract in vitro. Endocrinology 74:408–414

Dhariwal APS, Grosvenor CE, Antunes-Rodrigues J, McCann SM (1968) Studies on the purification of ovine prolactin-inhibiting factor. Endocrinology 82:1236–1241

Don Carlos LL, Ho RA, Berelowitz M, Finkelstein JA (1986) Immunoreactive somatostatin in the hypothalamus of genetically obese and non-obese Zucker rats. (to be published)

Dunn JD, Arimura A (1974) Serum growth hormone levels following ablation of the medial basal hypothalamus. Neuroendocrinology 15:189–199

Elde R, Hökfelt T, Johansson O, Efendic S, Luft R (1976) Somatostatin containing pathways in the nervous system. Soc Neurosci Abst 11:759

Epelbaum J, Willoughby JO, Brazeau P, Martin JB (1977) Effects of brain lesions and hypothalamic deafferentation on somatostatin distribution in the rat brain. Endocrinology 101:1495–1502

Esch FS, Boehlen P, Ling NC, Brazeau WB, Thorner MO, Cronin MJ, Guillemin R (1982) Characterization of a 40 residue peptide from a human pancreatic tumor with growth hormone-releasing activity. Biochem Biophys Res Commun 109:152–158

Everett JW (1954) Luteotrophic function of autografts of the rat hypophysis. Endocrinology 54:685–690

Falck B, Hillarp NA, Thieme G, Torp A (1962) Fluorescence of catecholamines and related compounds condensed with formaldehyde. J Histochem Cytochem 10:348–354

Fellmann C, Gouget A, Bugnon C (1983) Mise en envidence d'un nouveau systeme neuronal peptidergique immunoreactif à un immun serum anti-hpGRF 44 dans le cerveau du lerot (*Eliomys quercinus*). Cr Hebd Seanc Sci Paris Ser III 296:487–492

Fink G (1985) Has the prolactin inhibiting peptide at last been found? Nature 316:487–488

Finley JCW, Grossman GH, Dimeo P, Petrusz P (1978) Somatostatin-containing neurons in the rat brain: widespread distribution revealed by immunocytochemistry after pretreatment with pronase. Am J Anat 153:483–488

Forssmann WG (1978) A new somatostatinergic system in the mammalian spinal cord. Neurosci Lett 10:293–297

Fuxe K (1964) Cellular localization of monoamines in the median eminence and the infundibular stem of some mammals. Z Zellforsch 61:710–724

Fuxe K, Hökfelt T (1969) Catecholamines in the hypothalamus. In: Martini L, Ganong WF (eds) Frontiers in neuroendocrinology. Oxford University Press, New York

Gibbs DM (1984) Dissociation of oxytocin, vasopressin and corticotropin secretion during different types of stress. Life Sci 35:487–491

Gillies GE, Linton EA, Lowry PJ (1982) Corticotropin releasing activity of the new CRF is potentiated several times by vasopressin. Nature 299:355–357

Gomez-Pan A, Albinus M, Reed JJD, Shaw B, Hall R, Besser GM, Coy DH, Kastin AJ, Schally AV (1975) Direct inhibition of gastric acid and pepsin secretion by growth hormone in cats. Lancet 1:888–890

Grosz HJ, Rothballer AB (1961) Hypothalamic control of lactogenic function in the cat. Nature 190:349–350
Guillemin R, Rosenberg B (1955) Humoral hypothalamic control of anterior pituitary: a study with combined tissue cultures. Endocrinology 57:599–607
Guillemin R, Brazeau P, Boehlen P, Esch F, Ling N, Wehrenber W (1982) Growth hormone releasing factor from a human pancreatic tumor that caused acromegaly. Science 218:585–587
Halasz B, Pupp L, Uhlarik S (1962) Hypophysiotrophic area in the hypothalamus. J Endocrinol 25:147–154
Halasz B, Pupp L, Uhlarik S, Tima L (1965) Further studies on the hormone secretion of the anterior pituitary transplanted into the hypophysiotrophic area of the rat hypothalamus. Endocrinology 77:343–355
Halasz B, Rethelyi M, Rohlich P, Koritsanszky S, Bodoky M, Nagy L (1979) Recent observations on the structural organization of the rat medial basal hypothalamus. In: Proceedings of the International Symposium on Neuroendocrine Regulatory Mechanisms, vol VI. Scientific Assemblies, Serbian, Academic Scientific Arts, Belgrade
Haymaker W, Anderson E, Nauta WJ (1969) The hypothalamus. Thomas, Springfield
Hoffman GE (1983) LHRH neurons and their projections. In: Sano Y, Ibata Y, Zimmerman EA (eds) Structure and function of peptidergic and aminergic neurons. Japan Sci, Tokyo
Hoffman GE, Finch CE (1986) LHRH neurons in the female C57BL/6J mouse brain during reproductive aging: no loss up to middle age. Neurobiol Aging 7:45–48
Hoffman GE, Gibbs FP (1982) Deafferentation spares a subchiasmatic LHRH projection to the median eminence. Neuroscience 8:1979–1993
Hoffman GE, Hayes T (1979) Somatostatin neurons and their projections in dog diencephalon. J Comp Neurol 186:371–392
Hoffman GE, Sladek JR Jr (1980) Age-related changes in dopamine, LHRH and somatostatin in the rat hypothalamus. Neurobiol Aging 1:27–37
Hoffman GE, Melnyk V, Hayes T, Bennett-Clarke C, Fowler E (1978) Immunocytology of LHRH neurons. In: Scott DE, Kozlowski GP, Weindl A (eds) Brain-endocrine interactions, vol 3. Karger, Basel
Hoffman GE, Wray S, Goldstein M (1982) Relationship of catecholamines and LHRH: light microscopic study. Brain Res Bull 9:417–430
Hökfelt T, Efendic S, Hellerstrom C, Johansson O, Luft L, Arimura A (1975a) Cellular localization of somatostatin in endocrine-like cells and neurons of the rat with special reference to the A1 cells of the pancreatic islets and to the hypothalamus. Acta Endocrinol 200:5–41
Hökfelt T, Fuxe K, Johansson O, Jeffcoate S, White N (1975b) Distribution of thyrotropin-releasing hormone (TRH) in the central nervous system as revealed with immunohistochemistry. Eur J Pharmacol 34:389–392
Hökfelt T, Elde R, Fuxe K, Johansson O, Ljungdahl A, Goldstein M, Luft R, Efendic S, Nilsson G, Terenius L, Ganten D, Jeffcoate SL, Rehfeld R, Said S, Perez de la Mora M, Possani L, Tapia R, Teran L, Palacois R (1978) Aminergic and peptidergic pathways in the nervous system with special reference to the hypothalamus. In: Reichlin S, Baldessarini RJ, Martin JB (eds) The hypothalamus. Raven, New York, pp 69–135
Hökfelt T, Fahrenkrug T, Tatemoto K, Mutt V, Werner S, Hulting AL, Terenius L, Chang KJ (1983) The PHI (PHI-27)/corticotropin releasing factor/enkephalin immunoreactive hypothalamic neuron: possible morphological basis for integrated control of prolactin, corticotropin, and growth hormone secretion. Proc Natl Acad Sci USA 80:895–898
Holzwarth-McBride MA, Horst EM, Knigge KM (1976) Monosodium glutamate induced lesions of the arcuate nucleus. I. Endocrine deficiency and ultrastructure of the median eminence. Anat Rec 186:371–392
Ibata Y, Watanabe K, Kinoshita H, Kubo S, Sano Y, Sin S, Hashimura E, Imagawa K (1979) The location of LHRH neurons in the rat hypothalamus and their pathways to the median eminence. Cell Tissue Res 198:381–395

Ibata Y, Tani N, Obata HL, Tanaka M, Kubo S, Fukui K, Fujimoto M, Kinoshita H, Watanabe K, Sano Y (1981) Correlative ontogenetic development of catecholamine- and LHRH-containing nerve endings in the median eminence of the rat. Cell Tissue Res 216:31–38
Ishikawa K, Inoue K, Tosaka H, Shimada O, Suzuki M (1984) Immunohistochemical characterization of thyrotropin-releasing hormone-containing neurons in rat septum. Neuroendocrinology 39:448–452
Jackson IMD, Reichlin S (1977) Brain thyrotropin releasing hormone is independent of the hypothalamus. Nature 267:853–854
Jacobowitz DM, Schulte H, Chrousos GP, Loriaux DL (1983) Localization of GRF-like immunoreactive neurons in the rat brain. Peptides 4:521–524
Jennes L, Stumpf WE (1980) LHRH-systems in the brain of the golden hamster. Cell Tissue Res 209:239–256
Jennes L, Stumpf WE, Tappaz ML (1983) Anatomical relationships of dopaminergic and GABAergic systems with the GnRH-systems in the septo-hypothalamic area. Exp Brain Res 50:91–99
Jennes L, Stumpf WE, Sheedy ME (1985) Ultrastructure of GnRH-like neurons. J Comp Neurol 232:534–546
Jonsson G, Fuxe K, Hökfelt T (1972) On the catecholamine innervation of the hypothalamus with special reference to the median eminence. Brain Res 40:271–281
Kamberi IA, Mical RS, Porter JC (1971) Effects of anterior pituitary perfusion and intraventricular injection of catecholamines on prolactin release. Endocrinology 88:1012–1020
Kavanagh A, Weisz J (1973) Localization of dopamine and norepinephrine in the medial basal hypothalamus of the rat. Neuroendocrinology 13:201–212
Kawata M, Hashimoto K, Takahara J, Sano Y (1982) Immunohistochemical demonstration of the localization of corticotropin releasing factor-containing neurons in the hypothalamus of mammals including primates. Anat Embryol (Berlin) 165:303–313
King JC, Anthony ELP (1984) LHRH neurons and their projections in humans and other mammals: species comparisons. Peptides 5 (Suppl 1): 195–207
King JC, Tobet SA, Snavely FL, Arimura A (1982) LH-RH immunopositive cells and their projections to the median eminence and organum vasculosum of the lamina terminalis. J Comp Neurol 209:287–300
King JC, Anthony ELP, Gustafson AW, Damassa DA (1984) Luteinizing hormone-releasing hormone (LH-RH) cells and their projections in the forebrain of the bat, *Myotis lucifugus lucifugus*. Brain Res 298:298–301
Kiss JZ, Mezey E, Skirboll L (1984) Corticotropin-releasing factor-immunoreactive neurons of the paraventricular nucleus become vasopressin positive after adrenalectomy. Proc Natl Acad Sci USA 81:1854–1858
Kizer JS, Palkovits M, Brownstein MJ (1976) The projections of the A8, A9, and A10 dopaminergic cell bodies: evidence for a nigral-hypothalamic-median eminence dopaminergic pathway. Brain Res 108:363–370
Koerker D, Ruch W, Chideckel E, Palmer J, Goodner CJ, Ensinck J, Gale CC (1974) Somatostatin: hypothalamic inhibitor of the endocrine pancreas. Science 184:482–483
Konig JFR, Klippel RA (1963) The rat brain: a stereotaxic atlas of the forebrain and lower parts of the brain stem. Williams and Wilkins, Baltimore
Krieger MS, Conrad LCA, Pfaff DW (1979) An autoradiographic study of the efferent connections of the ventromedial nucleus of the hypothalamus. J Comp Neurol 183:785–816
Krisch B (1978) Hypothalamic and extrahypothalamic distribution of somatostatin-immunoreactive elements in the rat brain. Cell Tissue Res 195:499–513
Krisch B, Leonhardt H (1979) Demonstration of a somatostatin-like activity in retinal cells of the rat. Cell Tissue Res 204:127–140
Lechan RM, Jackson IMD (1982) Immunohistochemical localization of thyrotropin-releasing hormone in the rat hypothalamus and pituitary. Endocrinology 111:55–65
Lechan RM, Nestler JL, Jacobson S (1980) The hypothalamic "tuberoinfundibular" system of the rat as demonstrated by horseradish peroxidase (HRP) microiontophoresis. Brain Res 195:13–27

Lichtensteiger W, Langemann H (1966) Uptake of exogenous catecholamines by monoamine-containing neurons of the central nervous system: uptake of catecholamines by arcuato-infundibular neurons. J Pharmacol Exp Ther 151:400–408

Liposits Z, Lengbari I, Vigh S, Schally AV, Flerko B (1983) Immunohistological detection of degenerating CRF-immunoreactive nerve fibers in the median eminence after lesion of paraventricular nucleus of the rat. A light and electron microscopic study. Peptides 4:941–953

Liposits Z, Paull WK, Setalo G, Vigh S (1985) Evidence for local corticotropin releasing factor (CRF)-immunoreactive neuronal circuits in the paraventricular nucleus of the rat hypothalamus. An electron microscopic immunohistochemical analysis. Histochemistry 83:5–16

Lu KH, Meites J (1971) Inhibition by I-dopa and monoamine oxidase inhibitors of pituitary prolactin release; stimulation by methyldopa and d-amphetamine. Endocrinology 91:868–872

MacLeod RM (1976) Regulation of prolactin secretion. In: Martini L, Ganong WF (eds) Frontiers in neuroendocrinology. Raven, New York

MacLeod RM, Lehmeyer JE (1974) Studies on the mechanism of the dopamine-mediated inhibition of prolactin secretion. Endocrinology 94:1077–1085

Makara GB, Stark E, Karteszi M, Palkovits M, Rappay GY (1981) Effects of paraventricular lesions on stimulated ACTH release and CRF in stalk-median eminence of the rat. Am J Physiol 240:E441–E446

Marshall PE, Goldsmith PC (1980) Neuroregulatory and neuroendocrine GnRH pathways in the hypothalamus and forebrain of the baboon. Brain Res 193:353–372

Martin JB, Reichlin S (1972) Plasma thyrotropin (TSH) response to hypothalamic electrical stimulation and to injection of synthetic thyrotropin releasing hormone (TRH). Endocrinology 90:1079–1085

Martin JB, Brazeau P, Tannenbaum GS, Willoughby JO, Epelbaum J, Terry LC, Durand D (1978) Neuroendocrine organization of growth hormone regulation. In: Reichlin S, Baldessarini RJ, Martin JB (eds) The hypothalamus. Raven, New York

Matsuo H, Baba Y, Nair R, Arimura A, Schally AV (1971) Structure of the porcine LH- and FSH-releasing hormone. I. The proposed amino acid sequence. Biochem Biophys Res Commun 43:1334–1339

McNeill TH, Sladek JR Jr (1978) Fluorescence-immunocytochemistry: simultaneous localization of catecholamines and gonadotropin-releasing hormone. Science 200:72–74

McNeill TH, Scott DE, Sladek JR Jr (1980) Simultaneous monoamine histofluorescence and neuropeptide immunocytochemistry. I. Localization of catecholamines and gonadotropin-releasing hormone in the rat median eminence. Peptides 1:59–68

Merchenthaler I, Vigh S, Petrusz P, Schally AV (1982) Immunocytochemical localization of corticotropin releasing factor (CRF) in the rat brain. Am J Anat 165:385–396

Merchenthaler I, Vigh S, Petrusz P, Schally AV (1983) The paraventriculo-infundibular corticotropin releasing factor (CRF) pathway as revealed by immunocytochemistry in long term hypophysectomized or adrenalectomized rats. Regul Pept 5:295–305

Merchenthaler I, Thomas CR, Arimura A (1984a) Immunocytochemical localization of growth hormone releasing factor (GHRF)-containing structures in the rat brain using anti-rat GHRF serum. Peptides 5:1017–1019

Merchenthaler I, Vigh S, Schally AV, Petrusz P (1984b) Immunocytochemical localization of growth hormone-releasing factor in the rat hypothalamus. Endocrinology 114:1082–1085

Mezey E, Kiss JZ (1985) Vasoactive intestinal peptide-containing neurons in the paraventricular nucleus may participate in regulating prolactin secretion. Proc Natl Acad Sci USA 82:245–247

Millard WJ, Martin JB Jr, Audet J, Sagar SM, Martin JB (1982) Evidence that reduced growth hormone secretion observed in monosodium glutamate-treated rats in the result of a deficiency in growth hormone-releasing factor. Endocrinology 110:540–550

Neill JD (1980) Neuroendocrine regulation of prolactin secretion. In: Ganong WF, Martini L (eds) Frontiers in neuroendocrinology. Raven, New York

Nikitovitch-Winer MB (1965) Effect of hypophysial stalk transection on luteotropic hormone secretion in the rat. Endocrinology 77:658–666

Nikitovitch-Winer M, Everett J (1958) Functional restitution of pituitary grafts retransplanted from kidney to median eminence. Endocrinology 63:916–930

Nikitovitch-Winer M, Everett J (1959) Histocytologic changes in grafts of rat pituitary on the kidney and upon re-transplantation under the diencephalon. Endocrinology 65:357–368

Nikolics K, Mason AJ, Szonyi E, Ramachandran J, Seeburg PH (1985) A prolactin-inhibiting factor within the precursor for human gonadotropin-releasing hormone. Nature 316:511–517

Olschowka JA, D'Donahue TL, Mueller GP, Jacobowitz DM (1982) Hypothalamic and extrahypothalamic distribution of CRF-like immunoreactive neurons in the rat brain. Neuroendocrinology 35:305–308

Palkovits M (1977) Neural pathways involved in ACTH regulation. Ann NY Acad Sci 297:455–476

Palkovits M, Brownstein MJ (1985) Distribution of neuropeptides in the central nervous system using biochemical micromethods. 2.2 Thyrotropin releasing hormone. In: Björklund A, Hökfelt T (eds) Handbook of chemical neuroanatomy, vol 4: GABA and neuropeptides in the CNS. Elsevier, Amsterdam

Palkovits M, Saavedra JM, Jacobowitz DM, Kizer JS, Zaborsky L, Brownstein MJ (1977) Serotonergic innervation of the forebrain: effect of lesions on serotonin and tryptophan hydroxylase levels. Brain Res 130:121–134

Palkovits M, Mezey E, Zaborsky L, Feminger A, Versteeg DHG, Wijnen HLJM, DeJong W, Fekete MIK, Herman JP, Kanyicska B (1980 a) Adrenergic innervation of the rat hypothalamus. Neurosci Lett 18:237–243

Palkovits M, Zaborsky L, Feminger A (1980 b) Noradrenergic innervation of the rat hypothalamus: experimental biochemical and electron microscopic studies. Brain Res 191:160–171

Paull WK, Scholer J, Arimura A, Meyers CA, Chang J, Chang D, Shimizu M (1982) Immunocytochemical localization of CRF in the ovine hypothalamus. Peptides 1:183–191

Phelps CJ, Hoffman GE (1986) Immunocytochemical localization of growth hormone releasing factor in normal and dwarf mouse brain. Endocrinology 118 (Suppl):396 (abstract)

Phelps CJ, Hymer WC (1986) Pituitary lactotroph sedimentation profiles and in vitro secretory activity after ablation of the medial hypothalamus. Anat Rec 215:365–373

Richoux JP, Dubois MP (1980) Neuronal systems immunologically related to the somatostatin system in the garden dormouse. Cell Tissue Res 209:455–472

Rivier C, Vale W (1984) Interaction of corticotropin releasing factor (CRF) and arginine vasopressin (AVP) on ACTH secretion in vivo. Endocrinology 113:939–942

Rivier J, Spiess J, Thorner MO, Vale W (1982) Characterization of a growth hormone-releasing factor from a human pancreatic islet tumor. Nature 30:276–278

Rivier C, Rivier J, Mormede P, Vale W (1984) Studies of the nature of the interaction between vasopressin and corticotropin releasing factor on adrenocorticotropin release in the rat. Endocrinology 115:882–886

Roth KA, Weber E, Barchas JD (1982) Immunoreactive corticotropin releasing factor (CRF) and vasopressin are colocalized in a subpopulation of the immunoreactive vasopressin cells in the paraventricular nucleus of the hypothalamus. Life Sci 31:1857–1860

Roth KA, Weber E, Barchas JD, Chang D, Chang J-K (1983) Immunoreactive dynorphin-(1-8) and corticotropin-releasing factor in subpopulation of hypothalamic neurons. Science 219:189–190

Saffran M, Schally AV (1955) The release of corticotrophin by anterior pituitary tissue in vitro. Can J Biochem Physiol 33:408–415

Saper CB, Swanson LW, Cowan WM (1978) The efferent connections of the anterior hypothalamic areas of the rat, cat and monkey. J Comp Neurol 183:575–600

Sarkar DK, Gottschall PE, Meites J (1981) Damage to hypothalamic dopamine neurons is associated with development of prolactin-secreting tumors. Science 218:684–686
Sawchenko P, Swanson LW, Vale W (1984a) Co-expression of corticotropin-releasing factor and vasopressin immunoreactivity in parvocellular neurosecretory neurons of the adrenalectomized rat. Proc Natl Acad Sci USA 81:1883–1887
Sawchenko P, Swanson LW, Vale W (1984b) Corticotropin releasing factor: coexpression within distinct subsets of oxytocin-, vasopressin-, and neurotensin-immunoreactive neurons in the hypothalamus of the male rat. J Neurosci 4:1118–1129
Schafer MK-H, Burke S, Sherman TG, Watson SJ (1985) Co-existence of opioid peptides with oxytocin and vasopressin in the rat hypothalamic magnocellular nuclei. International Narcotic Res Conference Abst 133:47
Schally AV, Meites J, Bowers CY, Ratner R (1964) Identity of prolactin inhibitory factor (PIF) and luteinizing hormone-releasing factor (LRF). Proc Soc Exp Biol Med 117:252–254
Schally AV, Redding TW, Bowers CY, Barrett JF (1969) Isolation and properties of porcine thyrotropin-releasing hormone. J Biol Chem 244:4077–4088
Scott PM, Knigge KM (1981) Immunocytochemistry of luteinizing hormone-releasing hormone, vasopressin and corticotropin following deafferentation of the basal hypothalamus of the male rat brain. Cell Tissue Res 219:393–402
Seifert H, Perrin M, Rivier J, Vale W (1985) Binding sites for growth hormone releasing factor on rat anterior pituitary cells. Nature 313:487–488
Selmanoff M (1981) The lateral and medial median eminence: distribution of dopamine, norepinephrine, and luteinizing hormone-releasing hormone and the effect of prolactin on catecholamine turnover. Endocrinology 108:1716–1722
Silverman AJ, Renaud L (1985) Which LHRH neurons project to the median eminence: a combined retrograde tracing and immunocytochemical analysis. Soc Neurosci Abst 11:20
Silverman AJ, Antunes JL, Abrams GM, Nilaver G, Thau R, Robinson JA, Ferin M, Krey LC (1982) The luteinizing hormone-releasing hormone pathways in rhesus (*Macaca mulatta*) and pigtailed (*Macaca nemestrina*) monkeys: new observations on thick, unembedded sections. J Comp Neurol 211:309–317
Sladek JR Jr, Sladek CD, McNeill TH, Wood JG (1978) New sites of monoamine localization in the endocrine hypothalamus as revealed by new methodological approaches. In: Scott DE, Kozlowski GP, Weindl A (eds) Brain-endocrine interactions III. Neural hormones and reproduction, 3rd Int Symp, Würzburg, Karger, Basel
Smith GC, Fink G (1972) Experimental studies on the origin of monoamine-containing fibres in the hypothalamo-hypophyseal complex of the rat. Brain Res 43:37–51
Smith PE (1927) The disabilities caused by hypophysectomy and their repair. J Am Med Assoc 88:158–161
Sofroniew MV (1985) Vasopressin, oxytocin and their related neurophysins. In: Björklund A, Hökfelt T (eds) Handbook of chemical neuroanatomy, vol 4: GABA and neuropeptides in the CNS. Elsevier, Amsterdam
Soper BD, Weick RF (1980) Hypothalamic and extrahypothalamic mediation of pulsatile discharages of luteinizing hormone in the ovariectomized rat. Endocrinology 106:348–355
Spiess J, Rivier J, Thorner MO, Vale W (1982) Sequence analysis of a growth hormone releasing factor from a human pancreatic islet tumor. Biochemistry 21:6037–6040
Spiess J, Rivier J, Vale W (1983) Characterization of rat hypothalamic growth hormone releasing factor. Nature 303:532–535
Stillman MA, Recht LD, Rosario SL, Seif SM, Robinson AG, Zimmerman EA (1977) The effect of adrenalectomy and glucocorticoid replacement on vasopressin and vasopressin-neurophysin in the zona externa of the median eminence of the rat. Endocrinology 101:42–49
Swanson LW, Kuypers JM (1980) The paraventricular nucleus of the hypothalamus: cytoarchitectonic subdivisions and organization of projections to the pituitary, dorsal vagal complex and spinal cord, as demonstrated by retrograde fluorescence double-labeling methods. J Comp Neurol 194:555–570

Swanson LW, Sawchenko P, Wiegand SJ, Price JL (1980) Separate neurons in the paraventricular nucleus project to the median eminence and to the medulla or spinal cord. Brain Res 198:190–195

Swanson LW, Sawchenko PE, Rivier J, Vale W (1983) Organization of ovine corticotropin-releasing factor immunoreactive cells and fibers in the rat brain: an immunohistochemical study. Neuroendocrinology 36:165–186

Szentagothai J (1964) The parvocellular neurosecretory system. Prog Brain Res 5:135–146

Takatsuke K, Shiosada S, Dakanaka M, Inagaki S, Senba E, Takagi H, Tohyama M (1981) Somatostatin in the auditory system of the rat. Brain Res 213:211–216

Terasawa E, Davis GA (1983) The LHRH neuronal system in female rats: relation to the medial preoptic nucleus. Endocrinol Jpn 30:405–417

Thorner MO, Flückinger E, Caine DB (1980) Bromocriptine: a clinical and pharmacological review. Raven, New York

Thorner MO, Perryman RL, Cronin MJ, Rogol AD, Draznin M, Johanson A, Vale W, Horvath E, Kovacs K (1982) Successful treatment of acromegaly by removal of a pancreatic islet tumor secreting a growth hormone-releasing factor. J Clin Invest 70:965–977

Thorner MO, Rivier J, Spiess JL, Borges J, Vance ML, Bloom SR, Rogol AD, Cronin MJ, Kaiser DL, Evans WS, Webster RM, MacLeod FM, Vale W (1983) Human pancreatic growth-hormone-releasing factor selectively stimulates growth-hormone secretion in man. Lancet 1:24

Tilders F, Tatemoto K, Berkenbosch F (1984) The intestinal peptide PHI-27 potentiates the action of corticotropin-releasing factor on ACTH release from rat pituitary fragments in vitro. Endocrinology 115:1633–1635

Tixier-Vidal A, Gourdji D (1981) Mechanism of action of synthetic hypothalamic peptides on anterior pituitary cells. Physiol Rev 61:974–1011

Tramu G, Croix C, Pillez A (1983) Ability of the CRF immunoreactive neurons of the paraventricular nucleus to produce a vasopressin-like material. Neuroendocrinology 37:467–469

Turpen C, Dunn JD (1976) The effect of surgical isolation or ablation of the medial basal hypothalamus on serum prolactin levels in male rats. Neuroendocrinology 20:224–234

Turpen C, Sladek JR Jr (1978) Localization of glyoxylic acid-induced histofluorescence in surgically isolated medial basal hypothalamus. Cell Tissue Res 187:449–456

Vale W, Rivier C, Brazeau P, Guillemin R (1974) Effects of somatostatin on the secretion of thyrotropin and prolactin. Endocrinology 95:968–977

Vale W, Spiess J, Rivier C, Rivier J (1981) Characterization of a 41-residue ovine hypothalamic peptide that stimulates secretion of corticotropin and B-endorphin. Science 213:1394–1397

Vale W, Rivier C, Brown MR, Spiess J, Koob G, Swanson L, Binezikjian L, Bloom F, Rivier J (1983) Chemical and biological characterization of corticotropin releasing factor. Recent Prog Horm Res 39:245–270

Vijanian E, Samson W, Said S, McCann SM (1979) Vasoactive intestinal polypeptide: evidence for a hypothalamic site of action to release growth hormone, luteinizing hormone and prolactin in conscious ovariectomized rats. Endocrinology 104:53–57

Watson RE, Wiegand SJ, Clough RW, Hoffman GE (1986) Use of cryoprotectant to maintain longterm peptide immunoreactivity and tissue morphology. Peptides 7:155–159

Watson SJ, Richard CW, Barchas JD (1978) Adrenocorticotropin in rat brain: immunocytochemical localization in cells and axons. Science 200:1180–1182

Weiner RI, Shryne JE, Gorski RA, Sawyer CH (1972) Changes in the catecholamine content of the rat hypothalamus following deafferentation. Endocrinology 90:867–873

Whitnall MH, Mezey E, Gainer H (1985) Co-localization of corticotropin-releasing factor and vasopressin in median eminence neurosecretory vesicles. Nature 317:248–250

Wiegand SJ, Price JL (1980) Cells of origin of the afferent fibers to the median eminence. J Comp Neurol 192:1–19

Wiegand SJ, Terasawa E (1978) Persistent estrus and blockade of progesterone induced LH release follows lesions which do not damage the suprachiasmatic nucleus. Endocrinology 102:1645–1648

Wiegand SJ, Watson RE, Hoffman GE, Tang L (1985) Differential effects of lesions of the LHRH system or the medial preoptic nucleus (MPN) on phasic gonadotropin release in the rat. Biol Reprod 32 [Suppl 1]:110

Winokur A, Utiger RD (1974) Thyrotropin-releasing hormone: regional distribution in rat brain. Science 185:265–266

Wolfson B, Manning RW, Davis LG, Arentzen R, Baldino F (1985) Co-localization of corticotropin releasing factor and vasopressin mRNA in neurons after adrenalectomy. Nature 315:59–61

Wray S, Hoffman GE (1986) Postnatal morphological changes in rat LHRH neurons correlated with sexual maturation. Neuroendocrinology 43:93–97

Zaborszky L (1982) Afferent connections of the medial basal hypothalamus. Adv Anat Embryol Cell Biol 69:1–107

Zimmerman EA, Liotta A, Krieger DT (1978) β-lipotropin in brain: localization in hypothalamic neurons by immunoperoxidase technique. Cell Tissue Res 186:393–398

Afferents onto Neuroendocrine Cells

M. Palkovits

Contents

1 Neuroendocrine Cells in Hypothalamic Nuclei

Several substances, mainly peptides, are produced by neurons in the central nervous system and have their effects outside the brain, transported by the circulation. This is consistent with their classification as neurohormones. The hormone-producing cells, generally termed neuroendocrine cells, are present mainly but not exclusively in the hypothalamus. Neurohormones produced by neuroendocrine cells are conveyed directly by axonal transport to the posterior pituitary, or indirectly by a neurovascular route to the anterior pituitary. Substances which are closely involved in the regulation of pituitary functions are called releasing or release-inhibiting hormones, while others (e.g., vasopressin and oxytocin), after axonal transport into and storage in the posterior lobe, act as neurohormones on the periphery.

Although several substances have been identified during the last two decades as releasing or release-inhibiting hormones, there are only five which have been identified, characterized, and topographically localized in the central nervous system. These are luteinizing hormone releasing hormone (LHRH), thyrotropin-releasing hormone (TRH), corticotropin-releasing hormone (CRF), growth hor-

First Department of Anatomy, Semmelweis University Medical School, Tűzoltó-u 58, H-1450 Budapest, IX.

Current Topics in Neuroendocrinology, Vol. 7

mone-releasing hormone (GRF), and growth hormone release-inhibiting hormone (somatostatin). Evidence based on morphological, physiological, and pharmacologic experiments indicates that the role of these neuropeptides is not restricted to that of a neurohormone. Their presence in nerve terminals in extrahypothalamic brain regions indicates that they can also serve as neurotransmitters or -modulators.

Those releasing hormone- or release inhibiting hormone-producing cells which project to the median eminence, however, are localized exclusively within the hypothalamus and preoptic area: LHRH cells inside or in the vicinity of the medial preoptic nucleus (and in the arcuate nucleus in certain mammals and in man; TRH and CRF cells in the paraventricular nucleus; GRF cells in the arcuate nucleus; and somatostatin cells in the hypothalamic periventricular nucleus [Palkovits 1984a–c; 1986a, b]). Data obtained from retrograde tract-tracing studies indicate (Wiegand and Price 1980; Lechan et al. 1982) that cells from all these hypothalamic nuclei project directly into the median eminence (Fig. 1). There are

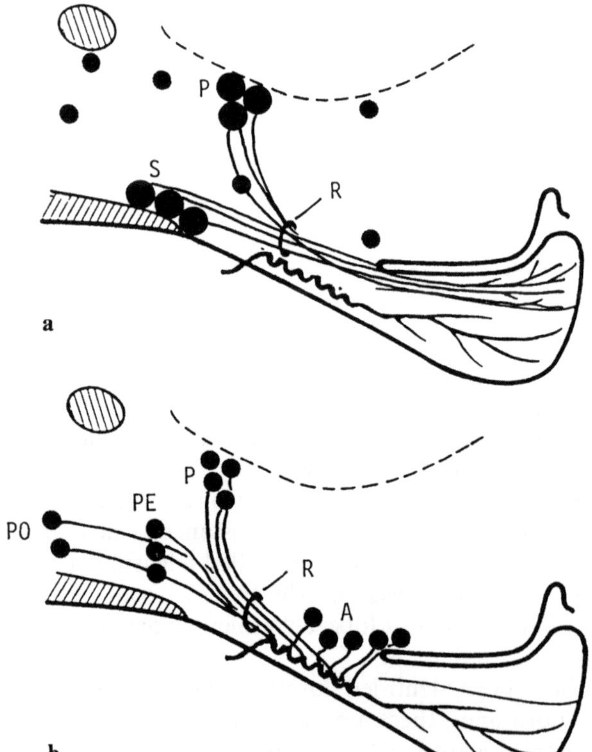

Fig. 1 a, b. Neuroendocrine cells in the rat hypothalamus: projections to the posterior pituitary (**a**), and the median eminence (**b**). Fibers other than of arcuate nucleus origin (tuberoinfundibular neurons) enter the median eminence through a common gate in the lateral retrochiasmatic area (*R*). *A*, arcuate nucleus; *P*, paraventricular nucleus; *PE*, hypothalamic periventricular nucleus; *PO*, medial preoptic nucleus; *S*, supraoptic nucleus other, scattered neurons throughout the hypothalamus with median eminence projections, but the afferent connections of these single cells are unknown; they are therefore beyond the scope of this review.

Vasopressin and oxytocin were the first neurohormones isolated in the hypothalamus. In the past few years, evidence has been furnished of the existence of vasopressin not only in the magnocellular supraoptic and paraventricular nuclei, but also in parvocellular hypothalamic (suprachiasmatic) and extrahypothalamic neurons. According to current knowledge, only magnocellular neurons from the supraoptic and paraventricular nuclei, from accessory magnocellular cell groups in the hypothalamus and from scattered cells from the preoptic area, and from the bed nucleus of the stria terminalis (NIST), from the zona incerta, and from the arcuate nucleus project to the posterior pituitary (Kelly and Swanson 1980). Afferents to the above scattered neurons, including the magnocellular accessory cell groups, which consitute a fairly high percentage (500–1800 cells) of the hypothalamic magnocellular neurosecretory system (Palkovits et al. 1974; Rhodes et al. 1980), are unknown.

In this chapter, neuronal afferents to the five major hypothalamic neuroendocrine cell groups (the supraoptic, paraventricular, medial preoptic, periventricular, and arcuate nuclei) will be summarized. This summary is offered as a guide, with the following points:

1. Neuroendocrine cells from the above five nuclei occur in the hypothalamus. Afferents to these scattered cells, however, have not yet been established.
2. Only a certain percentage of neurons in these five nuclei project to the median eminence/posterior pituitary, while many others have no neuroendocrine characteristics. Consequently, neuronal afferents to hypothalamic nuclei may innervate neurons other than the neuroendocrine ones. Only a combined use of tract-tracing techniques with immunocytochemistry at electron microscopic level (see Sect. 2.4) can provide direct evidence for the existence of specific and characteristic afferent inputs to neuroendocrine cells.
3. Several aminergic, cholinergic, and gamma-aminobutyric acid-(GABA)ergic, and certain peptidergic fibers terminate in the median eminence, and some of them also in the intermediate and posterior pituitary lobes. Even so, they are not considered to be neuroendocrine cells. They may produce their effect within the median eminence and the posterior pituitary by stimulating or inhibiting the release of neurohormones into the portal or pituitary circulations. Therefore, afferent innervation of these neurons is beyond the scope of this chapter.
4. To date, only five neuropeptides have been recognized as releasing or release-inhibiting hormones. There are, however, at least 30 other peptides in the hypothalamus which are synthesized by neurons. Many of these peptidergic neurons also project to the median eminence or the posterior pituitary, and may also act as neurohormones. It is noteworthy that these neurons are present almost exclusively in the five releasing hormone-containing hypothalamic cell groups. Furthermore, some of these neuropeptides are colocalized with one of the releasing hormone-, vasopressin-, or oxytocin-containing neurons (Palkovits 1984a–c).

5. The neuronal connections of the hypothalamus have been studied in a great number of species (Palkovits and Záborszky 1979). This chapter is based on data obtained only from the rat.

2 Neuroanatomical Techniques for Visualization of Neuronal Afferents

Neuroanatomical techniques for the identification and localization of neuronal connections have constantly improved. Several powerful methods have been introduced, and as a result a growing number of neuronal projections to the neuroendocrine cells have been identified in the past 5 years. In this short review, only morphological data by which afferents could be visualized and topographically localized, both at light and electron microscopic levels, are considered. Data obtained from physiological (mainly electrophysiological), biochemical, and pharmacologic studies are not summarized here.

No details, only the basic principles of some neuroanatomical techniques used for identification and localization of fiber connections, are mentioned here. The technical merits and the advantages and disadvantages of their application have been dealt with by several reports (Palkovits and Záborszky 1979; Záborszky 1982; Palkovits 1983). It should also be noted that critical evaluation of the specificity of certain techniques for identifying neuronal connections between brain areas is not the subject of this review.

Information obtained by the application of various methods can be classified in four categories: a) demonstration of the existence of certain neuronal afferents onto hypothalamic nuclei and localization of cells of origin; b) identification of the chemical nature of afferent fibers and projecting cells; c) type and numbers of afferent nerve terminals synapsing on hypothalamic cells; and d) chemical characterization of both presynaptic afferents and postsynaptic target neurons in the hypothalamic nuclei.

2.1 Tract-Tracing Techniques and Cells of Origin

The following are major neuroanatomical techniques for studying afferent connections:

1. *Nerve-Degenerating Technique*. This method is based on the fact that the distal portion of the axon, including the nerve terminals, will degenerate following transection of the axon or lesioning of the cell body. Degenerating nerve terminals or axon fragments can be recognized under both the light and the electron microscope.
2. *The Autoradiographic Fiber-Tracing Technique*. This method is based on the light or electron microscopic demonstration of anterogradely transported labeled amino acids in nerve terminals.
3. *Retrograde Tract-Tracing Techniques*. Horseradish peroxidase, various lectins, and fluorescent dyes are injected (or applied iontophoretically) into the inves-

tigated cell groups. The trace substance is transported retrogradely in the axons and accumulated in cells which project to the investigated area.

2.2 Chemical Identification of Afferent Neurons

Immunohistochemistry is widely used for identifying chemical substances in nerve cells. The terminal arborization of axons, but not the pathways, can be recognized on immunostained sections. Immunohistochemistry in combination with other techniques may provide further information. Transection of fibers or lesioning of projecting neurons may result in a disappearance of immunostained material distal to the lesion; this material may accumulate retrogradely in the transected fibers and cells of origin. A depletion or disappearance of substances after lesions can be quantitated by biochemical microassays in microdissected hypothalamic nuclei.

By combining tract-tracing techniques with immunohistochemistry, both afferents and target neurons can be recognized in the same section. Technically, retrograde labeling with fluorescent dyes and immunofluorescent staining has been improved to become a powerful tool for simultaneous localization and chemical characterization of projecting neurons.

2.3 Synaptic Contacts: Qualitative and Quantitative Investigations

Light microscopic techniques give considerable information about the origin of fibers and their terminal patterns in brain nuclei. The demonstration of nerve terminals synapsing with perikarya or dendrites is the subject of electron microscopy. Even if a clear pattern of terminals can be seen under a light microscope, the presence of afferent fibers in a brain nucleus does not provide direct evidence of the innervation of each particular neuron within the cell group investigated. Demonstration of anterogradely transported substances, labeled amino acids in presynaptic boutons, or degenerating nerve terminals following lesions in synaptic contact with cells or dendrites in the investigated nucleus are direct evidence of the existence of afferents onto an individual neuron in the brain.

The number of afferents to a single neuron or cell group can be quantitated. The synaptic bouton: cell ratio represents the best parameter for that kind of measurement. The number of presynaptic terminals can be directly counted on electron micrographs or under an electron microscope and expressed by their surface or volume density (number of boutons/1000 μm^2 or 1000 μm^3). That density divided by the cell-packing density (cell number/1000 μm^2 or 1000 μm^3, measured under a light microscope) gives us a bouton: cell ratio, which expresses the average number of boutons terminating on a single neuron (Palkovits 1976). The ratio is a useful measure for determining the number of local (intrinsic) nerve terminals and those arising from various brain regions to a particular nucleus (extrinsic nerve terminals). In such studies, the investigated brain nucleus is surgically deafferentated (isolated from the other parts of the brain), and several days later the intact and the degenerating synaptic boutons are counted. The intact boutons represent the intrinsic terminals, whereas the degenerating or absent nerve terminals (degenerating boutons disappear quickly owing to rapid phagocytosis by glial cells) are of extranuclear origin (Záborszky et al. 1975; Kiss et al. 1983b). In consecutive experiments, specific brain regions are lesioned or pathways transected, and degenerating bouton: cell ratios determined. Each such ratio represents the number of afferent terminals from a certain brain region, and they can finally be summarized in a mosaiclike manner (Záborszky et al. 1975).

2.4 Chemical Identification of Pre- and Postsynaptic Elements

Characterization of both pre- and postsynaptic elements is chemically possible by electron microscopic double or multiple immunostainings for simultaneous localization of two or more antigens in the same section. To investigate neuronal inputs to an immunologically characterized neuron, either autoradiography or a nerve-degenerating technique (or both) can be used in combination with electron microscopic immunocytochemistry. In the first case (autoradiography), anterogradely transported labeled amino acids can be visualized in nerve terminals synapsing with immunoreactive cells or dendrites (Oldfield and Silverman 1985), while in the second case (nerve-degenerating technique) degenerating nerve terminals are found in synaptic contact with perikarya or dendrites that can be characterized by immunostaining (Palkovits et al. 1982).

3 Supraoptic Afferents

The synaptic bouton: cell ratio is relatively low in the supraoptic nucleus; it concerns only 600 nerve terminals per cell. Two-thirds of these nerve terminals appear to be of intranuclear origin, as many remain intact following a total deafferentation of the nucleus (Léránth et al. 1975). (This number is unexpectedly high. Although connections with cells in the immediate vicinity of supraoptic cells remained intact and their projections were therefore judged to be intrinsic, the intranuclear connections are significant. This observation is consistent with those obtained from similar quantitative studies performed in other brain nuclei such as the suprachiasmatic, paraventricular, or deep cerebellar nuclei, or in larger brain regions like the hippocampus or cerebral cortex.) The intrinsic synapses may represent the nerve terminals of interneurons or collaterals of projecting supraoptic neurons.

No immunohistochemical data are available about the chemical character of the intrinsic supraoptic terminals. They may arise from vasopressin and oxytocin cells, but the existence of several other neuropeptides in supraoptic neurons (independently of or in colocalization with vasopressin or oxytocin cells) has recently been demonstrated (Table 1).

The major sources of supraoptic afferents are the lower brain stem, the hypothalamus, and the limbic system (Table 2). On the whole, about one-third of supraoptic afferents originate from the limbic system (Záborszky et al. 1975); fibers

Table 1. Neurohormone- and neurotransmitter-containing cells and nerve terminals in the neuroendocrine cell groups. (Palkovits 1984b, 1986a, b)

Neuropeptide	Nucleus				
	Medial preoptic	Supraoptic	Paraventricular	Periventricular	Arcuate
LH-RH	**C** F				
TRH	C F		**C** F	C F	C F
CRF	C F	C F	**C** F	C F	F
GRF	F	F	C F	F	**C** F
Somatostatin	C F	F	C F	**C** F	C F
Vasopressin	F	**C** F	**C** F	F	F
Oxytocin		**C** F	**C** F		F
Prolactin	C F	C F	C F	F	F
ACTH	F		F	F	**C** F
β-Endorphin	F	F	F	**F**	**C** F
α-MSH	C **F**	F	F	**F**	**C** F
Enkephalins	C F	C F	C F	F	C F
Dynorphins	F	C F	C F	F	C F
Substance P	C F	F	C F	F	C F
NPY	C F	F	C F	C F	**C** F
VIP	F	F	C F	F	F
PHI 27			C F		
CCK	C F	C F	C F	**C** F	C F
Bombesin	C F		**C** F		F
Motilin	F		F	F	C F
Galanin	C F	C F	C F	C F	C F
Glucagon		C F	C F	F	F
Neurotensin	**C** F		C F	C F	C F
Angiotensin II	F	**C** F	C F	C F	
Ranatensin	C F		C F	C F	
FMRFamide	F	F	C F	**C** **F**	C F
Bradykinin	F		C F	F	
Atriopeptin (ANF)	C F		C F	C F	C F
GABA	F	F	F	C F	C F
Dopamine	F	F	F	C F	**C** F
Noradrenaline	F	F	F	F	F
Adrenaline	F	F	F	F	F
Serotonin	F	F	F	F	F

Abbreviations: LH-RH, luteinizing hormone-releasing hormone; TRH, thyrotropin releasing hormone; CRF, corticotropin releasing factor; GRF, growth hormone releasing factor; ACTH, adrenocorticotropin; α-MSH, α-melanocyte stimulating hormone; NPY, neuropeptide Y; VIP, vasoactive intestinal polypeptide; PHI-27, peptide histidine isoleucine amide-27; CCK, cholecystokinin; GABA, γ-aminobutyric acid.

C, perikarya; F, nerve fibres and terminals; **C/F**, very dense; ACTH, adrenocorticotropic hormone; α-MSH, alpha melanocyte-stimulating hormone; NPY, neuropeptide Y; VIP, vasoactive intestinal polypeptide; PHI-27, ANF, antinuclear factor; CCK, cholecystokinin; FMRFamide

Table 2. Neuronal afferents to the supraoptic nucleus in the rat

Cells of origin	Chemical nature of fibres	Techniques applied	References
Telencephalon			
Ventral subiculum		Lesions + synaptic degenerations	Záborszky et al. (1975)
		Tract-tracing	Tribollet et al. (1985)
Septum		Lesions + synaptic degenerations	Záborszky et al. (1975)
		Autoradiography	Garris (1979)
		Autoradiography + EM immunocytochemistry	Silverman and Oldfied (1984)
Olfactory tubercle		Lesions + synaptic degenerations	Záborszky et al. (1975)
Amygdala		Lesions + synaptic degenerations	Záborszky et al. (1975)
		Retrograde tract-tracing	Iijima and Ogawa (1981)
		Autoradiography + EM immunocytochemistry	Silverman and Oldfield (1984)
OVLT		Tract-tracing	Tribollet et al. (1985)
Subfornical organ		Autoradiography	Miselis et al. (1979)
		Autoradiography	Miselis (1981)
		Retrograde tract-tracing	Lind et al. (1982)
		Lesions + synaptic degenerations	Renaud et al. (1983)
		Autoradiography + tract-tracing	Sawchenko and Swanson (1983)
		Tract-tracing	Tribollet et al. (1985)
Nucleus accumbens	GABA	Lesions + biochemistry	Meyer et al. (1980)
NIST		Autoradiography + tract-tracing	Sawchenko and Swanson (1983)
Hypothalamus			
Preoptic area		Autoradiography	Swanson (1976)
		Autoradiography	Conrad and Pfaff (1976b)
Median nucleus		Lesions + synaptic degenerations	Carithers et al. (1980)
		Autoradiography + tract-tracing	Sawchenko and Swanson (1983)
		Tract-tracing	Tribollet et al. (1985)
Paraventricular nucleus		Autoradiography	Conrad and Pfaff (1976b)
Medial basal hypothalamus		Lesions + synaptic degenerations	Záborszky et al. (1975)
Arcuate nucleus	ACTH	Tract-tracing + immunohistochemistry	Sawchenko et al. (1982)
Dorsomedial nucleus		Autoradiography + tract-tracing	Sawchenko and Swanson (1983)
Lateral hypothalamus	Acetylcholine	Tract-tracing + histochemistry	Mason et al. (1983)
A11–A14 cell groups	Dopamine	Histofluorescence	Lindvall et al. (1984)

Midbrain and pons			
Dorsal raphe nucleus	Serotonin	Lesions + biochemistry	Palkovits et al. (1977)
Raphe nuclei		Autoradiography	Moore et al. (1978)
		Autoradiography	Azmitia and Segal (1978)
	Serotonin	Tract-tracing + immunohistochemistry	Sawchenko et al. (1983)
Locus ceruleus	Noradrenaline	Lesions + biochemistry	Záborszky et al. (1977)
		Autoradiography	Jones and Moore (1977)
		Tract-tracing + immunohistochemistry	Sawchenko and Swanson (1982)
Medulla oblongata			
Lateral reticular nucleus (A1)		Autoradiography	McKellar and Loewy (1981)
	Noradrenaline	Tract-tracing + immunohistochemistry	Sawchenko and Swanson (1982)
		Tract-tracing	Tribollet et al. (1985)
Nucleus tractus solitarii (A2)	Noradrenaline	Tract-tracing + immunohistochemistry	Sawchenko and Swanson (1982)
		Tract-tracing	Tribollet et al. (1985)

OVLT, organum vasculosum of lamina terminalis; ACTH, adrenocorticotropic hormone; EM, electron microscope may arise from the hippocampus, septum, olfactory tubercle, and amygdala. Recent studies indicate that limbic afferents may not terminate inside the nucleus (Silverman and Oldfield 1984; Silverman et al. 1985; Tribollet et al. 1985), but around it, synapsing with vasopressinergic dendrites which extend beyond the nuclear border. The neurotransmitter content of limbic afferents has not been characterized yet.

Lower brain stem fibers constitute the other third of the supraoptic afferents (Záborszky et al. 1975). Most of these afferents are aminergic (Table 2). The noradrenergic innervation of the supraoptic nucleus is complex. The cells of noradrenergic fibers are localized mainly in the A1 cell group (lateral reticular nucleus in the medulla oblongata), but the locus ceruleus and the noradrenergic cells in the nucleus of the solitary tract (A2 cell group) also participate in the innervation of the supraoptic nucleus (Záborszky et al. 1977; Swanson et al. 1981; Sawchenko and Swanson 1982). Serotonin-containing afferents to the supraoptic nucleus arise from the midbrain and dorsal raphe nuclei (Palkovits et al. 1977; Sawchenko et al. 1983). Both the catecholaminergic (Lindvall and Björklund 1974; Sawchenko and Swanson 1982) and the serotonergic fibers (Palkovits et al. 1977) travel in the lateral hypothalamus among the fibers of the medial forebrain bundle. Some of the noradrenergic fibers decussate at the retrochiasmatic area and merge with fibers of the supraoptic decussations (Lindvall and Björklund 1974), reaching the supraoptic nucleus from the medial direction.

About 20% of supraoptic afferents are represented by those arising from the hypothalamic nuclei (Záborszky et al. 1975). Dopaminergic projections appear to originate from each of the dopamine-containing cell groups (A11–A14 cell groups) of the hypothalamus (Buijs et al. 1984; Lindvall et al. 1984). Cholinergic and GABAergic neurons projecting to the supraoptic nucleus were found in the lateral hypothalamic area, immediately dorsal and lateral to the supraoptic nucleus (Meyer et al. 1980; Mason et al. 1983). Adrenocorticotropin (ACTH)-containing nerve terminals in the supraoptic nucleus are of arcuate nucleus origin (Sawchenko et al. 1982).

Afferent projections to the supraoptic nucleus from the subfornical organ, the organum vasculosum laminae terminalis (OVLT), and the median preoptic nucleus have been established by various techniques (Table 2). They may serve as a morphological link in angiotensin–vasopressin interrelationships in the brain which control body fluid homeostasis.

A number of peptidergic fibers and nerve terminals in the supraoptic nucleus have been visualized immunohistochemically (Table 1). Cells of origin of these fibers are partly intrinsic; several others – GRF, somatostatin, β-endorphin (β-END), alpha melanocyte-stimulating hormone (α-MSH), substance P, neuropeptide Y (NPY), and vasoactive intestinal polypeptide (VIP) – appear to be extrinsic, but the exact location of projecting neurons has not been established yet.

4 Paraventricular Afferents

Quantitative analysis of the magnocellular paraventricular nucleus has revealed that approximately 2800 axon terminals are associated with a single magnocellular neurosecretory neuron in the rat (Kiss et al. 1983 a). No data are available regarding the synaptic bouton: cell ratio of parvocellular paraventricular neurons. A rich intrinsic organization of the paraventricular nucleus has recently been reported (van der Pol 1982), which is consistent with quantitative data obtained from rats with a "deafferentated" paraventricular nucleus. At least 43% of the total number of presynaptic boutons in the magnocellular subdivisions of the nucleus are intrinsic (Kiss et al. 1983 b). More than 20 neuropeptides have been demonstrated in paraventricular cells by immunohistochemistry (Table 1). They may project down to the median eminence, posterior pituitary, lower brain stem, and spinal cord, but most probably they also give rise to short axon collaterals or some of them serve as intrinsic neurons. Recently, the existence of a local CRF-containing neuronal circuit has been elegantly demonstrated by electron microscopic immunohistochemistry (Liposits et al. 1985).

Afferent connections to the paraventricular nucleus have been traced by several investigators. The most extensive direct inputs to the nucleus arise from the lower brain stem, the limbic system, and cell groups in the hypothalamus (Table 3).

The vast majority of brain stem afferents to the paraventricular nucleus are aminergic. The sites of origin of aminergic inputs are complex, and terminal patterns of these fibers are restricted to cytoarchitectonically defined subdivisions of the nucleus. The ascending fibers may reach the nucleus through either a lateral (ventral tegmental bundle/medial forebrain bundle system) or a medial (periventricular system) route. Three aminergic cell groups – the A1–D1 cell group of the ventral medulla, the A2–C2 cell group in the dorsomedial medulla (inside and around the nucleus of the solitary tract), and the locus ceruleus (A6 cell group) – provide almost all of the noradrenergic, and at least a part of the adrenergic inputs to the hypothalamus, including those to the paraventricular nucleus. The A1 cell group projects to all parvocellular subdivisions of the nucleus, and mainly to the vasopressin-containing portion of the magnocellular subdivisions. Axons from cells in the A2 cell group innervate the medial, periventricular, and dorsal parvocellular subdivisions, but not the magnocellular cells. The locus ceruleus projects to the most medial aspects of the paraventricular nucleus (Swanson et al. 1981; Sawchenko and Swanson 1982). Noradrenergic synapses on vasopressin (Silverman and Oldfield 1984; Silverman et al. 1985) and oxytocin (Yamano et al. 1985) cells, or dopamine-containing nerve terminals on magnocellular neurons (Buijs et al. 1984) have recently been demonstrated by electron microscopic immunocytochemistry. It appears more likely that serotonin inputs to the paraventricular nucleus arise primarily from midbrain and dorsal raphe nuclei (Table 3).

Further brain stem projections arise from the parabrachial (mainly from the dorsal) nuclei, projecting mainly to the medial parts of the parvocellular subdivisions of the paraventricular nucleus. The biochemical specificity of these neural afferents is unknown.

Retrograde axonal transport studies on limbic afferents to the paraventricular nucleus (Tribollet and Dreifuss 1981; Berk and Finkelstein 1981; Silverman et al. 1981; Sawchenko and Swanson 1983) suggest that the septum (primarily the lateral nucleus), the amygdala (mainly the medial nucleus), and the hippocampal

Table 3. Neuronal afferents to the paraventricular nucleus

Cells of origin	Chemical nature of fibres	Techniques applied	References
Telencephalon			
Ventral subiculum		Retrograde tract-tracing	Tribollet and Dreifuss (1981)
		Retrograde tract-tracing	Berk and Finkelstein (1981)
		Retrograde tract-tracing	Silverman et al. (1981)
		Autoradiography + tract-tracing	Sawchenko and Swanson (1983)
		Lesions + synaptic degenerations	Kiss et al. (1983b)
		Autoradiography + EM immunocytochemistry	Silverman and Oldfield (1984)
Lateral septal nucleus		Retrograde tract-tracing	Tribollet and Dreifuss (1981)
		Retrograde tract-tracing	Berk and Finkelstein (1981)
		Retrograde tract-tracing	Silverman et al. (1981)
		Autoradiography + tract-tracing	Sawchenko and Swanson (1983)
		Lesions + synaptic degenerations	Kiss et al. (1983b)
		Autoradiography + EM immunocytochemistry	Silverman and Oldfield (1984)
Medial amygdala		Retrograde tract-tracing	Tribollet and Dreifuss (1981)
		Retrograde tract-tracing	Berk and Finkelstein (1981)
		Retrograde tract-tracing	Silverman et al. (1981)
		Autoradiography + tract-tracing	Sawchenko and Swanson (1983)
		Autoradiography + EM immunocytochemistry	Silverman et al. (1985)
NIST		Autoradiography	Swanson (1976)
		Retrograde tract-tracing	Silverman et al. (1981)
		Autoradiography + tract-tracing	Sawchenko and Swanson (1983)
OVLT		Retrograde tract-tracing	Tribollet and Dreifuss (1981)
		Retrograde tract-tracing	Silverman et al. (1981)
		Autoradiography + tract-tracing	Sawchenko and Swanson (1983)
Subfornical organ		Retrograde tract-tracing	Tribollet and Dreifuss (1981)
		Autoradiography	Miselis (1981)
		Retrograde tract-tracing	Silverman et al. (1981)
		Retrograde tract-tracing	Lind et al. (1982)
		Autoradiography + tract-tracing	Sawchenko and Swanson (1983)
	Angiotensin II	Immunohistochemistry	Lind et al. (1984)
Diencephalon			
Medial preoptic nucleus		Retrograde tract-tracing	Tribollet and Dreifuss (1981)
		Retrograde tract-tracing	Silverman et al. (1981)
		Autoradiography + tract-tracing	Sawchenko and Swanson (1983)

Periventricular (median) preoptic nucleus		Autoradiography + tract-tracing	Sawchenko and Swanson (1983)
		Lesions + synaptic degenerations	Carithers and Johnson (1985)
Lateral preoptic area		Retrograde tract-tracing	Silverman et al. (1981)
Anterior hypothalamic nucleus		Retrograde tract-tracing	Tribollet and Dreifuss (1981)
		Autoradiography + tract-tracing	Sawchenko and Swanson (1983)
Periventricular nucleus		Retrograde tract-tracing	Tribollet and Dreifuss (1981)
Suprachiasmatic nucleus		Retrograde tract-tracing	Berk and Finkelstein (1981)
		Retrograde tract-tracing	Silvermann et al. (1981)
		Autoradiography + tract-tracing	Sawchenko and Swanson (1983)
	Vasopressin	Lesions + immunohistochemistry	Eaton-Hanna et al. (1984)
Arcuate nucleus		Retrograde tract-tracing	Tribollet and Dreifuss (1981)
		Retrograde tract-tracing	Silverman et al. (1981)
	ACTH	Tract-tracing + immunohistochemistry	Sawchenko et al. (1982)
		Autoradiography + tract-tracing	Sawchenko and Swanson (1983)
	ACTH/β-E/α-MSH	Immunohistochemistry	Kiss et al. (1984b)
	NPY	Lesions + immunohistochemistry	Bai et al. (1985)
Ventromedial nucleus		Retrograde tract-tracing	Tribollet and Dreifuss (1981)
		Retrograde tract-tracing	Silverman et al. (1981)
		Autoradiography + tract-tracing	Sawchenko and Swanson (1983)
Dorsomedial nucleus		Retrograde tract-tracing	Tribollet and Dreifuss (1981)
		Autoradiography + tract-tracing	Sawchenko and Swanson (1983)
Premamillary nuclei		Retrograde tract-tracing	Tribollet and Dreifuss (1981)
		Autoradiography + tract-tracing	Sawchenko and Swanson (1983)
Supramamillary nucleus		Autoradiography + tract-tracing	Sawchenko and Swanson (1983)
Posterior hypothalamic nucleus		Autoradiography + tract-tracing	Sawchenko and Swanson (1983)
Lateral hypothalamic area		Retrograde tract-tracing	Silverman et al. (1981)
		Autoradiography + tract-tracing	Sawchenko and Swanson (1983)
Zona incerta		Autoradiography + tract-tracing	Silverman et al. (1981)
		Autoradiography + tract-tracing	Sawchenko and Swanson (1983)
A11 and A13 cell groups		Retrograde tract-tracing	Tribollet and Dreifuss (1981)
	Dopamine	Histofluorescence	Lindvall et al. (1984)
Dorsal thalamic nucleus		Autoradiography + tract-tracing	Sawchenko and Swanson (1983)
Mesencephalon			
Central gray matter		Retrograde tract-tracing	Berk and Finkelstein (1981)
Raphe nuclei (B8, B9)		Retrograde tract-tracing	Tribollet and Dreifuss (1981)
		Retrograde tract-tracing	Berk and Finkelstein (1981)
	Serotonin	Tract-tracing + immunohistochemistry	Sawchenko et al. (1983)
Peripeduncular		Retrograde tract-tracing	Berk and Finkelstein (1981)

Table 3 (continued)

Cells of origin	Chemical nature of fibres	Techniques applied	References
Pons			
Dorsal raphe nucleus	Serotonin	Lesions + biochemistry	Palkovits et al. (1977)
		Retrograde tract-tracing	Tribollet and Dreifuss (1981)
		Retrograde tract-tracing	Berk and Finkelstein (1981)
	Serotonin	Tract-tracing + immunohistochemistry	Sawchenko et al. (1983)
Dorsolateral tegmental area		Retrograde tract-tracing	Berk and Finkelstein (1981)
Parabrachial nuclei		Autoradiography	Norgren (1976)
		Autoradiography	Saper and Loewy (1980)
		Autoradiography	McKellar and Loewy (1981)
		Retrograde tract-tracing	Tribollet and Dreifuss (1981)
		Retrograde tract-tracing	Berk and Finkelstein (1981)
Locus ceruleus	Noradrenaline	Lesions + synaptic degradation + biochemistry	Záborszky et al. (1977)
		Autoradiography	Jones and Moore (1977)
		Retrograde tract-tracing	Tribollet and Dreifuss (1981)
		Autoradiography	McKellar and Loewy (1981)
		Retrograde tract-tracing	Berk and Finkelstein (1981)
	Noradrenaline	Tract-tracing + immunohistochemistry	Sawchenko and Swanson (1982)
Medulla oblongata			
Lateral reticula nucleus (C1)	Adrenaline	Lesions + synaptic degradation + biochemistry	Palkovits et al. (1980b)
		Autoradiography	McKellar and Loewy (1981)
(A1)	Noradrenaline	Lesions + synaptic degradation + biochemistry	Palkovits et al. (1980a)
		Retrograde tract-tracing	Tribollet and Dreifuss (1981)
		Autoradiography	McKellar and Loewy (1981)
		Retrograde tract-tracing	Berk and Finkelstein (1981)
	Noradrenaline	Tract-tracing + immunohistochemistry	Sawchenko and Swanson (1982)
Nucleus tractus solitarii (A2)		Autoradiography + tract-tracing	Ricardo and Koh (1978)
		Retrograde tract-tracing	Tribollet and Dreifuss (1981)
		Autoradiography	McKellar and Loewy (1981)
		Retrograde tract-tracing	Berk and Finkelstein (1981)
	Noradrenaline	Tract-tracing + immunohistochemistry	Sawchenko and Swanson (1982)

EM, electron microscope

formation (specifically the ventral subiculum) all project to the nucleus. However, neither light (Sawchenko and Swanson 1983) nor electron microscopic (Silverman and Oldfield 1984) studies confirm the existence of direct projections from these limbic areas to the paraventricular nucleus. Axons from the septum, amygdala, or ventral subiculum do not terminate within the nucleus, but around it. Ultrastructural examinations of this perinuclear area showed that limbic afferents make synaptic contacts with vasopressinergic dendrites which extend beyond the nuclear boundaries (Silverman and Oldfield 1984). According to retrograde tract-tracing and autoradiographic studies, the NIST is the only limbic nucleus with direct paraventricular projections. It appears that certain limbic inputs to the paraventricular nucleus may be relayed by neurons of the NIST.

Both antero- and retrograde tract-tracing studies have reported that direct inputs to the paraventricular nucleus arise from almost all of the hypothalamic and preoptic nuclei (Table 3). Most of these nuclei project preferentially to the parvocellular subdivisions of the nucleus, wherein each input has a unique distribution pattern (Sawchenko and Swanson 1983). Only the projections from the median preoptic and dorsomedial hypothalamic nuclei appear to terminate in the magnocellular subdivisions of the paraventricular nucleus. Projections from the median preoptic nucleus distribute uniformly in areas rich in both oxytocin- and vasopressin-containing neurons, while dorsomedial afferents innervate preferentially oxytocinergic parts of the magnocellular paraventricular nucleus (Sawchenko and Swanson 1983).

Our present knowledge of the biochemical specificity of hypothalamic inputs to the paraventricular nucleus is rather limited. ACTH- or pro-opiomelanocortin-immunoreactive fibers from the arcuate nucleus are found in most parts of the parvocellular subdivisions and in magnocellular regions rich in oxytocin-containing cells (Sawchenko et al. 1982; Kiss et al. 1984b). Recently, neuropeptide Y (NPY) projections have been reported from the arcuate nucleus to the paraventricular nucleus (Bai et al. 1985).

5 Medial Preoptic Afferents

Rather sparse inputs to the median eminence arise from the preoptic region. Although several neuropeptide-containing neurons have been verified in this area (Table 1), only the LHRH neurons are accepted as neuroendocrine cells. They distribute mainly in the medial preoptic nucleus and its vicinity (periventricular preoptic nucleus, nucleus tractus diagonalis, medial septal nucleus, and NIST). Less than 10% of total medial preoptic neurons are stained with LHRH. This fact indicates that only a low percentage of the medial preoptic afferents constitute specific afferent inputs to neuroendocrine cells. Therefore, studies based on double or multiple immunocytochemistry or combined use of autoradiography and immunostaining at electron microscopic level have considerable importance. By these techniques, chemically characterized presynaptic terminals can be demonstrated in synaptic contact with LHRH neurons: serotonin- (Kiss and Halász 1985), tyrosine hydroxylase- (Léránth et al. 1986), GABA- (Léránth et al. 1985b), and ACTH-containing (Léránth et al. 1985a) nerve terminals synapse with LHRH neurons in the medial preoptic nucleus of the rat. By immunocytochemistry, the presence of an intrinsic LHRH neuronal circuit has been reported in this nucleus, on the basis of electron microscopically observed LHRH-containing presynaptic terminals in synaptic contact with LHRH-immunostained dendrites (Léránth et al. 1985d).

Afferents have been verified from all major brain areas to the medial preoptic nucleus using various types of tract-tracing technique (Table 4). A significant number of fibers arise from the limbic system and the hypothalamus (Chiba and Murata 1985), and aminergic fibers, like other neuroendocrine cells, are of lower brain stem origin.

The number of data on the biochemical specificity of neuronal input from well-localized cell groups are increasing: noradrenergic fibers arise from the A1 and A2 cell groups in the medulla; dopaminergic fibers from the periventricular dopamine-synthesizing neurons (A14 cell groups); serotonergic axons from both midbrain and dorsal raphe nuclei; GABA projections from cells in the lateral preoptic and hypothalamic areas; vasopressin-containing fibers from the suprachiasmatic nucleus; substance P-containing fibers from the anteroventral hypothalamic nucleus; and ACTH-containing axons from the arcuate nucleus (Table 4). As has been mentioned above, these afferents may not terminate exclusively on LHRH neurons in the medial preoptic nucleus.

6 Hypothalamic Periventricular Afferents

The periventricular nuclei consist of a few layers of cells and a delicate fiber network of extreme density on both sides of the third ventricle throughout the rostrocaudal extent of the preoptic and anterior hypothalamic areas. The hypothalamic periventricular nucleus is the caudal continuation of the preoptic periventricular nucleus. (The latter includes a relatively neuron-dense area, called the nucleus preopticus medianus.) Among several neuropeptide-containing cells localized in the periventricular nucleus (Table 1), the somatostatin-containing neurons represent the neuroendocrine cells in this nucleus. Rather sparse somatostatinergic cells in the preoptic periventricular nucleus project to the median eminence (Jew et al. 1984), which is consistent with data obtained from retrograde tract-tracing experiments (Lechan et al. 1982).

Investigating the presence of specific afferents to the periventricular nucleus is rather difficult. The nucleus, a narrow stripe with several traversing fibers, is not an easy target for the application of tract-tracing techniques. Although several neuronal inputs have been reported from various brain areas to the periventricular nucleus (Table 5), some of them are probably traversing rather than terminating fibers. Specifically, direct evidence (electron microscopically verified synaptic contacts) of the innervation of periventricular somatostatin-containing cells is still missing.

Table 4. Neuronal afferents to the medial preoptic nucleus in the rat

Cells of origin	Chemical nature of fibres	Techniques applied	References
Telencephalon			
Ventral subiculum		Autoradiography	Krieger et al. (1979)
Septum (medial)		Autoradiography	Swanson and Cowan (1979)
(lateral)		Autoradiography	Garris (1979)
		Retrograde tract-tracing	Berk and Finkelstein (1981)
Amygdala		Retrograde tract-tracing	Berk and Finkelstein (1981)
Nucleus accumbens		Autoradiography	Conrad and Pfaff (1976c)
NIST		Autoradiography	Conrad and Pfaff (1976a)
Subfornical organ		Retrograde tract-tracing	Lind et al. (1982)
Hypothalamus			
Periventricular nucleus (A14)	Dopamine	Histofluorescence	Björklund et al. (1975)
		Retrograde tract-tracing	Day et al. (1980)
Anterior hypothalamic nucleus		Autoradiography	Conrad and Pfaff (1976b)
Paraventricular nucleus		Autoradiography	Conrad and Pfaff (1976b)
Suprachiasmatic nucleus	Vasopressin	Lesions + immunohistochemistry	Eaton-Hanna et al. (1984)
Anteroventral hypothalamic nucleus	Substance P	Lesions + immunohistochemistry	Takatsuki et al. (1983)
Arcuate nucleus		Lesions + synaptic degeneration	Záborszky and Makara (1979)
	ACTH	Tract-tracing + EM immunohistochemistry	Léránth et al. (1985)
Ventromedial nucleus		Autoradiography	Krieger et al. (1979)
Lateral hypothalamus	GABA	EM immunocytochemistry	Léránth et al. (1985b)
Mesencephalon			
Raphe nuclei		Lesions + degeneration + autoradiography	Conrad et al. (1974)
	Serotonin	Autoradiography	Moore et al. (1978)
		Lesions + autoradiography	Azmitia and Segal (1978)
		Retrograde tract-tracing	Sakumoto et al. (1978)
	Serotonin	Tract-tracing + immunohistochemistry	Simersly et al. (1984)

Table 4 (continued)

Cells of origin	Chemical nature of fibres	Techniques applied	References
Pons			
Dorsal raphe nucleus		Lesions + degeneration + autoradiography	Conrad et al. (1974)
		Retrograde tract-tracing	Sakumoto et al. (1978)
		Retrograde tract-tracing	Berk and Finkelstein (1981)
	Serotonin	Tract-tracing + immunohistochemistry	Simersly et al. (1984)
Parabrachial nuclei		Retrograde tract-tracing	Sakumoto et al. (1978)
		Retrograde tract-tracing	Berk and Finkelstein (1981)
Medulla Oblongata			
Lateral reticular nucleus (A1)		Retrograde tract-tracing	Sakumoto et al. (1978)
	Catecholamines	Tract-tracing + histofluorescence	Day et al. (1980)
		Retrograde tract-tracing	Berk and Finkelstein (1981)
Nucleus tractus solitarii (A2)		Tract-tracing + autoradiography	Ricardo and Koh (1978)
		Retrograde tract-tracing	Sakumoto et al. (1978)
	Catecholamines	Tract-tracing + histofluorescence	Day et al. (1980)
		Retrograde tract-tracing	Berk and Finkelstein (1981)

EM, electron microscope

Table 5. Neuronal afferents to the periventricular hypothalamic nucleus in the rat

Cells of origin	Chemical nature of fibres	Techniques applied	References
Telencephalon			
Septum		Autoradiography	Conrad and Pfaff (1976c)
Subfornical organ		Autoradiography	Miselis (1981)
Hypothalamus			
Preoptic area		Autoradiography	Swanson (1976)
Suprachiasmatic nucleus	Vasopressin	Lesions + immunohistochemistry	Eaton-Hanna et al. (1984)
Anterior hypothalamic nucleus		Autoradiography	Conrad and Pfaff (1976b)
Paraventricular nucleus		Autoradiography	Conrad and Pfaff (1976b)
Arcuate nucleus		Lesions + synaptic degeneration	Záborszky and Makara (1979)
Ventromedial nucleus		Autoradiography	Saper et al. (1976)
Mesencephalon			
Central gray matter		Retrograde tract-tracing	Záborszky (1982)
Midbrain raphe nucleus		Retrograde tract-tracing	Záborszky (1982)
Pons			
Dorsal raphe nucleus		Retrograde tract-tracing	Záborszky (1982)
Locus ceruleus		Autoradiography	Jones and Moore (1977)
	Noradrenaline	Lesions + biochemistry	Palkovits et al. (1980a)
		Retrograde tract-tracing	Záborszky (1982)
Medulla oblongata			
Lateral reticular nucleus (A1–C1)	Noradrenaline	Lesions + biochemistry	Palkovits et al. (1980a)
	Adrenaline	Lesions + biochemistry + synaptic degeneration	Palkovits et al. (1980b)

7 Arcuate Afferents

The arcuate nucleus is the third largest hypothalamic nucleus containing a number of peptidergic, dopaminergic, and GABAergic neurons (Table 1). A fairly high percentage of arcuate neurons project to the median eminence and further down to the intermediate and posterior pituitary lobes. These cells constitute a separate group of neuroendocrine cells: the tuberoinfundibular system.

Almost all of the brain neuropeptides discovered so far are present in the neuronal meshwork of the arcuate nucleus (Table 1). Some of them are intrinsic, like substance P- (Tsuruo et al. 1984), ACTH- (Chen and Pelletier 1983; Kiss and Williams 1983) or prolactin- (Nishizuka et al. 1984)-containing neurons, while others are extrinsic. The cells of origin of the latter group appear to be widely distributed in several brain areas (Table 6), but their chemical nature remains to be characterized.

Rather sparse afferents to the arcuate nucleus arise from the limbic system, but probably none from other telencephalic regions. The intrahypothalamic input seems to be more numerous (Záborszky 1982), mainly from the neighboring nuclei, and especially from the ventromedial nucleus (Table 6).

A significant proportion of arcuate afferents derive from the lower brain stem, and, like brain stem afferents to other neuroendocrine cells, they are almost exclusively aminergic. Most of the noradrenergic nerve terminals present in the arcuate nucleus arise from the A1 cell group in the lateral reticular nucleus in the medulla and from the locus ceruleus. Only a minor percentage, if any, derive from the A2 cell group (Table 6). Dopamine-containing fibers appear to be intrinsic: cell bodies are present inside the arcuate nucleus (A12 cell group) and tyrosine hydroxylase-like immunoreactivity can be demonstrated on both pre- and postsynaptic elements by electron microscopy (Léránth et al. 1985c). Serotonin-containing axons and terminals in the arcuate nucleus are, at least for the major part, of extrahypothalamic origin: serotonergic neurons may be localized in midbrain and dorsal raphe nuclei (Table 6). Recently, serotonin nerve terminals synapsing with ACTH neurons in the arcuate nucleus have been demonstrated by immunoelectron microscopy in combination with autoradiography (Kiss et al. 1984b).

8 Concluding Remarks

The central nervous system mediates a wide range of endocrine, behavioral, and automatic responses that maintain body homeostasis. A great many studies indicate that hypothalamic neuroendocrine cells play a substantial role in these mechanisms. It seems clear that neuroendocrine cells are effector neurons in a rather broad sense, receiving neuronal inputs from various other sources and having only a single output channel. This output is a straightforward mechanism: neuroendocrine cells produce hormonal substances which exert their actions on the pituitary or the periphery.

Table 6. Neuronal afferents to the arcuate nucleus in the rat

Cells of origin	Chemical nature of fibres	Techniques applied	References
Telencephalon			
Septum		Autoradiography	Conrad and Pfaff (1976c)
		Autoradiography	Garris (1979)
		Lesions + degenerations	Palkovits and Záborszky (1980)
NIST		Autoradiography	Conrad and Pfaff (1976a)
		Autoradiography	Swanson (1976)
Olfactory tubercle		Lesions + degenerations	Palkovits and Záborszky (1980)
Hypothalamus			
Preoptic area		Autoradiography	Conrad and Pfaff (1976a)
		Autoradiography	Swanson (1976)
		Lesions + synaptic degenerations	Köves and Réthelyi (1976)
Anterior hypothalamic nucleus		Autoradiography	Conrad and Pfaff (1976b)
Paraventricular nucleus		Autoradiography	Conrad and Pfaff (1976b)
Ventromedial nucleus		Autoradiography	Krieger et al. (1979)
		Lesions + synaptic degenerations	Záborszky and Makara (1979)
		Retrograde tract-tracing	Wiegand and Price (1980)
Premamillary nuclei		Lesions + synaptic degenerations	Záborszky and Makara (1979)
Mesencephalon			
Ventral tegmental area		Retrograde tract-tracing	Záborszky (1982)
Midbrain raphe nucleus		Autoradiography	Moore et al. (1978)
		Autoradiography	Azmitia and Segal (1978)
Pons			
Dorsal raphe nucleus	Serotonin	Lesions + synaptic degeneration + biochemistry	Palkovits et al. (1977)
Parabrachial nuclei		Autoradiography	Saper and Loewy (1980)
Locus ceruleus	Noradrenaline	Lesions + synaptic degeneration + biochemistry	Palkovits et al. (1980a)
		Retrograde tract-tracing	Záborszky (1982)
Medulla oblongata			
Lateral reticular nucleus (A1)	Noradrenaline	Lesions + synaptic degeneration + biochemistry	Palkovits et al. (1980a)
(C1)	Adrenaline	Lesions + biochemistry	Palkovits et al. (1980b)
Nucleus tractus solitarii		Tract-tracing + autoradiography	Ricardo and Koh (1978)

There is a large variety of kinds and sources of fibers that may influence the activity of neuroendocrine cells. Evidence suggests that neuroendocrine cells receive feedback signals from the pituitary as well as from peripheral endocrine organs. It has long been very widely believed that such feedback is almost exclusively humoral. Neuronal feedback signals may reach the neuroendocrine cells either directly or be relayed by other neurons. Relay neurons in that sense occur in several brain areas, including the limbic and autonomic brain centers. Signals influencing the activity of an individual neuroendocrine neuron may arise from a number of hypothalamic, limbic, and lower brain stem nuclei. The large number of neuronal afferents terminating on a single cell gives at first an impression of chaos rather than the feeling of better understanding of a neuronal mechanism. However, taking into consideration the great variability of brain signals influencing the hypothalamohypophyseal magno- or parvocellular systems, the thousands of axon terminals on a single neuron (which phenomenon is, in fact, a common feature of central nervous system neurons) become a more feasible proposition. Although the neuroendocrine cells having efferent projections to the median eminence/posterior pituitary are effector neurons, they may be capable of informing other neurons in the central nervous system about their neurohormonal activity. All those neuropeptides which act as releasing or inhibiting hormones in the pituitary may exert neurotransmitter actions if they are transported by axons or axon collaterals of neuroendocrine cells into the nerve terminals synapsing with an other neuron.

The full extent of the neuroendocrine mechanisms can be understood only after detailed neuroanatomical studies are carried out to determine their afferent and efferent neuronal projections at light microscopic level and their neuronal connections are verified by electron microscopy. The neuromorphological technical arsenal has indeed been developed considerably during the past few years, and several powerful methods are available for extended and detailed studies on the morphological aspects of brain mechanisms. Even so, the most elegant double or multiple immunohistochemical procedures in combination with the most powerful tract-tracing techniques may provide us with a great deal of information about the existence and chemical nature of neuroendocrine input-output channels; morphological data without functional significance are only food for speculation. On the other hand, there is no doubt that such food is fundamental for any kind of further study performed for the better understanding of brain mechanisms.

References

Azmitia EC, Segal M (1978) An autoradiographic analysis of the differential ascending projections of the dorsal and median raphe nuclei in the rat. J Comp Neurol 179:641–668

Bai FL, Yamano M, Shiotani Y, Emson PC, Smith AD, Powell JF, Tohyama M (1985) An arcuato-paraventricular and -dorsomedial hypothalamic neuropeptide Y-containing system which lacks noradrenaline in the rat. Brain Res 331:172–175

Berk ML, Finkelstein JA (1981) Afferent projections to the preoptic area and hypothalamic regions in the rat brain. Neuroscience 6:1601–1624

Björklund A, Lindvall O, Nobin A (1975) Evidence of an incerto-hypothalamic dopamine neurone system in the rat. Brain Res 89:29–42

Buijs RM, Geffard M, Pool CW, Hoorneman EMD (1984) The dopaminergic innervation of the supraoptic and paraventricular nucleus. A light and electron microscopic study. Brain Res 323:65–72

Carithers J, Johnson AK (1985) Lesions of the tissue surrounding the preoptic recess (AV3V region) affect neurosecretory cells in the paraventricular nuclei in the rat. Brain Res 337:233–243

Carithers J, Bealer SL, Brody MJ, Johnson AK (1980) Fine structural evidence of degeneration in supraoptic nucleus and subfornical organ of rats with lesions in the anteroventral third ventricle. Brain Res 201:1–12

Chen YY, Pelletier G (1983) Demonstration of contacts between proopiomelanocortin neurons in the rat hypothalamus. Neurosci Lett 43:271–276

Chiba T, Murata Y (1985) Afferent and efferent connections of the medial preoptic area in the rat: a WGA-HRP study. Brain Res Bull 14:261–272

Conrad LCA, Pfaff DW (1976a) Efferents from medial basal forebrain and hypothalamus in the rat. I. An autoradiographic study of the medial preoptic area. J Comp Neurol 169:185–220

Conrad LCA, Pfaff DW (1976b) Efferents from medial basal forebrain and hypothalamus in the rat. II. An autoradiographic study of the anterior hypothalamus. J Comp Neurol 169:221–262

Conrad LCA, Pfaff DW (1976c) Autoradiographic tracing of nucleus accumbens efferents in the rat. Brain Res 113:589–596

Conrad LCA, Leonard CM, Pfaff PW (1974) Connections of the median and dorsal raphe nuclei in the rat: an autoradiographic and degeneration study. J Comp Neurol 156:179–205

Day TA, Blessing W, Willoughby JO (1980) Noradrenergic and dopaminergic projections to the medial preoptic area of the rat. A combined horseradish peroxidase/catecholamine fluorescence study. Brain Res 193:543–548

Eaton-Hanna A, Brownfield MS, Hostetter G, Kozlowski GP (1984) Vasopressinergic projections from the suprachiasmatic nucleus in rats. Anat Rec 208:50A

Garris DR (1979) Direct septo-hypothalamic projections in the rat. Neurosci Lett 13:83–90

Iijima K, Ogawa T (1981) An HRP study on the distribution of all nuclei innervating the supraoptic nucleus in the rat brain. Acta Histochem (Jena) 69:274–295

Jew JY, Léránth CS, Arimura A, Palkovits M (1984) Preoptic LH-RH and somatostatin in the rat median eminence. An experimental light and electron microscopic immunocytochemical study. Neuroendocrinology 38:169–175

Jones BJ, Moore RY (1977) Ascending projections of the locus coeruleus in the rat. II. Autoradiographic study. Brain Res 127:23–53

Kelly J, Swanson LW (1980) Additional forebrain regions projecting to the posterior pituitary: preoptic region, bed nucleus of the stria terminalis, and zona incerta. Brain Res 197:1–9

Kiss J, Halász B (1985) Demonstration of serotoninergic axons terminating on luteinizing hormone-releasing hormone neurons in the preoptic area of the rat using a combination of immunocytochemistry and high resolution autoradiography. Neuroscience 14:69–78

Kiss JZ, Williams TH (1983) ACTH-immunoreactive boutons from synaptic contacts in the hypothalamic arcuate nucleus of rat: evidence for local opiocortin connections. Brain Res 263:142–146

Kiss JZ, Palkovits M, Záborszky L, Tribollet E, Szabó D, Makara GB (1983a) Quantitative histological studies on the hypothalamic paraventricular nucleus in rats. I. Number of cells and synaptic boutons. Brain Res 262:217–224

Kiss JZ, Palkovits M, Záborszky L, Tribollet E, Szabó D, Makara GB (1983b) Quantitative histological studies on the hypothalamic paraventricular nucleus in rats. II. Number of local and certain afferent nerve terminals. Brain Res 265:11–20

Kiss J, Léránth CS, Halász B (1984a) Serotoninergic endings on VIP-neurons in the suprachiasmatic nucleus and on ACTH-neurons in the arcuate nucleus of the rat hypothalamus. A combination of high resolution autoradiography and electron microscopic immunocytochemistry. Neurosci Lett 44:119–124

Kiss JZ, Cassell MD, Palkovits M (1984b) Analysis of the ACTH/β-END/α-MSH-immunoreactive afferent input to the hypothalamic paraventricular nucleus of rat. Brain Res 324:91–99

Köves K, Réthelyi M (1976) Direct neural connection from the medial preoptic area to the hypothalamic arcuate nucleus of the rat. Exp Brain Res 25:529–539

Krieger MS, Conrad LCA, Pfaff DW (1979) An autoradiographic study of the efferent connections of the ventromedial nucleus of the hypothalamus. J Comp Neurol 183:785–816

Lechan RM, Nestler JL, Jacobson S (1982) The tuberoinfundibular system of the rat as demonstrated by immunohistochemical localization of retrogradely transported wheat germ agglutinin (WGA) from the median eminence. Brain Res 245:1–15

Léránth CS, Záborszky L, Marton J, Palkovits M (1975) Quantitative studies on the supraoptic nucleus in the rat. I. Synaptic organization. Exp Brain Res 22:509–523

Léránth CS, MacLusky MJ, Naftolin F (1985a) Proopiomelanocortin-derived neuropeptide immunoreactive cells of the ventromedial arcuate nucleus establish direct synaptic connections with LH-RH neurons of the medial preoptic area in the rat. Proc Annu Meet Neurosci Soc 15:146 (abstract 47.10)

Léránth CS, MacLusky NJ, Sakamoto H, Shanabrough M, Naftolin F (1985b) Glutamic acid decarboxylase-containing axons synapse on LHRH neurons in the rat medial preoptic area. Neuroendocrinology 40:536–539

Léránth CS, Sakamoto H, MacLusky NJ, Shanabrough M, Naftolin F (1985c) Intrinsic tyrosine hydroxylase (TH) immunoreactive neurons synapse with TH immunopositive neurons in the rat arcuate nucleus. Brain Res 331:371–375

Léránth CS, Segura LMG, Palkovits M, MacLusky NJ, Naftolin F (1985d) The LHRH-containing neuronal network in the preoptic area of the rat: demonstration of LHRH-containing nerve terminals in synaptic contact with LHRH neurons. Brain Res 345:332–336

Léránth CS, Naftolin F, MacLusky NJ (1986) TH immunoreactive axons establish direct synaptic connections on LHRH and GAD immunopositive neurons in the rat medial preoptic area.

Lind RW, van Hoesen GW, Johnson AK (1982) An HRP study of the connection of the subfornical organ of the rat. J Comp Neurol 210:265–277

Lind RW, Swanson LW, Ganten D (1984) Angiotensin II immunoreactive pathways in the central nervous system of the rat: evidence for a projection from the subfornical organ to the paraventricular nucleus of the hypothalamus. Clin Exp Hypertens A6:1915–1920

Lindvall O, Björklund A (1974) The organization of the ascending catecholamine neuron system in the rat brain. Acta Physiol Scand 412 [Suppl]:1–48

Lindvall O, Björklund A, Skagerberg G (1984) Selective histochemical demonstration of dopamine terminal systems in rat di- and telencephalon: new evidence for dopaminergic innervation of hypothalamic neurosecretory nuclei. Brain Res 306:19–30

Liposits ZS, Paull WK, Sétáló G, Vígh S (1985) Evidence for local corticotropin releasing factor (CRF)-immunoreactive neuronal circuits in the paraventricular nucleus of the rat hypothalamus. An electron microscopic immunohistochemical analysis. Histochemistry 83:5–16

Mason WT, Ho YW, Eckenstein F, Hatton GI (1983) Mapping of cholinergic neurones associated with rat supraoptic nucleus: combined immunocytochemical and histochemical identification. Brain Res Bull 11:617–626

McKellar S, Loewy AD (1981) Organization of some brain stem afferents to the paraventricular nucleus of the hypothalamus of the rat. Brain Res 217:351–357

Meyer DK, Oertel WH, Brownstein MJ (1980) Deafferentation studies on the glutamic acid decarboxylase content of the supraoptic nucleus of the rat. Brain Res 200:165–168

Miselis RR (1981) The efferent projections of the subfornical organ of the rat: a circumventricular organ within a neural network subserving water balance. Brain Res 230:1–23

Miselis RR, Shapiro RE, Hand PJ (1979) Subfornical organ efferents to neural system for control of body water. Science 207:1022–1025

Moore RY, Halaris AE, Jones BE (1978) Serotonin neurons of the midbrain raphe: ascending projections. J Comp Neurol 180:417–438

Nishizuka M, Shivers BD, Pfaff DW, Léránth Cs (1984) Electron-microscopic identification of prolactin-like immunoreactivity in medial basal hypothalamus. Proc Annu Meet Neurosci Soc 14:434 (abstract 128.18)

Norgren R (1976) Taste pathways to hypothalamus and amygdala. J Comp Neurol 166:17–30

Oldfield BJ, Silverman A-J (1985) A light microscopic HRP study of limbic projections to the vasopressin-containing nuclear groups of the hypothalamus. Brain Res Bull 14:143–157

Palkovits M (1976) Determination of axon terminal density in the central nervous system. Brain Res 108:413–417

Palkovits M (1983) Neuroanatomical techniques. In: Krieger DT, Brownstein MJ, Martin JB (eds) Brain peptides. Wiley, Chichester

Palkovits M (1984a) Topography of chemically identified neurons in the central nervous system: progress in 1981–1983. In: Müller EE, MacLeod RM (eds) Neuroendocrine perspectives, vol 3. Elsevier, Amsterdam

Palkovits M (1984b) Distribution of neuropeptides in the central nervous system: a review of biochemical mapping studies. Prog Neurobiol 23:151–189

Palkovits M (1984c) Neuropeptides in the hypothalamo-hypophyseal system: lateral retrochiasmatic area as a common gate for neuronal fibers towards the median eminence. Peptides 5 [Suppl 1]:35–39

Palkovits M (1986a) Neuropeptides in the median eminence. Neurochem Int

Palkovits M (1986b) Distribution of neuropeptides in the brain. In: Negro-Vilar A, Conn PM (eds) Peptide hormones: effects and mechanisms of action. CRC Press, Boca Raton

Palkovits M, Záborszky L (1979) The anatomy of the hypothalamus. In: Morgane PJ, Panksepp J (eds) Handbook of the hypothalamus, vol 1. Dekker, New York

Palkovits M, Záborszky L, Ambach G (1974) Accessory neurosecretory cell groups in the rat hypothalamus. Acta Morphol Hung 22:21–33

Palkovits M, Saavedra JM, Jacobowitz DM, Kizer JS, Záborszky L, Brownstein MJ (1977) Serotoninergic innervation of the forebrain: effect of lesions on serotonin and tryptophan hydroxylase levels. Brain Res 130:121–134

Palkovits M, Záborszky L, Feminger A, Mezey É, Fekete MIK, Herman JP, Kanyicska B, Szabó D (1980a) Noradrenergic innervation of the rat hypothalamus: experimental biochemical and electron microscopic studies. Brain Res 191:161–171

Palkovits M, Mezey É, Záborszky L, Feminger A, Versteeg DHG, Wijnen HJLM, de Jong W, Fekete MIK, Herman JP, Kanyicska B (1980b) Adrenergic innervation of the rat hypothalamus. Neurosci Lett 18:237–243

Palkovits M, Léránth Cs, Jew JY, Williams TH (1982) Simultaneous characterization of pre- and postsynaptic neuron contact sites in brain. Proc Natl Acad Sci USA 79:2705–2708

Renaud LP, Rogers J, Sgro S (1983) Terminal degeneration in supraoptic nucleus following subfornical organ lesions: ultrastructural observations in the rat. Brain Res 275:365–368

Rhodes CH, Morrell JI, Pfaff DW (1981) Immunohistochemical analysis of magnocellular elements in rat hypothalamus: distribution and number of cells containing neurophysin, oxytocin and vasopressin. J Comp Neurol 198:45–64

Ricardo JA, Koh ET (1978) Anatomical evidence of direct projections from the nucleus of the solitary tract to the hypothalamus, amygdala, and other forebrain structures in the rat. Brain Res 153:1–26

Sakumoto T, Tohyama M, Satoh K, Kimoto Y, Kinugasa T, Tanizawa O, Kurachi K, Shimizu N (1978) Afferent fiber connections from the lower brainstem to hypothalamus studied by the horseradish peroxidase method with special reference to noradrenaline innervation. Exp Brain Res 31:81–94

Saper CB, Loewy AD (1980) Efferent connections of the parabrachial nucleus in the rat. Brain Res 197:291–317

Saper CB, Swanson LW, Cowan WM (1976) The efferent connections of the ventromedial nucleus of the hypothalamus of the rat. J Comp Neurol 169:409–422

Sawchenko PE, Swanson LW (1982) The organization of noradrenergic pathways from the brainstem to the paraventricular and supraoptic nuclei in the rat. Brain Res Rev 4:275–325
Sawchenko PE, Swanson LW (1983) The organization of forebrain afferents to the paraventricular and supraoptic nuclei of the rat. J Comp Neurol 218:121–144
Sawchenko PE, Swanson LW, Joseph SA (1982) The distribution and cells of origin of ACTH(1–39)-stained varicosities in the paraventricular and supraoptic nuclei. Brain Res 232:365–374
Sawchenko PE, Swanson LW, Steinbusch HWM, Verhofstad AAJ (1983) The distribution and cells of origin of serotonergic inputs to the paraventricular and supraoptic nuclei of the rat. Brain Res 277:355–360
Silverman A-J, Oldfield BJ (1984) Synaptic input to vasopressin neurons of the paraventricular nucleus (PVN). Peptides 5 [Suppl 1]:139–150
Silverman A-J, Hoffman DL, Zimmerman EA (1981) The descending afferent connections of the paraventricular nucleus of the hypothalamus (PVN). Brain Res Bull 6:47–61
Silverman A-J, Oldfield B, Hou-Yu A, Zimmerman EA (1985) The noradrenergic innervation of vasopressin neurons in the paraventricular nucleus of the hypothalamus: an ultrastructural study using radioautography and immunocytochemistry. Brain Res 325:215–229
Simerly RB, Swanson LW, Gorski RA (1984) The cells of origin of a sexually dimorphic serotonergic input to the medial preoptic nucleus of the rat. Brain Res 324:185–189
Swanson LW (1976) An autoradiographic study of the efferent connections of the preoptic region in the rat. J Comp Neurol 167:227–256
Swanson LW, Cowan WM (1979) The connections of the septal region in the rat. J Comp Neurol 186:621–656
Swanson LW, Sawchenko PE, Berod BK, Hartman BK, Helle KB, van Orden DE 81981) An immunohistochemical study of the organization of catecholaminergic cells and terminal fields in the paraventricular and supraoptic nuclei of the hypothalamus. J Comp Neurol 196:271–285
Takatsuki K, Sakanaka M, Takagi H, Tohyama M, Shiotani Y (1983) Experimental immunohistochemical studies on the distribution and origins of substance P in the medial preoptic area of the rat. Exp Brain Res 53:183–192
Tribollet E, Dreifuss JJ (1981) Localization of neurones projecting to the hypothalamic paraventricular nucleus area of the rat: a horseradish peroxydase study. Neuroscience 6:1315–1328
Tribollet E, Armstrong WE, Dubois-Dauphin M, Dreifuss JJ (1985) Extra-hypothalamic afferent inputs to the supraoptic nucleus area of the rat as determined by retrograde and anterograde tracing techniques. Neuroscience 15:135–148
Tsuruo Y, Hisano S, Daikoku S (1984) Morphological evidence for synaptic junctions between substance P-containing neurons in the arcuate nucleus of the rat. Neurosci Lett 46:65–69
van den Pol AN (1982) The magnocellular and paravocellular paraventricular nucleus of rat: intrinsic organization. J Comp Neurol 206:317–345
Wiegand SJ, Price JL (1980) Cells of origin of the afferent fibers to the median eminence in the rat. J Comp Neurol 192:1–19
Yamano M, Bai F-L, Tohyama M, Shiotani Y (1985) Ultrastructural evidence of direct synaptic contact of catecholamine terminals with oxytocin-containing neurons in the parvocellular portion of the rat hypothalamic paraventricular nucleus. Brain Res 336:176–179
Záborszky L (1982) Afferent connections of the medial basal hypothalamus. Adv Anat Embryol Cell Biol 69:1–107
Záborszky L, Makara GB (1979) Intrahypothalamic connections: An electron microscopic study in the rat. Exp Brain Res 34:201–215
Záborszky L, Léránth Cs, Makara GB, Palkovits M (1975) Quantitative studies on the supraoptic nucleus in the rat. II. Afferent fiber connections. Exp Brain Res 22:525–540
Záborszky L, Brownstein M, Palkovits M (1977) Ascending brain stem pathways from the dorsolateral pontine tegmentum: a biochemical and electron microscopic study. Acta Morphol Hung 25:175–189

Interconnectedness of Steroid Hormone-Binding Neurons: Existence and Implications

S. L. Cottingham and D. Pfaff

Contents

1 Introduction

Neurons in the central nervous system concentrate estrogen, progesterone, testosterone or dihydrotestosterone, and corticosterone. The location of these neurons in the brain has been determined by using autoradiographic (Pfaff 1968; Pfaff and Keiner 1973; Stumpf et al. 1975; Morrell and Pfaff 1981) and steroid receptor-

Laboratory of Neurobiology and Behavior, The Rockefeller University, 1230 York Avenue, New York, NY 10021-6399, USA binding techniques (McEwen et al. 1979). Steroid sex hormone-binding cells are found in the medial preoptic area, medial hypothalamus, and limbic forebrain areas in the vertebrate brain; in general, these neurons are found in brain regions that participate in behavioral and pituitary reproductive functions (McEwen et al. 1979). In contrast corticosterone-concentrating neurons are found primarily in the hippocampus.

Several authors who have studied steroid-concentrating cells have noted the network formed by their many interconnections (Pfaff and Keiner 1973; Pfaff and Conrad 1978; Rees et al. 1980). This characteristic may be important for integrating complex physiological and environmental signals to produce reproductive behaviors and regulate the pituitary. This chapter will review the locations of steroid-concentrating neurons, describe connections between brain regions that concentrate estradiol; and suggest possible functions for the network formed by estrogen-concentrating neurons.

2 Steroid-Concentrating Cells – Anatomy

The distribution of estrogen (E)-, progesterone (P)-, testosterone (T)-, dihydrotestosterone (DHT)-, and corticosterone-concentrating neurons in the brains of diverse vertebrate species has been studied (Pfaff and Keiner 1973; Wade et al. 1973; Sar and Stumpf 1973a, b; Stumpf et al. 1975; Warembourg 1975a–c; Stumpf and Sar 1976; Lieberburg et al. 1977; Lieberburg and McEwen 1977; Pfaff et al. 1976, 1978; MacLuskey and McEwen 1980; Parsons et al. 1982; Rainbow et al. 1982a, b; McEwen et al. 1982) using steroid autoradiography and biochemical techniques. E, T, and DHT are related biochemically; testosterone is converted to DHT by 5a-reductase and to estradiol by an aromatase. Both of these enzymes are found in the brain (McEwen et al. 1979). Thus, studies of testosterone may identify cells that concentrate T, E, or DHT. However, if labelling of cells with T is seen in regions where E-concentrating cells are not found, the cells must be T- or DHT-concentrating neurons.

Across vertebrate species, a limbo-hypothalamic distribution of sex steroid-concentrating neurons holds for mammals (Pfaff 1968; Pfaff and Keiner 1973; Krieger et al. 1976; Morrell et al. 1977; Pfaff et al. 1976), birds (Zigmond et al. 1973; Arnold et al. 1976; Barfield et al. 1978), amphibians (Morrell et al. 1975b; Kelley et al. 1975), fish (Morrell et al. 1975a; Davis et al. 1977), and reptiles (Morrell et al. 1979; Halpern et al. 1982). Variation in hormone binding correlates with the sensitivity of a species to hormone, and location of cells correlates with brain regions that function in reproductive behavior (reviewed by Morrell et al. 1975a; Kelley and Pfaff 1978). For example, estradiol binding in the hamster is less than in the rat, and hamsters are less sensitive behaviorally to estrogen administration (Feder et al. 1974; Gorzalka and Whalen 1977). In male songbirds brain nuclei important for testosterone-dependent song production contain T-concentrating neurons, in addition to the septal-preoptic-hypothalamic axis (Zigmond et al. 1973; Arnold et al. 1976; Kelley and Pfaff 1978).

Corticosterone receptors and corticosterone-binding neurons have also been characterized in rat and monkey brains (Warembourg 1975a, b; Pfaff et al. 1976). Their distribution, in contrast to that of the sex steroid-concentrating neurons, is largely hippocampal.

2.1 Estrogen-Concentrating Neurons

The pattern of estrogen binding in the brain does not obey nuclear boundaries precisely, but does allow description by nuclear groups. The discussion in this section will review brain regions of E-concentrating neurons from highest to lowest numbers of cells, focusing on the rat (Pfaff and Keiner 1973), cat (Rees et al. 1980), and monkey (Pfaff et al. 1976), since Sect. 3 will refer to these species.

2.1.1 Hypothalamus

The hypothalamus contains the highest concentration of E-binding neurons in the brain for all species studied. The distribution of these neurons will be considered in sequence from anterior to posterior hypothalamic cell groups.

2.1.1.1 Medial Preoptic Area

The medial preoptic area (mPOA) contains a high percentage of E-binding neurons throughout vertebrate phylogeny. In the rat (Pfaff and Keiner 1973), cells in the mPOA are heavily labelled with radioactive estradiol. The highest percentage of E-concentrating neurons are found near the midline, below the anterior commissure, and in the suprachiasmatic POA. The lateral POA contains scattered labelled cells, including some cells in the medial forebrain bundle (mfb), this level of labelling being constant at most levels of the mfb. A similar pattern is seen in the rhesus monkey (Pfaff et al. 1976), cat (Rees et al. 1980), mink (Morrell et al. 1977), mouse (Stumpf and Sar 1975b), and hamster (Krieger et al. 1976), although labelling in the hamster is lighter throughout (less heavily labelled cells and fewer cells) than in the rat.

2.1.1.2 Anterior Hypothalamic Area

A high percentage of cells in the anterior hypothalamic area (AHA) concentrate radioactive estradiol. This cell group, at a level behind the anterior commissure and anterior to the paraventricular nucleus (PVN), runs from a ventromedial point near the third ventricle dorsolaterally toward the stria medullaris, nucleus of the stria terminalis, and fornix. Between the PVN and the suprachiasmatic nucleus (SCN), E-concentrating cells are found near the third ventricle.

The same distribution is seen in the cat and monkey, the most heavily labelled cells being found ventrally and medially in the AHA.

2.1.1.3 Suprachiasmatic Nucleus

In the rat, cells at the dorsomedial and dorsolateral borders of the SCN are labelled, but very few labelled cells are found in the SCN. The SCN is unlabelled in the cat and monkey.

2.1.1.4 Paraventricular Nucleus

Few cells in the PVN are labelled in the rat. Those cells that do concentrate estrogen are found in the magnocellular part of the PVN. Cells lateral to the PVN are also labelled. In the cat light labelling is seen over the PVN, and in the monkey a few cells in the magno- and parvocellular regions are labelled. Also in the monkey, a few large cells between the PVN and supraoptic nucleus (SON) show intense labelling.

2.1.1.5 Supraoptic Nucleus

No label was seen in the SON for any mammalian species.

2.1.1.6 Arcuate Nucleus

The arcuate (infundibular) nucleus extends from the anterior hypothalamus to the mamillary nuclei. High numbers of intensely labelled estrogen cells are found throughout its extent in the cat, rat, and monkey. In the monkey, cells are spread in dorsal and lateral directions from the arcuate nucleus.

2.1.1.7 Ventromedial Nucleus

The ventromedial nucleus (VMN) of the hypothalamus is an example of a region where label is restricted to a specific area of the nucleus and not confined to the border of the nucleus. In mammals E-concentrating neurons are found throughout the VMN, but are densest in the ventrolateral, posterior part. In the rat and cat, labelled cells are also found outside the VMN, scattered toward the arcuate nucleus, the dorsomedial nucleus (DMN) and fornix, and the lateral hypothalamic area.

2.1.1.8 Dorsomedial Nucleus

Weak labelling is seen in the DMN in all species.

2.1.1.9 Premamillary Region

In the rat a high percentage of E-concentrating neurons are found in and around the ventral premamillary nucleus (VPM). As with the VMN, the cells are not restricted to the VPM proper, but scatter medially to the arcuate nucleus, laterally into the lateral hypothalamic area (LHA), and dorsally toward the fornix. A group of cells is regularly seen lateral to the VPM at the ventral surface of the brain. A few labelled cells are seen at the lateral tip of the dorsal premamillary nucleus (DPM). Denser labelling of cells has been described lateral to (toward the fornix) and medial to (toward the ventricle) the DPM.

2.1.1.10 Lateral Hypothalamus

It has already been noted that in the rat labelled cells are mingled with mfb fibers in the lateral hypothalamus. Two fields of cells have also been noted: First, well-labelled cells just medial to the tip of the cerebral peduncle; secondly, the field described above, which lies lateral to the VPM at the bottom of the brain, extending from the level of the VMN to the VPM.

Very few labelled cells are found in the lateral hypothalamus of the cat, and in the monkey it is virtually unlabelled.

2.1.1.11 Posterior Hypothalamus

In the rat no labelled cells are seen in the mamillary bodies, the supramamillary nucleus, or the fields of Forel. Continuing caudally, the interpeduncular nucleus and the ventral tegmental area also remain unlabelled. In the cat a few unlabelled cells are distributed in the premamillary, anterior mamillary, tuberomamillary, and supramamillary nuclei. The mamillary bodies are unlabelled. In the monkey the posterior hypothalamus is largely unlabelled. A few lightly labelled cells are found near the medial and posterior borders of the medial mamillary nucleus, but the mamillary bodies are unlabelled.

2.1.2 Limbic System

2.1.2.1 Septum and Adjacent Structures

In the rat the lateral septum, the nucleus triangularis septi, and the subfornical organ contain many well-labelled cells. In the lateral septum fewer cells are seen at posterior than anterior levels, and overall, labelling never reaches the density seen in the mPOA or medial hypothalamus. The ventral part of the lateral septum contains more E-concentrating cells than the dorsal part, with especially well-labelled cells just medial to the dorsal tip of the nucleus accumbens. The nucleus accumbens contains a small number of well-labelled cells located dorsal and medial to the anterior limb of the anterior commissure.

Moderate to heavy labelling is seen in the lateral septum and posterior septum of the cat. The medial septum is rarely labelled. Light labelling of a few cells is seen in the subfornical organ.

The same general pattern holds for the rhesus monkey. Labelling is seen in the ventrolateral septum and subfornical organ, and the nucleus triangularis septi is lightly labelled. Sparse labelling of cells has been noted in the medial septum.

E-concentrating cells are labelled in the diagonal bands of Broca in the rat and cat, but not in the monkey. Cells in the bed nucleus of the stria terminalis are also labelled throughout in the rat, cat, and monkey.

The olfactory tubercle is labelled in the cat and monkey. There are a few lightly labelled cells in the rat. The islands of Calleja are labelled in the rat and monkey; in the cat the islands of Celleja are unlabelled, but cells immediately adjacent to this structure are labelled.

The nucleus accumbens has scattered cells that are labelled in the rat and cat, but is unlabelled in the monkey.

2.1.2.2 Amygdala

In the rat fields of labelled cells are seen ventrolateral to the cerebral peduncle and dorsolateral to the SON, in the anterior amygdaloid area. The nucleus of the lateral olfactory tract contains few labelled neurons. The medial amygdaloid nucleus is densely labelled. At anterior levels some labelled cells are found along the dorsomedial edge; at more posterior levels cells are found reliably throughout the nucleus. Labelling is densest opposite the hypothalamus from the level of the VMN to the VPM, especially along the dorsomedial edge. The dorsal part of the nucleus is usually more heavily labelled than the ventral part. The central amygdaloid nucleus has many E-concentrating neurons along its medial edge at anterior levels.

At more posterior levels well-labelled cells are found in the ventromedial part of this nucleus, just lateral to the stria terminalis at the point where it separates from the amygdala, and between the cortical and medial nuclei, just ventral to the stria terminalis. The cortical amygdaloid nucleus is labelled throughout, with many more labelled cells at posterior than anterior levels. The rostro-caudal level of the heaviest labelling is opposite the VMN and arcuate nuclei. A few labelled cells are seen in the basolateral complex, above the cortical nucleus.

In the cat the anterior amygdaloid area contains a few labelled cells. Numerous cells are seen in the amygdalohippocampal area; and the anterior and posterior cortical nuclei, the medial nucleus, and the basomedial nucleus are moderately labelled. There are a few labelled cells in the central division.

In the monkey the medial nucleus of the amygdala is labelled. A small number of labelled cells has been reported in, or lateral to, the posterior aspect of the cortical nucleus, near the stria terminalis. The nucleus basalis accessory medialis is moderately labelled. There is no label in other amygdaloid nuclei.

2.1.2.3 Hippocampus

Pyramidal cells in Ammon's horn and granule cells in the dentate gyrus are labelled in the rat. The ventral hippocampus has more label than the dorsal hippocampus, especially at the level of the posterior amygdala. In the cat granule cells in the ventral dentate gyrus are weakly labelled. There is no label in the entorhinal cortex, the subiculum, or the piriform cortex. In the monkey the hippocampus, the entorhinal cortex, and the piriform cortex are unlabelled.

2.1.3 Thalamus

A very small number of cells is labelled in the midline thalamic nuclei of the rat, and in the parafascicular nucleus of the cat. There is no label in the thalamus in the monkey.

2.1.4 Midbrain

The majority of E-concentrating neurons in the rat mesencephalon are found in the midbrain central gray (MCG). The greatest number of cells are located lateral and ventrolateral to the cerebral aqueduct, in the central gray and just lateral to it. There are a few labelled neurons near the third ventricle, where it forms the cerebral aqueduct, and near the fourth ventricle, just behind the aqueduct. There are also a few labelled cells in and near the posterior part of the nucleus of the tractus solitarius. The reticular formation is unlabelled throughout its extent, from the mesencephalon through the medulla. In the cat there are numerous E-concentrating cells in the MCG, and some cells are seen in the marginal nucleus of the superior cerebellar peduncle, the nucleus incertus, the medial and lateral nucleus of the solitary tract, the spinal nucleus of the trigeminal nerve, and interfascicular nuclei. In the monkey the periventricular stratum (which projects to the MCG) contains a few lightly labelled cells. There is a small number of labelled cells in the dorsolateral and ventrolateral MCG.

2.1.5 Spinal Cord

E-concentrating neurons have been described in the dorsal horn and intermediate gray of the spinal cord, but not in motoneurons, of the rat (Morrell et al. 1982). Cat and monkey spinal cords have not been examined for E-concentrating neurons.

2.2 Progesterone-Concentrating Neurons

Progesterone receptors and progesterone-concentrating neurons have been identified in rat and guinea pig (Sar and Stumpf 1973b; Wade et al. 1973; Kato and Onouchi 1977; Warembourg 1977, 1979; MacLuskey and McEwen 1978, 1980; Kato 1985, reviewed by Etgen 1984). After estrogen treatment progesterone receptors are found in the medial preoptic area, the ventromedial nucleus of the hypothalamus, and the arcuate nucleus-median eminence area.

2.3 Testosterone-Concentrating Neurons

In general, there were fewer T-concentrating neurons than E-concentrating neurons in all species studied, but T-concentrating regions of the brain overlap almost completely with E-concentrating regions (Pfaff 1968; Sar and Stumpf 1973a, 1977; Kelley et al. 1975; Arnold et al. 1976; Stumpf and Sar 1976; Lieberburg and McEwen 1977; Lieberburg et al. 1977; Selmanoff et al. 1977; Barfield et al. 1978; Sheridan 1979; Morrell et al. 1979; Halpern et al. 1982). In the rat ^{3}H-testosterone-concentrating neurons are found in the lateral septum and nucleus of the stria terminalis, medial preoptic area, medial nucleus of the amygdala, hippocampus, and periventricular, ventromedial, arcuate, and ventral premamillary nuclei of the hypothalamus. Unlike estrogen, testosterone labels motoneurons in the spinal cord.

2.4 Corticosterone-Concentrating Neurons

Corticosterone binding has been studied in the rat and monkey, using autoradiographic and biochemical techniques (McEwen et al. 1969, 1970; Gerlach and McEwen 1972; Warembourg 1975a, b; Gerlach et al. 1976; Pfaff et al. 1976). In the hippocampus pyramidal cells of Ammon's horn and granule cells of the dentate gyrus concentrate ^{3}H-corticosterone. Cells in the subiculum also concentrate corticosterone. Outside the hippocampus corticosterone-binding neurons are found in the lateral septum and the cortical nucleus of the amygdala. There is relatively little accumulation of corticosterone in the hypothalamus, where estradiol-concentrating neurons are found in great numbers.

3 Connections of Steroid–Binding Cells

This section describes the connections of estrogen-concentrating neurons, based on results using the techniques of amino acid autoradiography and retrograde transport of horseradish peroxidase (HRP) or fluorescent dyes (see Fig. 1). The results of these experiments have confirmed and extended earlier fiber degeneration studies. As in the previous section the summary is organized by brain region, roughly from areas of highest to lowest concentration of estrogen labelling. Each brain region is discussed in terms of its efferent projections.

3.1 Hypothalamus

3.1.1 Medial Preoptic Area

The mPOA has ascending dorsal and lateral, and descending projections (Nauta 1958; Valverde 1965; Wolf and Sutin 1966; Enoch and Kerr 1967; McClure and

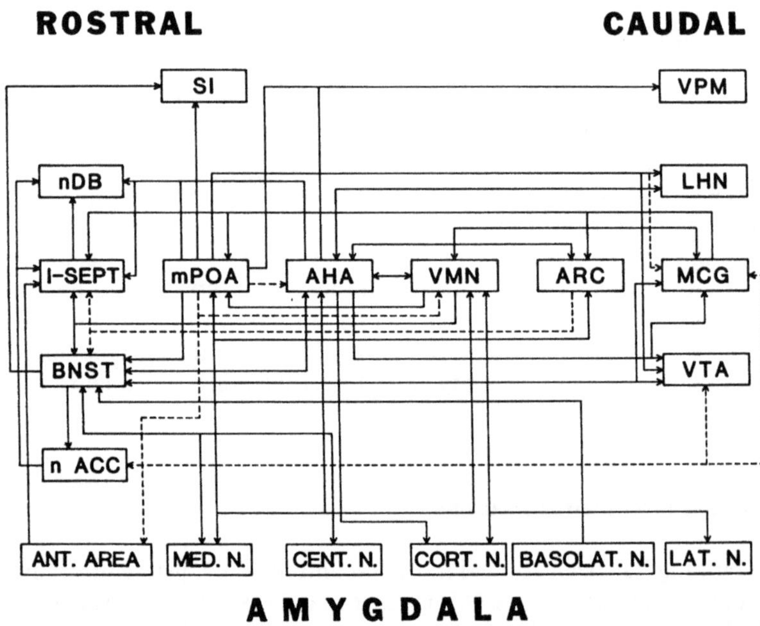

Fig. 1. Schematic showing the high degree of interconnectedness among estrogen (E)-concentrating regions in the rat. Virtually all E-concentrating areas are included. (See text for references) *Solid lines* represent prominent projections, and *dotted lines* represent weak projections; *intersections* of lines are not branch points. *Arrows* indicate the direction of the projection, and lines with arrows at both ends represent reciprocal connections. *n. ACC*, nucleus accumbens; *AHA*, anterior hypothalamic area; *ANT. AREA*, anterior area; *MED. N.*, medial nucleus; *CENT. N*, central nucleus; *CORT. N.*, cortical nucleus; *BASOLAT. N.*, basolateral nucleus; *LAT. N.*, lateral nucleus; *ARC*, arcuate nucleus; *BNST*, bed nucleus of the stria terminalis; *nDB*, nucleus of the diagonal bands of Broca; *LHN*, lateral habenular nucleus; *MCG*, midbrain central gray; *mPOA*, medial preoptic area; *l-SEPT*, lateral septum; *SI*, substantia innominata; VMN, ventromedial nucleus of the hypothalamus; *VPM*, ventral premamillary nucleus of the hypothalamus; *VTA*, ventral tegmental area

Clark 1968; Millhouse 1969; Mizuno et al. 1969; Chi and Flynn 1971; Conrad and Pfaff 1976b; Swanson 1976; Anderson and Shen 1980; Chiba and Murata 1985; reviewed by Palkovits and Zaborszky 1979). Fibers from the mPOA project substantially to the vertical and horizontal nuclei of the Diagonal Bands of Broca (nDB), and to the lateral septum, except the midlateral septum. There is a lighter projection to the nucleus triangularis septi and the bed nucleus of the stria terminalis (BNST). Dorsally the mPOA projects to the periventricular and paratenial thalamic nuclei. The mPOA projects laterally to the medial amygdaloid nucleus and more lightly to the anterior amygdaloid area, but not to the cortical nor basolateral nuclei. There is also a lateral projection to the substantia innominata.

Within the hypothalamus the mPOA projects lightly to the AHA, the ventrolateral subdivision of the VMN (where E-concentrating neurons are located), and the dorsomedial and dorsal hypothalamic nuclei. Farther caudal the mPOA projection to the arcuate nucleus and median eminence continues laterally into the ventral premamillary nucleus. There is also a projection to the supramamillary nuclei. The dorsal premamillary nucleus receives a light projection, compared to the VPM. The (caudal) projection to the lateral habenular nucleus (LHN) is of note with regard to the location of E-concentrating neurons in the LHN. The mPOA projects to all parts of the anterior LHN, but at more posterior levels, the projection is almost entirely to the medial part of the LHN, where E-concentrating neurons are located.

The mPOA has a diffuse projection to the ventral tegmental area of Tsai, and a light projection to the mesencephalic central gray, particularly the midlateral, medial, and ventral parts. The mPOA also projects to the nucleus cuneiformis and parabrachial nucleus in the mesencephalic reticular formation, and the median raphe nucleus, especially its dorsal half.

3.1.2 Anterior Hypothalamic Area

The AHA has dorsal, caudal, and intrahypothalamic projections (Powell et al. 1963; Nauta and Haymaker 1969; Millhouse 1973b; Swanson and Cowan 1975b; Conrad and Pfaff 1976c; Saper et al. 1978, 1979; Shen and Anderson 1980; reviewed by Palkovits and Zaborszky 1979, Zaborszky 1982). The anterior hypothalamic area sends dorsal projections to the BNST, nDB, and lateral septum, particularly the midlateral part. More caudally AHA axons project to the medial half of the lateral habenula, and the cortical and medial amygdaloid nuclei.

Intrahypothalamic connections are to the median eminence, the arcuate nucleus, and the contralateral hypothalamus. The AHA projects heavily to the dorsomedial nucleus, a region of light E labelling, and to all subdivisions of the ventromedial nucleus, including E- and non-E-concentrating regions. The AHA also projects to the dorsal and ventral premamillary nuclei, with fewer axons going to the supramamillary nucleus and diffuse projections to the mamillary bodies.

In the midbrain the AHA projects to the central gray, to all but its dorsolateral and most ventral parts. AHA also projects to the ventral tegmental area of Tsai.

The most caudal AHA projection is to the lateral pontine central gray, around the level of the locus ceruleus.

3.1.3 Arcuate Nucleus

The arcuate nucleus projects most strongly to the median eminence (Dyer 1975; Makara and Hodacs 1975; Holtzwarth-McBride et al. 1976; Yagi and Sawaki 1977; Renaud 1977; reviewed by Palkovits and Zaborszky 1979; Zaborszky 1982). It also has weak connections to the AHA and mPOA, continuing into the BNST and lateral septum. There are very light lateral projections to the zona incerta, fields of Forel, SOC, and medial amygdaloid nucleus. Finally, there is a weak descending projection to an area dorsal to the mamillary bodies, and lateral central gray (Krieger et al. 1978).

3.1.4 Ventromedial Nucleus

Projections of the VMN are ascending, lateral, and descending (Wolf and Sutin 1966; Arees and Mayer 1967; Kaelber and Leeson 1967; Szentagothai et al. 1968; Sutin and Eager 1969; Chi 1970a; Millhouse 1973a, b; Wakefield and Hall 1974; Renaud and Martin 1975; Saper et al. 1976, 1979; Meibach 1977; Krieger et al. 1979; reviewed by Palkovits and Zaborsky 1979; Zaborszky 1982). Ascending projections of the VMN are from all parts of this nucleus, and go to the anterior hypothalamus, particularly the periventricular region, the preoptic area, the BNST, and the lateral septum. There is also a light projection to the periventricular thalamus.

One of the major lateral projections from the VMN is to the amygdala. This projection is mainly through the supraoptic commissure, although some fibers reach to amygdala via the stria terminalis. The VMN projects to the medial, cortical, and lateral nuclei, and to a capsule around the central nucleus. These sites correspond well to the location of E-concentrating cells in the amygdala. The VMN also projects laterally to the zona incerta and up and around the cerebral peduncle, crossing the thalamus and subthalamus, and terminating in the area lateral and dorsal to the posterior commissure.

The VMN projects to the median eminence in the rat (Krieger et al. 1979) but not in the cat or the monkey (Saper et al. 1979). Longer medial descending tracts go to the midbrain central gray, the dorsal projection being stronger than the ventral projection.

The ventrolateral part of the VMN contains significantly more E-concentrating neurons than the dorsomedial part. Krieger et al. (1979) state that the patterns of projection from these two parts of the VMN are similar, but differ in strength; the ventrolateral subdivision had greater connections to the amygdala and thalamus than the dorsomedial subdivision. Saper et al. (1979) also suggest that the ventrolateral VMN is responsible for rostral projections to the POA, the BNST, and the lateral septum. It may be, then, that the ventrolateral part of the VMN projects more strongly than the dorsomedial subdivision to the other brain regions that concentrate estradiol.

3.2 Lateral Septum

The lateral septum may be divided into dorsal, intermediate, and ventral subdivisions (Swanson and Cowan 1979). There are three major efferent projections from the lateral septal nucleus (Nauta 1956, 1958; Andy and Stephan 1964; Raisman 1966; DeFrance 1976; Meibach and Siegel 1977; Poulain 1977; Swanson and Cowan 1976, 1979). All parts of the lateral septum project to the medial septal-diagonal band system. The intermediate and ventral parts, which contain E-concentrating neurons, project to the lateral preoptic and lateral hypothalamic areas. This projection, through the mfb, also sends fibers to the dorsomedial nucleus of the hypothalamus, the supramamillary area, the capsule of the lateral mamillary nucleus, and the ventral tegmental area. The third projection is from the ventral subdivision to the paratenial and paraventricular nuclei of the thalamus, and to the medial habenular nucleus.

3.3 Posterior Septum

The posterior septum contains the nucleus triangularis septi and the septofimbrial nucleus. The nucleus triangularis septi, unlike the septofimbrial nucleus, contains well-labelled E-binding neurons. This nucleus projects to both the medial and lateral habenular nuclei, whereas the septofimbrial nucleus projects only to the medial habenular nucleus (Herkenham and Nauta 1977, 1979; Swanson and Cowan 1979). (The lateral, but not the medial, habenular nucleus contains E-concentrating cells.) Some fibers from the triangular nucleus do not terminate in the habenular nuclei, but project through the fasciculus retroflexus to the interpeduncular nucleus.

3.4 Bed Nucleus of the Stria Terminalis

Neurons in the bed nucleus of the stria terminalis (BNST) project to the hippocampus, and to regions that contain E-concentrating neurons (Valverde 1965; Millhouse 1969; Turner and Knapp 1976; Conrad and Pfaff 1976b; Swanson and Cowan 1979). Just as the BNST receives projections from the amygdaloid complex, it sends projections back (via the stria terminalis) to the central nucleus and the posterior part of the medial nucleus. The BNST sends a large projection through the mfb, laterally to the nucleus accumbens and substantia innominata, and ventrocaudally to the hypothalamus. The BNST projects to all of the hypothalamus except the VMN and mamillary nuclei, but as Swanson and Cowan (1979) point out, these nuclei may be innervated by the BNST through dendrites that reach the mfb. Finally, the BNST innervates the ventral tegmental area, the central tegmental fields, the central gray, and the median and dorsal raphe nuclei. The most caudal projections are to the parabrachial nucleus, and the locus ceruleus.

3.5 Nucleus Accumbens

The nucleus accumbens projects to limbic and motor areas (Swanson and Cowan 1975a; Conrad and Pfaff 1976a; Powell and Leman 1976; Williams et al. 1977;

Nauta et al. 1978). The nucleus accumbens projects anteriorly to the anterior and lateral septum, frontal cortex, and anterior end of the caudatoputamen. Descending projections innervate the nDB, globus pallidus, and medial and lateral preoptic areas. The nucleus accumbens does not project to the amygdaloid or habenular nuclei. The paratenial and paraventricular nuclei of the thalamus are labelled. The anterior and dorsomedial hypothalamus and the zona incerta are labelled, but the ventromedial, arcuate, premamillary and mamillary nuclei are unlabelled. Moving caudally, the posterior hypothalamus, supramamillary area, and medial anterior central gray receive light projections. Finally, some fibers were followed into the ventral tegmental area, substantia nigra, and ventrolateral tegmental reticular formation.

3.6 Amygdala

The projections of the cell groups of the amygdala that contain E-concentrating neurons (the cortical and medial nuclei, the medial part of the central nucleus, the posterior division of the lateral nucleus, and the anterior amygdala), in general, lead directly to the medial hypothalamus. Cells in the amygdala may also influence the hypothalamus indirectly through connections to the BNST, which then project to the hypothalamus (Hall 1963; Valverde 1965; Heimer and Nauta 1969; Leonard and Scott 1972; de Olmos and Ingram 1972; de Olmos 1972; reviewed by Gloor et al. 1972; Millhouse 1973 b; McBride and Sutin 1977; Krettek and Price 1978 a, b; Ottersen 1980; Price and Amaral 1981; Kelley et al. 1982; Kita and Oomura 1982).

The corticomedial and basolateral nuclei project to the preoptic area and ventromedial nucleus of the hypothalamus. The basolateral, basomedial, and central nuclei project to the lateral division of the BNST. The basomedial, medial, and posterior cortical nuclei, and the amygdalohippocampal area project to the medial division of the BNST. (The BNST contains E-concentrating neurons throughout its mediolateral extent.)

The amygdalo-hippocampal area and/or the posterior cortical nucleus project to the lateral septum. This projection is confined to the ventral lateral septum (where there are E-containing neurons) just dorsal to the BNST and nucleus accumbens.

3.7 Habenular Nuclei

Herkenham and Nauta (1977, 1979) divided the habenular complex into compartments based on their afferent and efferent connections (Cragg 1961; Mitchell 1963; Akagi and Powell 1968; Yamadori 1969; Kataoka et al. 1973; Kuhar et al. 1975; Leranth et al. 1975; Lenn 1976; Aghajanian and Wang 1977; Iwahori 1977; Herkenham and Nauta 1977, 1979; Marchand et al. 1980; Contestabile and Flumerfelt 1981). The medial habenular nucleus is a compartment that receives projections from the supracommissural septum and projects to the interpeduncular nucleus. The medial part of the lateral habenular nucleus (LHN), the second compartment, is the only part of the habenular nuclei that contains E-concentrating neurons. The lateral part of the LHN is the third compartment. The LHN receives afferents from the medial pallidum, projecting more strongly to the lateral than the medial part of the LHN. The LHN also receives an orderly projection from the lateral preoptic-hypothalamic regions, with more laterally placed neurons in the preoptic-hypothalamic area projecting to the lateral part of the LHN, and more medially placed neurons projecting to the medial part of the LHN. The medial, but not the lateral part of the LHN, has projections from the ventrolateral septum, the nDB, and the medial part of the nucleus accumbens (all regions of light to moderate E labelling). The medial part of the LHN projects bilaterally, via the fasciculus retroflexus, to ventral parts of the septum, lateral preoptic area, medial and lateral hypothalamic regions, and median and dorsal raphe nuclei with adjacent parts of the midbrain central gray.

3.8 Midbrain Central Gray

In the rat there are substantial reciprocal connections between the midbrain central gray (MCG) and the basomedial hypothalamus and limbic structures (Kuypers 1956; Nauta 1958; Hamilton and Skultety 1970; Chi 1970b; Hamilton 1973; Ruda 1976; Morrell et al. 1981; Berk and Finkelstein 1981; Morrell et al. 1981; Beitz 1982; Eberhart 1985; Marchand and Hagino 1983). Their largest hypothalamic input is from the ventromedial nucleus of the hypothalamus. In the rat MCG neurons project bilaterally to the olfactory tubercle, dorsal nucleus accumbens, BNST, medial and lateral septum, central nucleus of the amygdala, midline thalamic nuclei, preoptic area, the ventromedial, dorsomedial, and arcuate nuclei of the hypothalamus, the substantia nigra, zona incerta, and caudate putamen. Many of these connections are reciprocal. MCG neurons have descending efferents to the medullary reticular formation. These patterns are, for the most part, the same in cats and monkeys.

3.9 Electrophysiological Evidence on Function and Connections

Electrical activity of hypothalamic neurons changes during the estrous cycle of female rats (Cross and Dyer 1971; Dyer et al. 1972; Dyer 1973; Kawakami et al. 1970; Moss and Law 1971; Teresawa and Sawyer 1969; reviewed by Pfaff 1973) owing to the effect of estrogen and progesterone, and/or pituitary hormones. The effects of the gonadal steroids estrogen and progesterone on neural activity have been studied. Ovariectomized, estrogen-treated female rats have fewer spontaneously active neurons in the medial preoptic region and BNST than do untreated females, and more spontaneously active neurons in and near the ventromedial nucleus (Bueno and Pfaff 1976). A similar pattern has been seen in cats (Alcaraz et al. 1969). In another study lower spontaneous activity was also seen in units in the preoptic area and anterior border of the medial anterior hypothalamus, and the lateral septum (Lincoln 1967; Yagi 1970, 1973; Yagi and Sawaki 1973; Whitehead and Ruf 1974). Microelectrophoresed estradiol inhibits medial preoptic neurons in female rats (Kelly et al. 1976, 1977, 1978). That E-concentrating neurons do project from the areas where they are found has been established by means of combined retrograde tracer-steroid autoradiography: E-concentrating neurons in the ventrolateral VMN project to the dorsal midbrain; those in the POA and BNST project to the amygdala; and those in the POA project to the VTA (Morrell and Pfaff 1982; Morrell et al. 1984). VMN neurons can indeed be antidromically activated from CG (Sakuma and Pfaff 1982).

While there is no truly "central point" in the highly interconnected limbo-hypothalamic network described above, there is convergence from many regions to VMN neurons, and VMN neurons project to the median eminence and central gray, two regions that may be important "output lines" of the network. For these reasons, this discussion on functional interactions of the network will begin with the ventromedial nucleus and expand from there.

Tuberoinfundibular and unidentified neurons in the ventromedial nucleus of the hypothalamus are influenced by stimulation of the amygdala, stria terminalis, medial preoptic area, anterior hypothalamic area, mediodorsal nucleus of the thalamus, median eminence, and dorsal hippocampus (reviewed by Renaud 1979). With the idea of integration of signals in a network in mind, it is especially interesting that in one study in rats, 70%–80% of VMN neurons were affected by more than one of the regions listed above (Renaud and Martin 1975). The amygdaloid projection, primarily from the basomedial and basolateral nuclei, may be inhibitory or excitatory (Dreifuss et al. 1968; Dreifuss et al. 1968; Murphy et al. 1968 a, b; Murphy and Renaud 1969; Dreifuss 1972; Murphy 1972; Renaud 1976 a, b). The effect is frequency dependent: at a frequency of 1 Hz excitatory responses predominate, but at frequencies above 30 Hz responses are largely inhibitory (Renaud 1976 b).

Consideration of the AHA-POA is complex. VMN neurons that are excited by amygdaloid stimulation are also directly excited by stimulation of the AHA-POA (Renaud 1977). The amygdala may influence VNM neurons indirectly, through the AHA-POA: AHA-POA neurons receive largely excitatory inputs from the amygdala (Fenske et al. 1975; Demaine et al. 1976; Dyer et al. 1976 a, b; Carrer et al. 1978) and may then excite VMN neurons (Dyer 1973). Finally, in the rat VMN neurons project *back* to the basomedial and basolateral amygdala (Renaud and Hopkins 1977) and the AHA-POA (Dyer 1973, 1974, 1975; Renaud 1977), as shown by orthodromic activation.

The lateral septum also interacts in this amygdalo-mediobasal hypothalamic network. Amygdala and septal stimulation converge on VMN neurons in the cat (Murphy et al. 1968 a), in the rat (Renaud et al. 1977), and guinea pig (Poulain 1977). VMN neurons receive septal input and have a weak reciprocal connection back to the lateral septum. In contrast, stimulation of the lateral septum increases firing in mPOA neurons (Kawakami et al. 1979). VMN neurons may also influence neurons in the mediodorsal nucleus of the thalamus, and a small number are excited or inhibited by electrical stimulation of this same nucleus of the thalamus (Renaud and Martin 1975).

Finally, VMN neurons are influenced by stimulation of the dorsal hippocampus in the rat, an area where few E-concentrating neurons are found. Renaud and Martin (1975) reported excitatory and inhibitory responses, and noted that cells that responded to dorsal hippocampal stimulation did not respond to stimulation of the amygdala, i.e., there was no convergence of hippocampal and amygdaloid influence.

Electrophysiological studies in the monkey (Sandrew and Poletti 1984) and rat (Sakuma and Pfaff 1982) have shown that neurons from limbic areas converge on MCG neurons and may be facilitatory. In addition, medial hypothalamic neurons that project to the midbrain are influenced by extrahypothalamic, limbic areas (Pittman et al. 1979). MCG neurons have a low rate of spontaneous activity, but spontaneous rates are higher in E-treated female rats than in ovariectomized females (Sakuma and Pfaff 1980). Those MCG neurons that are antidromically activated from the reticular formation show a higher percentage of somatodendritic spikes when the VMN is electrically stimulated. *It has been suggested that massive recruitment of limbic inputs is required to activate MCG neurons (Sandrew and Poletti 1984)*. The MCG, in turn, projects to motor and autonomic control sites in the brainstem (Chi 1970 b; Ruda 1976; Eberhart 1986, in preparation). In fact, MCG can facilitate a motor system that activates muscles responsible for lordosis behavior (Cottingham et al. 1986, in prepraration). It is likely, then, that the MCG is important for integrating signals from areas of the forebrain that receive steroid hormone inputs and for activating systems that lead to coordinated motor patterns.

4 Functional Implications

The nature of synaptic inputs to a hormone-concentrating cell may influence the effects of the hormone on that cell. Estrogen effects on resting discharge of VMN neurons, obvious in vivo (Bueno and Pfaff 1976), can only be seen in vitro (where most inputs have been cut off) if an electrically-stimulated input is provided (Kow and Pfaff 1985). An estrogen-induced protein in hypothalamus (Mobbs et al. 1985), discovered with in vivo labelling, was not seen with in vitro labelling (Jones et al. 1985). Synaptic input during estrogen priming of hypothalamic neurons may even be important for eventual estrogen-stimulated behavior (Roy et al. 1985). Therefore, if synaptic inputs to a steroid-concentrating cell come largely from other steroid-concentrating cells, the network formed could be of great physiological importance.

4.1 Theoretical Importance of Steroid–Neuron Networking

Network theorists have studied the relation of equilibrium conditions, connectedness, and stability in complex systems, many times in the context of foodweb networks (MacDonald 1983). Studies of neural networks, especially theoretical ones, are very often concerned with learning and memory (John 1972; Little and Shaw 1975, 1978); others are concerned with pattern processing in a sensory system such as vision (Fuster and Jervey 1981).

This chapter has reviewed anatomical evidence that establishes the existence of connections between different areas of the limbic-hypothalamic axis; the complexity of connections in the limbic system has long been recognized (Papez 1937). The network formed by steroid-concentrating cells may be a subset of limbic-hypothalamic networks, with inputs and outputs being channeled through.

Several concepts about network functions may be relevant for the network formed by E-concentrating neurons. First, the steroid-concentrating cells in the input and output "lines" of the network may allow massive *amplification* of signals. Secondly, the network may permit greater *stability*. Thirdly, the network may impose a selective *filtering* upon inputs, channeling some to the hypothalamus or to an output line from the network (affecting pituitary or behavior).

The possibility of *amplification* by the network is obvious. If each hormone-concentrating neuron exerts a particular effect on an incoming signal, and if the signal must go through a path of two or three such neurons, the effect is *multiplied*. Similarly, if a hormone has a given percent effect on an individual neuron and if those hormone-sensitive neurons lie in a chain, the eventual hormone effect on the chain is *multiplied*. The concept of such a "cascade" has been recognized in immunology (Jerne 1974). Finally, where the connections are reciprocal and both directions are excitatory, the possibility of *self-reexcitation* arises, producing an extremely sharp rise in system-output. In terms of reproductive endocrinology, this is "positive feedback."

Extensive interconnectiveness may serve to add *stability*, as well as providing security through redundance. In endocrine terms this stability is "negative feedback." Direct reciprocal feedback is present in almost all connections in the E-concentrating network, e.g., VMN < = > AMYG, but may be modified by projections to either of these structures, e.g., BNST = > VMN and BNST = > AMYG. There is also indirect feedback, e.g., AMYG = > AHA-POA = > AMYG. Perhaps the indirect connections also provide *stability*. Finally, the redundancy of connections may make for more stable network performance (there are many ways to activate a particular group of neurons) by allowing for flexibility (as long as a certain form of the necessary conditions is met, the behavior or endocrine output will occur, but the conditions need not be met by the same part of the network every time).

Another possible role of the E-concentrating neuron network may be to *channel* (selectively filter) sensory input to facilitate reproductive behavior. Olfactory input is one sense that can affect reproductive behavior (Carr et al. 1965; Pfaff and Pfaffmann 1969). There are a few E-concentrating neurons in the olfactory tubercle and in the amygdala, which receives olfactory as well as vomeronasal input (Gloor 1960; Winans and Scalia 1970). A relay through the amygdala could "channel" (selectively transfer) this input to neurons in the POA. Somatosensory input that triggers E-dependent lordosis behavior activates neurons in the MCG and regions just lateral to it, and ultrasound which facilitates reproductive behavior also is registered here. Estradiol-binding MCG neurons have a massive projection to the medial hypothalamus (Eberhart et al. 1985) and selectively could transfer some of this information.

When a network is formed, interactive and cooperative phenomena can lead to an output that is a balance of the excitation and inhibition in the network, and the afferent input. Shaw and coworkers (1982) have described theoretical neural "nets" which have stable and unstable equilibrium states, and may make phase transitions that essentially change the gain of the input/output function. For example, neurons may interact cooperatively, and the output of the network may show hysteresis, sustaining a response to a brief stimulus. This could be a special case of the amplification function described above.

4.2 Comparisons with Other Neuronal Systems

Other chemically coded chains of neurons may exist, analogous to the E-binding network described. Notable are three systems of neurons, each containing a physiologically important peptide.

First, LHRH, the decapeptide, is crucial for control of reproduction by the pituitary and important for reproductive behavior. It is interesting, therefore, that the primary region of LHRH cell bodies in the preoptic area, diagonal band and septum receives an input from LHRH-containing cells in the nervus terminalis (Silverman and Krey 1978; Jennes and Stumpf 1980; Schwanzel-Fukuda and Silverman 1980).

Secondly, CCK is important for the peripheral and central control of appetite, as well as for its effects on the GI tract itself (reviewed by Kow and Pfaff 1986). A chain of CCK neurons links the likely peripheral sites of CCK action through the nucleus of the tractus solitarius and the parabrachial region, all the way to the ventromedial hypothalamus, where CCK-responsive neurons can be recorded (Kow and Pfaff 1985).

Thirdly, connections between angiotensin-binding neurons may be significant for osmotic balance and drinking behavior, as neurons in the subfornical region have receptors for angiotensin (Swanson et al. 1978) and may be part of a neural system that influences drinking (Miselis et al. 1979).

In contrast, corticosterone-concentrating neurons are found in the hippocampus, and do not appear to be highly interconnected. The major hippocampal outputs do not involve many other corticosterone-binding cell groups, and the only "network" would be comprised of septal-hippocampal connections. Also in contrast to the steroid and peptide-coded neuronal chains cited above, β-endorphin is a neuropeptide not distinguished by a chain or network of chemically-specific neurons (reviewed by Akil et al. 1984).

4.3 Conclusions

Sex steroid-binding neurons in the medial hypothalamus, preoptic area and limbic system are highly interconnected by a web of projections, often reciprocal. Such a network of hormone-binding neurons permits *amplification* of steroid hormone effects on physiological systems; *stability* of hormone-neuron system performance; and selective *channeling* of inputs to the network. Therefore, the high degree of interconnectedness of sex hormone neurons must influence the manner in which estrogen, androgen and progestin-binding hypothalamic and limbic neurons impact the median eminence (for control of anterior pituitary) and the midbrain central gray (for control of reproductive behavior).

Reference

Aghajanian GK, Wang RY (1977) Habenular and other midbrain raphe afferents demonstrated by a modified retrograde tracing technique. Brain Res 122:229–242

Akagi K, Powell EW (1968) Differential projections of habenular nuclei. J Comp Neurol 132:263–274

Akil H, Watson SJ, Young E, Lewis ME, Khachatorian H, Walker MJ (1984) Endogenous opioids: biology and function. Annu Rev Neurosci 7:223–255

Alcaraz M, Guzman-Flores C, Salas M, Beyer C (1969) Effect of estrogen on the responsivity of hypothalamic and mesencephalic neurons in the female cat. Brain Res 15:439–446

Anderson CH, Shen CL (1980) Efferents of the medial preoptic area in the guinea pig: an autoradiographic study. Brain Res Bull 5:257–265

Andy OJ, Stephan H (1964) The septum of the cat. Thomas, Springfield

Anschel S, Alexander M, Perachio AA (1982) Multiple connections of medial hypothalamic neurons in the rat. Exp Brain Res 46:383–392

Arees EA, Mayer J (1967) Anatomical connections between medial and lateral regions of the hypothalamus concerned with food intake. Science 157:1574–1575

Arnold A, Nottebohm F, Pfaff DW (1976) Hormone-concentrating cells in vocal control and other areas of the brain of the zebra finch (*Poephila guttata*). J Comp Neurol 165:487–512

Barfield R, Ronay G, Pfaff DW (1978) Autoradiographic localization of androgen-concentrating cells in the chicken brain. Neuroendocrinology 26:297–311

Beitz AJ (1982) The organization of afferent projections to the midbrain periaqueductal gray of the rat. Neuroscience 7:133–159

Berk ML, Finkelstein JA (1981) Afferent projections to the preoptic area and hypothalamic regions in the rat brain. Neuroscience 6:1601–1624

Berk ML, Finkelstein JA (1982) Efferent connections of the lateral hypothalamic area of the rat: an autoradiographic investigation. Brain Res Bull 8:511–526

Bueno J, Pfaff DW (1976) Single unit recording in hypothalamus and preoptic area of estrogen-treated and untreated ovariectomized female rats. Brain Res 101:67–78

Carr WJ, Loeb LS, Dissinger ML (1965) Response of rats to sex odors. J Comp Physiol Psychol 59:370–377

Carrer HF, Whitmoyer DI, Sawyer CH (1978) Effects of hippocampal and amygdaloid stimulation on the firing of preoptic neurons in the proestrous female rat. Brain Res 142:363–367

Chi CC (1970a) Afferent connections to the ventromedial nucleus of the hypothalamus in the rat. Brain Res 17:439–445

Chi CC (1970b) An experimental silver study of the ascending projections of the central gray substance and adjacent tegmentum in the rat with observations in the cat. J Comp Neurol 139:259–272

Chi CC, Flynn JP (1971) Neuroanatomic projections related to biting attack elicited from hypothalamus in cats. Brain Res 35:49–66

Chiba T, Murata Y (1985) Afferent and efferent connections of the medial preoptic area in the rat: a WGA-HRP study. Brain Res Bull 14:261–272

Conrad LCA, Pfaff DW (1976a) Autoradiographic tracing of nucleus accumbens efferents in the rat. Brain Res 113:589–596

Conrad LCA, Pfaff DW (1976b) Efferents from medial basal forebrain and hypothalamus in the rat. I. An autoradiographic study of the medial preoptic area. J Comp Neurol 169:185–220

Conrad LCA, Pfaff DW (1976c) Efferents from medial basal forebrain and hypothalamus in the rat. II. An autoradiographic study of the anterior hypothalamus. J Comp Neurol 169:221–262

Contestabile A, Flumerfelt BA (1981) Afferent connections of the interpeduncular nucleus and the topographic organization of the habenulo-interpeduncular pathway: an HRP study in the rat. J Comp Neurol 196:253–270

Cottingham SL, Femano PA, Pfaff DW (1986) Effects of electrical stimulation of micbrain central gray on reticulospinal activation of axial muscle emg in the rat. In preparation
Cragg BG (1961) The connections of the habenula in the rabbit. Exp Neurol 3:388–409
Cross BA, Dyer RG (1971) Cyclic changes in neurons of the anterior hypothalamus during the rat estrous cycle and the effect of anesthesia. In: Sawyer H, Gorski RA (eds) Steroid hormones and brain function. University of California Press, Los Angeles, pp 96–102 (UCLA Forum in medical sciences, no 15)
Davis RE, Morrell JI, Pfaff DW (1977) Autoradiographic localization of sex steroid-concentrating cells in the brain of the teleost *Macropodus opercularis* (*Osteichthyes: Belontiidae*). Gen Comp Endocrinol 33:496–505
DeFrance JF (ed) (1976) The septal nuclei. Plenum, New York
Demaine C, Perkins MN, Whitehead SA (1976) Amygdaloid-hypothalamic connections and catecholamines. J Physiol 260:57–58P
De Olmos JS (1972) The amygdaloid projection field in the rat as studied with the cupric-silver method. In: Eleftheriou BE (ed) The neurobiology of the amygdala. Plenum, New York, pp 145–204
De Olmos JS, Ingram WR (1972) The projection field of the stria terminalis in the rat brain. An experimental study. J Comp Neurol 146:303–334
Dreifuss JJ (1972) Effects of electrical stimulation of amygdaloid complex on the ventromedial hypothalamus. In: Eleftheriou BE (ed) Neurobiology of the amygdala. Plenum, New York, pp 295–317
Dreifuss JJ, Murphy JT, Gloor P (1968) Contrasting effets of two identified amygdaloid pathways on single hypothalamic neurons. J Neurophysiol 31:237–248
Dyer RG (1973) An electrophysiological dissection of the hypothalamic regions which regulate the pre-ovulatory secretion of luteinizing hormone in the rat. J Physiol 234:421–442
Dyer RG (1974) The electrophysiology of the hypothalamus and its endocrinological implications. Prog Brain Res 41:133–147
Dyer RG (1975) Characteristics of neurones projecting directly to the median eminence. In: Motta M, Crosignani PG, Martini L (eds) Hypothalamic hormones: chemistry, physiology, pharmacology and clinical uses. Academic, New York, pp 169–182
Dyer RG, Pritchett CJ, Cross BA (1972) Unit activity in the diencephalon of female rats during the oestrous cycle. J Endocrinol 53:151–160
Dyer RG, Ellendorff F, MacLeod NK (1976a) Non-random distribution of cell types in the preoptic and anterior hypothalamic areas. J Physiol 261:495–504
Dyer RG, MacLeod NK, Ellendorff F (1976b) Electrophysiological evidence for sexual dimorphism and synaptic convergence in the preoptic and anterior hypothalamic areas of the rat. Proc R Soc Lond 13:421–440
Eberhart JA, Morrell JI, Krieger MS, Pfaff DW (1985) An autoradiographic study of projections ascending from the midbrain central gray, and from the region lateral to it, in the rat. J Comp Neurol 241:285–310
Enoch DM, Kerr FWL (1967) Hypothalamic vasopressor and vesicopressor pathways: II. Anatomic study of their course and connections. Arch Neurol 16:307–320
Etgen AM (1984) Progestin receptors and the activation of female reproductive behavior: a critical review. Horm Behav 18:411–430
Feder H, Siegel H, Wade G (1974) Uptake of [6,7^3-H]estradiol-17β in ovariectomized rats, guinea pigs, and hamsters: correlation with species differences in behavioral responsiveness to estradiol. Brain Res 71:93–103
Fenske M, Ellendorff F, Wuttke W (1975) Response of medial preoptic neurons to electrical stimulation of the mediobasal hypothalamus, amygdala and mesencephalon in normal, serotonin or catecholamine deprived female rats. Exp Brain Res 22:495–507
Fuster JM, Jervey JP (1981) Inferotemporal neurons distinguish and retain behaviorally relevant features of visual stimuli. Science 212:952–955
Gerlach JL, McEwen BS (1972) Rat brain binds adrenal steroid hormone: radioautography of hippocampus with corticosterone. Science 175:1133–1136

Gerlach J, McEwen B, Pfaff D, Moskovitz S, Ferin M, Carmel P, Zimmerman E (1976) Cells in regions of rhesus monkey brain and pituitary retain radioactive estradiol, corticosterone and cortisol differentially. Brain Res 103:603–612
Gloor P (1960) Amygdala. In: Field J, Magoun HW, Hall VE (eds) Neurophysiology. American Physiological Society, Washington (Handbook of physiology, vol II, sect I)
Gloor P, Murphy JT, Dreifuss JJ (1972) Anatomical and physiological characteristics of the two amygdaloid proection systems to the ventromedial hypothalamus. In: Hockman CH (ed) Limbic systems mechanisms and autonomic function. Thomas, Springfield, pp 60–77
Gorski RA, Gordon JH, Shryne JE, Southam AM (1978) Evidence for a morphological sex difference within the medial preoptic area of the rat brain. Brain Res 148:333–346
Gorzalka BB, Whalen RE (1977) The effects of progestins, mineralocorticoids, glucocorticoids, and steroid solubility on the induction of sexual receptivity in rats. Horm Behav 8:94–99
Gray P, Brooks PJ (1984) Effect of lesion location within the medial preoptic-anterior hypothalamic continuum on material and male sexual behaviors in female rats. Behav Neurosci 98:703–711
Hall E (1963) Efferent connections of the basal and lateral nuclei of the amygdala in cat. Am J Anat 113:139–145
Halpern M, Morrell JI, Pfaff DW (1982) Cellular [^{3}H]estradiol and [^{3}H]testosterone localization in the brains of garter snakes: an autoradiographic study. Gen Comp Endocrinol 46:211–224
Hamilton BL (1973) Projections of the nuclei of the periaqueductal gray matter in the cat. J Comp Neurol 152:45–58
Hamilton BL, Skultety FM (1970) Efferent connections of the periaqueductal gray matter in the cat. J Comp Neurol 139:105–114
Heimer L, Nauta WJH (1969) The hypothalamic distribution of the stria terminalis in the rat. Brain Res 13:284–297
Herkenham M, Nauta WJH (1977) Afferent connections of the habenular nuclei in the rat. A horseradish peroxidase study, with a note on the fiber-of-passage problem. J Comp Neurol 173:123–146
Herkenham M, Nauta WJH (1979) Efferent connections of the habenular nuclei in the rat. J Comp Neurol 187:19–48
Holzwarth-McBride M, Hurst EM, Knigge KM (1976) Monosodium glutamate induced lesions of the arcuate nucleus: endocrine deficiency and ultrastructure of the median eminence. Anat Rec 186:185–196
Iwahori N (1977) A golgi study on the habenular nucleus of the cat. J Comp Neurol 171:319–344
Jennes L, Stumpf WE (1980) LHRH-systems in the brain of the golden hamster. Cell Tissue Res 209:239–256
Jerne NK (1974) Towards a network theory of the immune system. Ann Immunol C125:373–389
John ER (1972) Switchboard versus statistical theories of learning and memory. Science 177:850–864
Jones KJ, McEwen BS, Pfaff DW (1985) Effects of estradiol on protein synthesis in vitro, in the ventromedial hypothalamic nucleus (VMN) and preoptic area (POA) of the female rat. Soc Neurosci Abstr 11:738
Kaelber WW, Leeson CR (1967) A degeneration and electron microscopic study of the nucleus hypothalamicus ventromedialis of the cat. J Anat 101:209–221
Kataoka K, Nakamura Y, Hassler R (1973) Habenulointerpeduncular tract: a possible cholinergic neuron in rat brain. BrainRes 62:264–267
Kato K (1985) Progesterone receptors in brain and hypophysis. In: Ganten E, Pfaff D (eds) Current topics in neuroendocrinology, vol 5. Springer, Berlin Heidelberg New York Tokyo, pp 31–81
Kato J, Onouchi T (1977) Specific progesterone receptors in the hypothalamus and anterior hypophysis of the rat. Endocrinology 101:920–928

Kawakami M, Terasawa E, Ibuki T (1970) Changes in multiple unit activity of the brain during the estrous cycle. Neuroendocrinology 6:30–48

Kawakami M, Akema T, Ando S (1979) Electrophysiological studies on the neural networks among estrogan and progesterone effective brain areas on lordosic behavior of the rat. Brain Res 169:287–301

Kelley DB, Pfaff DW (1978) Generalizations from comparative studies on neuroanatomical and endocrine mechanisms of sexual behavior. In: Hutchison JB (ed) Biological determinants of sexual behavior. Wiley, Chichester

Kelley DB, Morrell JI, Pfaff DW (1975) Autoradiographic localization of hormone-concentrating cells in the brain of an amphibian (Xenopus laevis), I. Testosterone. J Comp Neurol 164:47–61

Kelley AE, Domesick VB, Nauta WJH (1982) The amygdalostriatal projection in the rat – an anatomical study by anterograde and retrograde tracing methods. Neuroscience 7:615–630

Kelly MJ, Moss RL, Dudley CA (1976) Differential sensitivity of preoptic-septal neurons to microelectrophoresed estrogen during the estrous cycle. Brain Res 114:152–157

Kelly MJ, Moss RL, Dudley CA (1977) The effects of microelectrophoretically applied estrogen, cortisol and acetylcholine on medial preoptic-septal unit activity throughout the estrous cycle of the female rat. Exp Brain Res 30:53–64

Kelly MJ, Moss RL, Dudley CA (1978) The effects of ovariectomy on the responsiveness of preoptic-septal neurons to microelectrophoresed estrogen. Neurodendocrinology 25:204–211

Kita H, Domura Y (1982) An HRP study of the afferent connections to rat medial hypothalamic region. Brain Res Bull 8:53–62

Kow L-M, Pfaff DW (1985) Estrogen effects on neuronal responsiveness to electrical and neurotransmitter stimulation: an in vitro study on the ventromedial nucleus of the hypothalamus. Brain Res 347:1–10

Kow L-M, Pfaff DW (1986) Behavioral effects of neuropeptides: some conceptual considerations. In: Negro-Vilar A, Conn PM (eds) Peptide hormones: effects and mechanisms of action. CRC Press, Boca Raton (in press)

Krettek JE, Price JL (1978 a) Amygdaloid projections to subcortical structures within the basal forebrain and brainstem in the rat and cat. J Comp Neurol 178:225–254

Krettek JE, Price JL (1978 b) A description of the amygdaloid complex in the rat and cat with observations on intra-amygdaloid axonal connections. J Comp Neurol 178:255–280

Krieger MS, Morrell JI, Pfaff DW (1976) Autoradiographic localization of estradiol-concentrating cells in the female hamster brain. Neuroendocrinology 22:193–205

Krieger MS, Morrell JI, Pfaff DW (1978) Neuroanatomical connections of steroid concentrating cell groups. In: Scott DE, Kozlowski GP, Weindl A (eds) Brain-endocrine interaction, vol III: Neural hormones and reproduction. Karger, Basel, pp 197–211

Krieger MS, Conrad LCA, Pfaff DW (1979) An autoradiographic study of the efferent connections of the ventromedial nucleus of the hypothalamus. J Comp Neurol 183:785–816

Kuhar M, Dehaven RN, Yamamura HI, Rommelspacher H, Simon VR (1975) Further evidence for cholinergic habenulointerpeduncular neurons: pharmacologic and functional characteristics. Brain Res 97:265–275

Kuypers HGJM (1956) Certain fiber connections of the mesencephalic central gray matter. In: Kappers JA (ed) Progress in neurobiology. Elsevier, Amsterdam, pp 264–272

Lenn JN (1976) Synapses in the interpeduncular nucleus: electron microscopy of normal and habenula lesioned rats. J Comp Neurol 166:73–100

Leonard CM, Scott JW (1972) Origin and distribution of the amygdalofugal pathways in the rat: an experimental neuroanatomical study. J Comp Neurol 141:313–330

Leranth CS, Browstein MJ, Zaborsky L, Jaranyi ZS, Palkovitz M (1975) Morphological and biochemical changes in the rat interpeduncular nucleus following the transection of the habenulo-interpeduncular tract. Brain Res 99:124–128

Lieberburg I, McEwen BS (1977) Brain Cell nuclear retention of testosterone metabolites, 5a-dihydrotestosterone and estradiol-17β, in adult rats. Endocrinology 100:588–597

Lieberburg I, MacLusky NJ, McEwen BS (1977) 5a-dihydrotestosterone (DHT) receptors in rat brain and pituitary cell nuclei. Endocrinology 100:598–607
Lincoln DW (1967) Unit activity in the hypothalamus, septum and preoptic area of the rat: characteristics of spontaneous activity and the effect of oestrogen. J Endocrinol 37:177–189
Little WA, Shaw GL (1975) A statistical theory of short and long term memory. Behav Biol 14:115–133
Little WA, Shaw GL (1978) Analytic study of the storage capacity of a neural network. Math Biosci 39:281–290
MacDonald N (1983) Trees and networks in biological models. Wiley, New York
MacLeod NK, Mayer ML (1980) Electrophysiological analysis of pathways connecting the medial preoptic area with the mesencephalic central grey matter in rats. J Physiol 298:53–70
MacLusky NJ, McEwen BS (1978) Oestrogen modulates progestin receptor concentrations in some brain regions but not in others. Nature 274:276–277
MacLusky NJ, McEwen BS (1980) Progestin receptors in the rat brain: Distribution and properties of cytoplasmic progestin binding sites. Endocrinology 106:192–202
Maeda H, Mogenson GJ (1980) An electrophysiological study of inputs to neurons of the ventral tegmental area from the nucleus accumbens and medial preoptic-anterior hypothalamic areas. Brain Res 197:365–376
Makara GB, Hodacs L (1975) Rostral projections from the hypothalamic arcuate nucleus. Brain Res 84:23–29
Marchand ER, Riley JN, Moore RY (1980) Interpeduncular nucleus afferents in the rat. Brain Res 193:339–352
Marchand JE, Hagino N (1983) Afferents to the periaqueductal gray in the rat. A horseradish peroxidase study. Neuroscience 9:95–106
McBride RL, Sutin J (1977) Amygdaloid and pontine projections to the ventromedial nucleus of the hypothalamus. J Comp Neurol 174:377–396
McClure TD, Clark G (1968) Descending connections from hypothalamus. Exp Neurol 22:343–349
McEwen BS, Weiss JM, Schwartz LS (1969) Uptake of corticosterone by rat brain and its concentration by certain limbic structures. Brain Res 16:227–241
McEwen BS, Weiss JM, Schwartz LS (1970) Retention of corticosterone by cell nuclei from brain regions of adrenalectomized rats. Brain Res 17:471–482
McEwen BS, Davis PG, Parsons B, Pfaff DW (1979) The brain as a target for steroid hormone action. Annu Rev Neurosci 2:65–112
McEwen BS, Biegon A, Davis PG, Krey LC, Luine VN, McGinnis MY, Paden CM, Parsons B, Rainbow TC (1982) Steroid hormones: humoral signals which alter brain cell properties and function. Recent Prog Horm Res 38:41–92
Meibach RC (1977) Afferent connections of the amygdaloid complex in the cat. Anat Rec 187:653
Meibach RC, Siegel A (1977) Efferent connections of the septal area in the rat: an analysis utilizing retrograde and anterograde transport methods. Brain Res 119:1–20
Millhouse OE (1969) A Golgi study of the descending medial forebrain bundle. Brain Res 15:341–363
Millhouse OE (1973a) The organization of the ventromedial hypothalamic nucleus. Brain Res 55:71–87
Millhouse OE (1973b) Certain ventromedial hypothalamic afferents. Brain Res 55:89–105
Miselis RR, Shapiro RE, Hand PJ (1979) Subfornical organ efferents to neural systems for control of body water. Science 205:1022–1025
Mitchell R (1963) Connections of the habenula and of the interpeduncular nucleus in the cat. J Comp Neurol 121:441–457
Mizuno N, Clemente CD, Sauerland EK (1969) Fiber projections from rostral basal forebrain structures in the cat. Exp Neurol 25:220–237
Mobbs CV, Harlan RE, Pfaff DW (1985) An estradiol-induced protein synthesized in the ventral medial hypothalamus (VMH) and transported to the midbrain central gray (MCG). Soc Neurosci Abstr 11:1271

Mogenson GJ, Jones DL, Yim CY (1980) From motivation to action: functional interface between the limbic system and the motor system. Prog Neurobiol 14:69–97

Morrell JI, Pfaff DW (1981) Autoradiographic technique for steroid hormone localization. Application to the vertebrate brain. In: Adler NT (ed) Neuroendocrinology of reproduction. Plenum, New York, pp 519–531

Morrell JI, Pfaff DW (1982) Characterization of estrogen-concentrating hypothalamic neurons by their axonal projections. Science217:1273–1276

Morrell JI, Kelley DB, Pfaff DW (1975 a) Sex steroid binding in the brains of vertebrates. Studies with light-microscopic autoradiography. In: Knigge KM, Scott DE, Kobayashi H, Ishii S (eds) Brain-endocrine interaction, vol II. Karger, Basel, pp 230–256

Morrell JI, Kelley DB, Pfaff DW (1975 b) Autoradiographic localization of hormone-concentrating cells in the brain of an amphibian, Xenopus laevis. II. Estradiol. J Comp Neurol 164:63–78

Morrell JI, Ballin A, Pfaff DW (1977) Autoradiographic demonstration of the pattern of H-estradiol concentrating cells in the brain of a carnivore, the mink, *Mustela vison*. Anat Rec 189:609–624

Morrell JI, Crews D, Ballin A, Morgentaler A, Pfaff DW (1979) ^{3}H-Estradiol, ^{3}H-testosterone and ^{3}H-dihydrotestosterone localization in the brain of the lizard anolis carolinensis: an autoradiographic study. J Comp Neurol 188:201–223

Morrell JI, Greenberger LM, Pfaff DW (1981) Hypothalamic, other diencephalic, and telencephalic neurons that project to the dorsal midbrain. J Comp Neurol 201:589–620

Morrell JI, Wolinsky TD, Krieger MS, Pfaff DW (1982) Autoradiographic identification of estradiol-concentrating cells in the spinal cord of the female rat. Exp Brain Res 45:144–150

Morrell JI, Schwanzel-Fukuda M, Fahrbach SE, Pfaff DW (1984) Axonal projections and peptide content of steroid hormone concentrating neurons. Peptides 5, [Suppl] 1:227–239

Moss RL, Law OT (1971) The estrous cycle: its influence on single unit activity in the forebrain. Brain Res 30:435–438

Murphy JT (1972) The role of the amygdala in controlling hypothalamic output. In: Eleftheriou BE (ed) Neurobiology of the amygdala. Plenum, New York, pp 295–317

Murphy JT, Renaud LP (1969) Mechanisms of inhibition in the ventromedial nucleus of the hypothalamus. J Neurophysiol 32:85–102

Murphy JT, Dreifuss JJ, Gloor P (1968 a) Topographical differences in the responses of single hypothalamic neurons to limbic stimulation. Am J Physiol 214:1443–1453

Murphy JT, Dreifuss JJ, Gloor P (1968 b) Responses of hypothalamic neurons to repetitive amygdaloid stimulation. Brain Res 8:153–166

Nauta WJH (1956) An experimental study of the fornix system in the rat. J Comp Neurol 104:247–271

Nauta WJH (1958) Hippocampal projections and related neural pathways to the midbrain in the cat. Brain 81:319–340

Nauta WJH, Haymaker W (1969) Hypothalamic nuclei and fiber connections. In: Haymaker W, Anderson E, Nauta WJH (eds) The hypothalamus. Thomas, Springfield, pp 136–209

Nauta WJH, Smith GP, Faull RLM, Domesick VB (1978) Efferent connections and nigral afferents of the nucleus accumbens septi in the rat. Neuroscience 3:385–401

Ottersen OP (1980) Afferent connections to the amygdaloid complex of the rat and cat. II. Afferents from the hypothalamus and the basal telencephalon. J Comp Neurol 194:267–289

Palkovits M, Zaborszky L (1979) Neural connections of the hypothalamus. In: Morgane PJ, Panksepp J (eds) Anatomy of the hypothalamus. Dekker, New York, pp 379–509 (Handbook of the hypothalamus, vol 1)

Papez JW (1937) A proposed mechanism of emotion. Arch Neurol Psychiatr 38:725–743

Parsons B, Rainbow TC, MacLusky NJ, McEwen BS (1982) Progestin receptor levels in rat hypothalamic and limbic nuclei. J Neurosci 2:1446–1452

Pfaff DW (1968) Autoradiographic localization of radioactivity in the rat brain after injection of tritiated sex hormones. Science 161:1355–1356

Pfaff DW (1973) Interactions of steroid sex hormones with brain tissue: studies of uptake and physiological effects. In: Segal S et al. (eds) The regulation of mamalian reproduction. Springfield, Ill: Thomas, pp 5–22
Pfaff DW, Conrad LCA (1978) Hypothalamic neuroanatomy: steroid hormone binding and patterns of axonal projections. Int Rev Cytol 54:245–265
Pfaff DW, Keiner M (1973) Atlas of estradiol-concentrating cells in the central nervous system of the female rat. J Comp Neurol 151:121–158
Pfaff DW, Pfaffmann CP (1969) Behavioral and electrophysiological responses of male rats to female rat urine odor. In: Pfaffmann C (ed) Olfaction and taste. Rockefeller University Press, New York, pp 258–267
Pfaff DW, Gerlach JL, McEwen BS, Ferin M, Carmel P, Zimmerman EA (1976) Autoradiographic localization of hormone-concentrating cells in the brain of the female rhesus monkey. J Comp Neurol 170:279–294
Phillipson OT (1979) Afferent projections to the ventral tegmental area of Tsai and interfascicular nucleus: a horseradish peroxidase study in the rat. J Comp Neurol 187:117–144
Pittman QJ, Blume HW, Kearney RE, Renaud LP (1979) Influence of midbrain stimulation on the excitability of neurons in the medial hypothalamus of the rat. Brain Res 174:39–53
Poulain P (1977) Septal afferents to the arcuate-median eminence region in the guinea pig: correlative electrophysiological and horseradish peroxidase studies. Brain Res 137:150–153
Powell EW, Leman RB (1976) Connections of the nucleus accumbens. Brain Res 105:389–403
Powell TPS, Cowan WM, Raisman G (1963) Olfactory relationships of the diencephalon. Nature 199:710–712
Price JL, Amaral DG (1981) An autoradiographic study of the projections of the central nucleus of the monkey amygdala. J Neurosci 1:1242–1259
Rainbow TC, Parsons B, MacLusky NJ, McEwen BS (1982a) Estradiol receptor levels in rat hypothalamic and limbic nuclei. J Neurosci 2:1439–1445
Rainbow TC, Parsons B, McEwen BS (1982b) Sex differences in rat brain oestrogen and progestin receptors. Nature 300:648–649
Raisman G (1966) The connexions of the septum. Brain 89:317–348
Rees HD, Switz GM, Michael RP (1980) The estrogen-sensitive neural system in the brain of female cats. J Comp Neurol 193:789–804
Renaud LP (1976a) Electrophysiological evidence for axon collaterals in the tuberoinfundibular system of the rat. J Physiol 254:20–21P
Renaud LP (1976b) An electrophysiological study of the amygdalohypothalamic projections to the ventromedial nucleus of the rat. Brain Res 105:45–58
Renaud LP (1977) Influence of medial preoptic-anterior hypothalamic area stimulation on the excitability of mediobasal hypothalamic neurons in the rat. J Physiol 264:541–564Renaud LP (1979) Neurophysiology and neuropharmacology of medial hypothalamic neurons and their extrahypothalamic connections. In: Morgane PJ, Panksepp J (eds) Anatomy of the hypothalamus. Dekker, New York, pp 593–693 (Handbook of the hypothalamus, vol 1)
Renaud LP, Hopkins DA (1977) Amygdala afferents from the mediobasal hypothalamus: An electrophysiological and neuroanatomical study in the rat. Brain Res 121:201–213
Renaud LP, Martin JB (1975) Electrophysiological studies of connections of hypothalamic ventromedial nucleus neurons in the rat: evidence for a role in neuroendocrine regulation. Brain Res 93:145–151
Renaud LP, Blume HW, Pittman QJ, Kearney RE, MacKenzie BW (1977) Influence of lateral septum stimulation on the excitability of mediobasal hypothalamic neurons in the rat. Neurosci Abstr 3:204
Ricardo JA (1983) Hypothalamic pathways involved in metabolic regulatory functions, as identified by track-tracing methods. In: Szabo AJ (ed) Advances in metabolic disorders, vol 10. Academic, New York

Rodgers CH, Schneider VM (1979) Facilitatory influences on mating behavior in the female rat affected by lesions of the habenula or the basolateral amygdaloid regions. Psychoneuroendocrinology 4:237–244
Roy E, Lynn D, Clark AS (1985) Inhibition of sexual receptivity by anesthesia during estrogen priming. Brain Res 337:163–166
Ruda MT (1976) Autoradiographic study of the efferent projections of the midbrain central gray in the cat. PhD dissertation, University of Pennsylvania
Sakuma Y, Pfaff DW (1979) Facilitation of female reproductive behavior from mesencephalic central gray in the rat. Am Physiol Soc 237:R278–R284
Sakuma Y, Pfaff DW (1980) Convergent effects of lordosis-relevant somatosensory and hypothalamic influences on central gray cells in the rat mesencephalon. Exp Neurol 70:269–281
Sakuma Y, Pfaff DW (1982) Properties of ventromedial hypothalamic neurons with axons to midbrain central gray. Exp Brain Res 46:292–300
Sandrew BB, Poletti CE (1984) Limbic influence on the periaqueductal gray: a single unit study in the awake squirrel monkey. Brain Res 303:77–86
Saper CB, Swanson LW, Cowan WM (1976) The efferent connections of the ventromedial nucleus of the hypothalamus of the rat. J Comp Neurol 169:409–442
Saper CB, Swanson LW, Cowan WM (1978) The efferent connections of the anterior hypothalamic area of the rat, cat, and monkey. J Comp Neurol 182:575–600
Saper CB, Swanson LW, Cowan WM (1979) An autoradiographic study of the efferent connections of the lateral hypothalamic area in the rat. J Comp Neurol 183:689–706
Sar M, Stumpf WE (1973a) Autoradiographic localization of radioactivity in the rat brain after the injection of 1,2-^{3}H-testosterone. Endocrinology 92:251–256
Sar M, Stumpf WE (1973b) Neurons of the hypothalamus concentrate [^{3}H]progesterone or its metabolites. Science 182:1266–1268
Sar M, Stumpf WE (1977) Distribution of androgen target cells in rat forebrain and pituitary after [^{3}H]-dihydrotestosterone administration. J Steroid Biochem 8:1131–1135
Schwanzel-Fukuda M, Silverman AJ (1980) The nervus terminalis of the guinea pig: a new luteinizing hormone-releasing hormone (LHRH) neuronal system. J Comp Neurol 191:213–225
Selmanoff MK, Brodkin LD, Weiner RI, Suteri PK (1977) Aromatization and 5a-reduction of androgens in discrete hypothalamic and limbic regions of the male and female rat. Endocrinology 101:841–848
Shaw GL, Harth E, Scheibel AB (1982) Cooperativity in brain function: assemblies of approximately 30 neurons. Exp Neurol 77:324–358
Shen CL, Anderson CH (1980) Efferents from the medial anterior hypothalamic area in the guinea pig. Brain Res Bull 5:693–701
Sheridan PJ (1979) The nucleus interstitialis striae terminalis and the medial nucleus of the amygdala, prime targets for androgen in the rat forebrain. Endocrinology 104:130–136
Silverman AJ, Krey LC (1978) The luteinizing hormone-releasing hormone (LH-RH) neuronal networks of the guinea pig brain. I. Intra- and extra-hypothalamic projections. Brain Res 157:233–246
Stumpf WE, Sar M (1975a) Autoradiographic techniques for localizing steroid hormones. In: O'Malley BW, Hardman JG (eds) Hormone action, part A. Academic, New York, pp 135–156 (Methods in enzymology, vol XXXVI)
Stumpf WE, Sar M (1975b) Hormone-architecture of the mouse brain with UH-estradiol. In: Stumpf WE, Grant LD (eds) Anatomical neuroendocrinology. Karger, Basel, pp 83–103
Stumpf WE, Sar M (1976) Steroid hormone target sites in the brain: the differential distribution of estrogen, progestin, androgen, and glucocorticosteroid. J Steroid Biochem 7:1163–1170
Stumpf WE, Sar M, Keefer DA (1975) Atlas of estrogen target cells in rat brain. In: Stumpf WE, Grant LD (eds) Anatomical neuroanatomy. Karger, Basel, pp 104–119
Sutin J, Eager RP (1969) Fiber degeneration following lesions in the hypothalamic ventromedial nucleus. In: Morgane PJ (ed) Neural regulation of food and water intake. Ann NY Acad Sci 157:610–628

Swanson LW (1976) An autoradiographic study of the efferent connections of the preoptic region in the rat. J Comp Neurol 167:227–256

Swanson LW, Cowan WM (1975a) A note on the connections and development of the nucleus accumbens. Brain Res 92:324–330

Swanson LW, Cowan WM (1975b) The efferent connections of the suprachiasmatic nucleus of the hypothalamus. J Comp Neurol 160:1–12

Swanson LW, Cowan WM (1976) Autoradiographic studies of the development and connections of the septal area in the rat. In: DeFrance JF (ed) The septal nuclei. Plenum, New York, pp 37–64

Swanson LW, Cowan WM (1979) The connections of the septal region in the rat. J Comp Neurol 186:621–656

Swanson LW, Kuypers HGJM (1980) A direct projection from the ventromedial nucleus and retrochiasmatic area of the hypothalamus to the medulla and spinal cord of the rat. Neurosci Lett 17:307–312

Swanson LW, Kucharczyk J, Mogenson GJ (1978) Autoradiographic evidence for pathways from the medial preoptic area to the midbrain involved in the drinking response to angiotensin II. J Comp Neurol 178:645–659

Szentagothai J, Fleko B, Mess B, Halasz B (1968) Hypothalamic control of the anterior pituitary. Akademiai Kiado, Budapest

Terasawa E, Sawyer CH (1969) Changes in electrical activity in the rat hypothalamus related to electrochemical stimulation of adenohypophyseal function. Endocrinology 85:143–149

Turner BH, Knapp ME (1976) Projections of the nucleus and tracts of the stria terminalis following lesions at the level of the anterior commissure. Exp Neurol 51:468–479

Valverde F (1965) Studies on the piriform lobe. Harvard University Press, Cambridge

Wade GN, Harding CF, Feder HH (1973) Neural uptake of (1,2 H) progesterone in ovariectomized rats, guinea pigs and hamsters: correlation with species differences in behavioral responsiveness. Brain Res 61:357–364

Wakefield C, Hall E (1974) Hypothalamic projections to the amygdala in the cat. Cell Tissue Res 151:499–508

Warembourg M (1975a) Radioautographic study of the rat brain after injection of [1,2-^{3}H]corticosterone. Brain Res 89:61–70

Warembourg M (1975b) Radioautographic study of the rat brain and pituitary after injection of ^{3}H-dexamethasone. Cell Tissue Res 161:183–191

Warembourg M (1975c) Radioautographic study of the brain and pituitary after ^{3}H-progesterone injection into estrogen-primed ovariectomized guinea pigs. Neurosci Lett 7:1–5

Warembourg M (1977) Radioautographic study of the brain and pituitary after ^{3}H-progesterone in ovariectomized rats, guinea pigs and hamsters: correlation with species differences in behavioral responsiveness. Brain Res 61:357–364

Warembourg M (1979) Uptake of ^{3}H-labeled synthetic progestin by rat brain and pituitary. A radioautography study. Neurosci Lett 9:329–332

Whitehead SA, Ruf KB (1974) Responses of antidromically identified preoptic neurons in the rat to neurotransmitters and to estrogen. Brain Res 79:185–198

Williams DJ, Crossman AR, Slater P (1977) The efferent projections of the nucleus accumbens in the rat. Brain Res 130:217–227

Winans SS, Scalia F (1970) Amygdaloid nucleus: new afferent input from the vomeronasal organ. Science 170:330–332

Wolf G, Sutin J (1966) Fiber degeneration after lateral hypothalamic lesions in the rat. J Comp Neurol 127:137–156

Yagi K (1970) Effects of estrogen on the unit activity of the rat hypothalamus. J Physiol Soc Jpn 32:692–693

Yagi K (1973) Changes in firing rates of single preoptic and hypothalamic units following an intravenous administration of estrogen in the castrated female rat. Brian Res 53:343–352

Yagi K, Sawaki Y (1973) Feedback of estrogen in the hypothalamic control of gonadotrophin secretion. In: Yagi K, Yoshida S (eds) Neuroendocrine control. University of Tokyo Press, Tokyo, pp 297–325

Yagi K, Sawaki Y (1977) Medial preoptic nucleus neurones: inhibition and facilitation of spontaneous activity following stimulation of the median eminence in female rats. Brain Res 120:342–346

Yamadori T (1969) Efferent fibers of the habenula and stria medullaris thalami in rats. Exp Neurol 25:541–558

Zaborszky L (1982) Afferent connections of the medial basal hypothalamus. In: Hild W, van Limborgh GJ, Ortmann R, Pauly KJE, Schiebler TH (eds) Advances in anatomy, embryology and cell biology, vol 69. Springer, Berlin Heidelberg New York, pp 1–107

Zigmond RE, Nottebohm F, Pfaff DW (1973) Androgen-concentrating cells in the midbrain of a songbird. Science 179:1005–1007

Ultrastructure of Regulatory Neuroendocrine Neurons and Functionally Related Structures *

B. KRISCH

Contents

1 Introduction

Until five years ago, the concept of hypothalamic neurohormones (Bargmann 1949, 1966; Scharrer 1936; Scharrer and Scharrer 1954) seemed to be rather simple and plausible, and was based on their function as regulatory or effectory hormones (Vollrath 1974).

Regulatory neurohormones are releasing factors (or hormones) – liberins [1], or release-inhibiting factors (or hormones) – statins [1], which regulate the production and release of adenohypophyseal hormones in a stimulatory or inhibitory way. Regulatory neurohormones in the sense of this original concept (Motta and Martini 1972; Martini 1974) are luliberin (LRF), corticoliberin (CRF; Vale et al. 1981; Rivier et al. 1983b), thyroliberin (TRF), somatoliberin (Guillemin et al. 1982; Rivier et al. 1982; Spiess et al. 1983), somatostatin, and prolactin-inhibiting factor

Anatomisches Institut der Universität Kiel, Olshausenstr. 40, D-2300 Kiel

* The work on which this review is based was supported by the Deutsche Forschungsgemeinschaft (Grant Nr. 569/5-1).

1 According to the Nomenclature of Peptide Hormones of the Comission on Biochemical Nomenclature, 1974; some of the abbreviations accord with those of the Endorine Society.

Current Topics in Neuroendocrinology, Vol. 7

(dopamine; see Table 1). In this chapter, they will be referred to as direct regulatory neurohormones.

However, in the last few years a great number of neurohormones have been described whose function is not easy to characterize (see Table 2). In part, they belong among the peptides, in part among the monoamines or amino acids (Mulder 1982). Concerning the possible functions of these neurohormones it has to be taken into account, on the one hand, that a single regulatory neurohormone may influence the production and release of more than one adenohypophyseal hormone, but, on the other hand, that a single adenohypophyseal cell may also be influenced by several neurohormones (McCann et al. 1980). Moreover, the

Table 1. Direct regulatory neurohormones

Neurohormone	Location of perikarya projecting to the neurohypophysis	Main effects	Additional influences
CRF	Most numerous in the parvocellular part of the paraventricular nucleus (Bloom et al. 1982; Kawata et al. 1982a, b, 1983; Olschowka et al. 1982; Pelletier et al. 1982; Liposits et al. 1983; Merchenthaler et al. 1983; Paull and Gibbs 1983; Swanson et al. 1983; Vale et al. 1983a, b)	Stimulates ACTH release (Rivier and Vale 1983a, b; Rivier et al. 1983a; De Souza et al. 1985)	
LRF	Septal-preoptic region (Barry 1979; Krisch 1980a, b; Jennes et al. 1985)	Stimulates gonadotropin release	
TRF	Area dorsal of suprachiasmatic nucleus, dorsomedial and paraventricular nuclei (Johansson and Hökfelt 1980; Johansson et al. 1980)	Stimulates TSH release	Increases prolactin release during proestrus in the rat (Fink et al. 1982)
Somatoliberin	Arcuate nucleus (Sawchenko et al. 1985)	Stimulates growth hormone release	
Somatostatin	Periventricular nuclei (Krisch 1980a, c; Pelletier 1980; Johansson et al. 1984)	Inhibits growth hormone release	Decreases ACTH release (Richardson and Schönbrunn 1981); inhibits TSH release (Chihara et al. 1978); inhibits TRH- or VIP-induced prolactin release (Enjalbert et al. 1982)
Dopamine	Arcuate nucleus (Moore and Bloom 1978)	Inhibits prolactin release	Promotes release of LRF (Millan and Herz 1985), somatostatin (Maeda and Frohmann 1980), and TRF (Beaumont 1983)

ACTH, adrenocorticotropic hormone; TSH, thyroid-stimulating hormone; TRH, thyrotropin-releasing hormone; VIP, vasoactive intestinal polypeptide

Tabelle 2. Indirect regulatory neurohormones

Neurohormones	Location of perikarya projecting to the neurohypophysis	Possible effects
Angiotensin II	Paraventricular nucleus (Lind et al. 1985)	Stimulates ACTH release (Anhut et al. 1982; Spinedi and Negro-Vilar 1983); modulates ACTH releasing potency of CRF (Rivier and Vale 1983b)
Cholecystokinin (CCK)/gastrin	Periventricular nucleus, peripheral parts of paraventricular and supraoptic nuclei, collocated with oxytocin (Vanderhaeghen et al. 1980, 1981; Bai et al. 1984)	Stimulates β-endorphin release (Matsumara et al. 1983); stimulates growth hormone release (Matsumara et al. 1984); stimulates prolactin release (Matsumara et al. 1985)
ACTH and other proopiomelanocortin derivatives	Arcuate nucleus (Bloom et al. 1980; O'Donohue and Dorsa 1982; Bai et al. 1984; Liotta et al. 1984)	ACTH indirectly retards LH secretion (Mann et al. 1985) and ACTH (4–10) has specific binding sites on axons in the median eminence (Van Houten et al. 1985); pro-opiomelanocortin derivatives inhibit dopamine release (Reymond et al. 1983), influence the perikarya of the arcuate nucleus (Kiss et al. 1985), and increase prolactin release via suppression of dopamine release (Grandison and Guidotti 1977; Cheung 1984)
Dynorphin (Proenkephalin B)	Magnocellular parts of supraoptic and paraventricular nuclei (together with vasopressin; Khachaturian et al. 1982; Watson et al. 1982, 1983; Bai et al. 1984)	Modulating influence on LRF release (Beaumont 1983)
Enkephalins (Proenkephalin A)	Together with oxytocin in paraventricular and supraoptic nuclei (Watson et al. 1982; Martin et al. 1983; Bai et al. 1984); together with somatostatin in periventricular nucleus (Tramu and Leonardelli 1979; Tramu et al. 1981; Beauvillain et al. 1984)	Suppress the release of oxytocin, stimulate prolactin release, and inhibit TSH release via inhibition of dopamine release (Beaumont 1983); modulate somatostatin release (Tramu et al. 1981)
Neurotensin	Paraventricular and arcuate nuclei (Emson et al. 1982; Jennes et al. 1982b; Bai et al. 1984; Ibata et al. 1984)	Stimulates somatostatin release (Maeda and Frohman 1980; Shimatsu et al. 1982)
VIP/PHI	Parvocellular part of paraventricular nucleus (Mezey and Kiss 1985)	Stimulates ACTH release and prolactin release (Rostène 1984; Abe et al. 1985; Čarillo et al. 1985; Mezey and Kiss 1985); stimulates growth hormone secretion (Denef et al. 1985) by inhibitory effect on somatostatin release (Shimatsu et al. 1982; Rostène 1984); stimulates prolactin release (Abe et al. 1985); reduces inhibitory effect of somatostatin on growth hormone release (Enjalbert et al. 1982); sti-

Table 2 (continued)

Neuro-hormones	Location of perikarya projecting to the neurohypophysis	Possible effects
		mulates prolactin release by inhibition of dopamine (Samson et al. 1983); stimulates LRF release (Samson et al. 1981); and provokes vasodilatation (Edvinsson 1985). PHI stimulates prolactin release (Hökfelt et al. 1983) and potentiates CRF effects (Tilders et al. 1984)
Substance P	Together with vasopressin in magnocellular hypothalamic nuclei (mice; Stoeckel et al. 1982) and in arcuate nucleus (rat, rhesus monkey; Tsuruo et al. 1983; Bai et al. 1984; Rønnekleiv et al. 1984)	Stimulatory effect on prolactin and somatostatin release; inhibition of LRF-induced gonadotropin release (Stoeckel et al. 1982); inhibition of LH release together with high-affinity binding sites on gonadotropic cells (Kerdelhué et al. 1985). Influence on neurons of the arcuate nucleus projecting to the median eminence (Tsuruo et al. 1983)

ACTH, adrenocorticotropic hormone; VIP, vasoactive intestinal polypeptide; PHI, *p*eptide bearing *h*istidine and *i*soleucin as terminal amino acids; LH, luteinizing hormone

regulatory neurohormones might not have only a direct influence on the adenohypophyseal cell, but might also have an indirect effect mediated by the structures which participate in the regulation of neurohormone release and distribution. In this chapter, these regulatory neurohormones will be referred to as indirect regulatory neurohormones. The results referred to in this study were obtained mainly from the rat.

Only a small proportion of the perikarya producing regulatory neurohormones are concentrated in well-defined nuclei. Rather, they are located subependymally as more or less single elements in the narrow periventricular gray of the hypothalamus, the perikarya of each system occupying a different part of the hypophyseotropic area (see Table 1). The perikarya of several other systems are disseminated within specific hypothalamic nuclei. Neurohormonal projections to neurohemal regions arise from the perikarya, as do peptidergic projections into other parts of the brain (Krisch 1981 a; Swaab 1982). It is not yet known whether both forms of projection arise from the same perikaryon, or whether each form of projection comes from a different perikaryon within the same peptidergic system.The peptidergic projections are parts of neuronal circuits which mostly belong to structures of the limbic system, and they have been omitted from this chapter.

Regulatory neurohormones are released intermittently, as indicated by the level of hormones in the blood. This intermittent release is partly regulated by efferent projections from the suprachiasmatic nucleus in the preoptic region, which mediates circadian rhythms, shorter time intervals being dominant (Krieger 1981; Moore 1981; Weitzman et al. 1981).

Furthermore, the release of regulatory neurohormones is dependent upon neuronal afferents (Morris 1983) to the perikarya and upon humoral and hormonal efferents to the axon terminals of neuroendocrine neurons. Neurons influencing other regulatory neuroendocrine axon terminals at the site of their release will be subsumed under the group of indirect regulatory neuroendocrine neurons effecting manifold interactions. This group includes peptidergic, monoaminergic, and gamma-aminobutyric acid-(GABA)ergic neurons. These neurons frequently also mediate influences from other brain areas, mainly from the limbic system and suprachiasmatic nucleus.

2 Organization of the Neurohemal Region as a Prerequisite for Neurohormonal Function

The prerequisite for neurohormonal function is the release of hormones into neurohemal regions (Knowles and Bern 1966) of circumventricular organs which lack the blood–brain barrier, where neurohormones can gain access to the bloodstream (Leonhardt et al. 1983). Neurohormones have been demonstrated in only three neurohemal regions of the six known circumventricular organs, namely in the vascular organ of the lamina terminalis (OVLT), in the subfornical organ (SFO), and in the neurohypophysis with its proximal part (median eminence) and its distal part (neural lobe; cf. Leonhardt 1980).

2.1 Circumventricular Organs Involved

Judging by wealth of neurohormones, the neurohypophysis is the most important neurohemal region (Kobayashi et al. 1970), and there is more or less sound evidence about the function and the targets of neurohormones released from it (Krisch 1980a). The neurohypophysis is connected with the capillaries of the adenohypophysis by portal vessels, providing a short and direct transport route to the target cells in the adenohypophysis (Duvernoy 1972; Ambach et al. 1976).

After the neurohypophysis, the OVLT contains rather high amounts of substances known as regulatory neurohormones (Krisch 1980a). However, its special vascular architecture (Ambach et al. 1978) suggests that the targets (as yet unknown) of the substances released into the fenestrated capillaries are situated outside the brain; thus, in the OVLT the "regulatory" neurohormones would in fact be effectory hormones analogous to vasopressin and oxytocin in the neural lobe. Another, albeit functionally rather enigmatic, possibility is the drainage of the capillaries of the OVLT into vessels of the preoptic area (Ambach et al. 1978). These vessels, however, would be subject to the blood–brain barrier, and a direct influence of the neurohormones of the OVLT on the preoptic area would require hitherto unproven special affinities and receptors for regulatory neurohormones at the luminal endothelial cell membrane, thus permitting access to the preoptic neuropil (Pardridge et al. 1981; Ermisch et al. 1984, 1985; Zlokovic et al. 1985).

A similar problem arises in the SFO, in which, although in limited amounts, regulatory peptides are released into the perivascular spaces of fenestrated capillaries (Krisch and Leonhardt 1980a). Target regions might be the plexus choroidei or might be situated in the OVLT preoptic region; in the latter case they would influence structures involved in the regulation of gonadal activity or water metabolism.

As far as can be seen from the latest available literature, there is no further reliable information about the OVLT and the SFO. Hence, these two neurohemal regions will not be discussed as release sites of regulatory hormones in this chapter, which will mainly focus on the proximal and distal neurohypophysis.

2.2 Open Questions on and Shortcomings of the Original Concept of Neurosecretion

On the basis of the history of neurosecretion, it has become customary to attribute the "regulatory neurohormones" which regulate the activity of the adenohypophysis to the proximal neurohypophysis (median eminence), whereas the distal neurohypophysis (neural lobe), is regarded as the storage and release site for "effectory neurohormones" which exert their influence via the general circulation directly on their target tissues (Vollrath 1974).

In the median eminence, the unmyelinated efferent fibers of regulatory neuroendocrine neurons extending into the neurohemal region of the neurohypophysis form largely circumscribed fiber bundles within the tuberoinfundibular tract. This tract probably also contains efferents from the ventromedial nucleus dorsally adjacent to the arcuate nucleus (nucleus infundibularis).

According to current concepts, two main feedback pathways influence the activity of regulatory neuroendocrine neurons (Leonhardt et al. 1983).

1. Influences from the target cells of regulatory neurohormones may reach the neuroendocrine perikarya via neural inputs or via the blood-stream, provided that special properties of the capillary endothelial cell membrane exist in the different functional units of the brain, resulting in selectivity of the blood–brain barrier. This route is beyond the scope of this review, especially because our knowledge of the properties of the capillary endothelial membrane in different areas of the brain is rather limited.
2. The inner and the outer cerebrospinal fluid (CSF) are discussed as vehicles for feedback mechanisms. Concerning the inner CSF, fibers immunoreactive to regulatory neurohormones have frequently been observed between the cuboid ependymal cells (Krisch 1981b). Thus, they gain access to information mediated by the CSF or are enabled to release their peptides between the cuboid cells. This feedback pathway seems more probable than that via the tanycytes, i.e., the specialized ependymal cells covering the median eminence in the floor of the third ventricle (Knigge et al. 1975). So far, no convincing results substantiate the hypothesis that the tanycytes transport regulatory neurohormones intracellularly from the median eminence to the inner CSF (cf. Leonhardt 1980). The possible feedback pathway via the outer CSF will be referred to in Sect. 2.3.

In view of recent results, some points in the clear-cut and plausible concept of regulatory neurohormones have to be revised, since questions have arisen which cannot be answered within the limits of the original concept of neurosecretion.

The theory of a close association of a regulatory neurohormone with a specific adenohypophyseal target cell has to be revised; the number of regulatory neurohormones, a number of which coexist within one axon terminal, by far exceeds the number of adenohypophyseal cell types. The distinction between regulatory and effectory neurohormones should be used within reasonable limits: effectory neurohormones can participate in the regulation of adenohypophyseal activity (e.g., vasopressin potentiates CRF effects; Rivier and Vale 1983a; Vale et al. 1983b; Rivier et al. 1984; Spinedi and Negro-Vilar 1984). Furthermore, to return to the entity of the neurohypophysis, the question whether the proximal neurohypophysis, or median eminence, is the sole release site of regulatory neurohormones should be discussed. Both parts of the neurohypophysis possess vascular connections with the adenohypophysis, and there are indications that neurohormones reach the adenohypophysis from the distal neurohypophysis, the neural lobe (Froehlich and Ben-Jonathan 1984; Frawley et al. 1985). Finally, when considering the functionally highly specialized target cells in the adenohypophysis, their functional reserve (surplus of hormones stored in the different populations of adenohypophyseal cells) and the quickly changing demands of the periphery, the question arises whether the regulatory neurohormones flood the gland profusely, or whether structures exist which to a certain extent control the access of regulatory neurohormones to their target cells. Moreover, the already mentioned feedback pathway via the outer CSF has to be critically reviewed.

In order to understand the distinction between direct and indirect regulatory neurohormones and the accessibility of the adenohypophsis to regulatory neurohormones, it might be helpful to consider the general neuronal organization of the central nervous system; both regulatory and effectory neurohormones are final efferent neurons, the "final neuroendocrine pathway" (as the volume on the sixth International Symposium on Neurosecretion was entitled). They are comparable to the second efferent neuron of the autonomic nervous system or to the motoneurons of the brain or spinal nerves. In all these cases, neuronal function can be understood only in connection with the target structure which, as far as the regulatory neuroendocrine neuron is concerned, includes the vessels and structures belonging to or bordering the intercellular spaces.

The neuroendocrine neuron is characterized not only by its intracellular equipment and its cellular capabilities, but also especially by its particular integration into the surrounding tissue elements (including the targets of indirect regulatory neurohormones) and by special structures modulating regulatory function at the level of the final target cells.

Finally, in view of the plasticity of neurons in vivo (Purves and Hadley 1985), mutual influences of axon terminals in the neurohypophysis have to be taken into account (Hökfelt et al. 1983; Tizabi et al. 1985).

2.3 Blood–CSF and Blood–Blood Barriers

As already mentioned, the prerequisite for neurohormonal function is free access for regulatory neurohormones to the blood, which is provided by the lack of a blood–brain barrier in the release site of neurohormones, the neurohypophysis. However, the undisturbed functioning of the surrounding brain requires a CSF milieu, and borders are built up, separating the hemal milieu of the proximal neurohypophysis from the surrounding CSF milieu. The discussion of these borders has to focus on the borders against the inner, ventricular CSF, the intercellular CSF of the surrounding neuropil, and the outer CSF; all the more so in the latter case, since it has been thought to provide routes for ultrashort feedback pathways to the brain (Mezey et al. 1978; Mezey and Palkovits 1982 a, b). As morphological substrates of these borders, tanycytes have to be considered, as do glial and meningeal cells.

The border between the hemal milieu of the median eminence and the inner CSF is well-established as a characteristic of the neurohemal region; it is formed by apical occluding contacts between tanycytes (Brightman et al. 1975). However, the tanycytes were also considered to bridge the hemal and CSF milieus in the ventricle by transporting neurohormones intracellularly into the ventricle (cf. Leonhardt 1980). So far, no convincing results substantiate this hypothesis.

However, in accordance with their function as a blood–CSF barrier bordering the ventricle, the tanycytes also isolate the hemal milieu of the median eminence from the surrounding intercellular CSF (Fig. 1). The arcuate nucleus in the direct vicinity is one of the most versatile nuclei involved in neurosecretion, and its correct functioning requires the CSF milieu for neuronal inputs. Hence, there must be a functional border between its neuropil and the median eminence. This is provided by two elements: phagocytotically active glial cells, and tanycytes.

Fig. 1. The blood–brain, blood–CSF, and blood–blood barriers in the frontally sectioned median eminence in low and higher magnifications (compare Fig. 2). In the *lower part,* the CSF milieu, consequence of the blood–brain barrier, is indicated by clear coloration, the hemal milieu of the neurohypophysis by gray, and the hemal milieu of the dura mater by black staining. *1–20,* Structures establishing the blood–CSF barrier. *1,* Apical occluding contacts between tanycytes (*2*) providing the barrier between the hemal milieu (*3*) of the median eminence and the ventricular CSF (*4*). *5,* Occluding contacts between perivascular tanycytal endings surrounding perivascular clefts (*6*) of vessels (*7*) which are common to the neurohemal region and the CSF milieu of the arcuate nucleus (*8*). *9,* Phagocytosing glial cell providing further control of the intercellular clefts surrounding the median eminence. *5* and *9* establish the delineation of the hemal milieu of the neurohemal region against the intercellular CSF milieu. In the periphery of the neurohemal region, the outer arachnoid layers (*10*), including the neurothelium (*11*), fuse with the inner arachnoid layers (*12*), forming a secluded frame (*13*) which contains outer CSF (*14*). The space (*15*) between the pial layers ends as an occluded intercellular cleft. The possible communication (*16*) between CSF and hemal milieu along the cleft between the inner pial layers (*17*) and the marginal glia (*18*) is controlled by phagocytosing cells (*19*). *20* Hypothetical occluding contacts between processes of the marginal glia in direct vicinity of the neurohemal region. *21–23,* Structures establishing the blood–blood barrier. From the arachnoid frame (*13*) a continuous layer of meningeal cells (*21*) arises, establishing the blood–blood barrier between the neurohormone-rich milieu (*3*) of the neurohypophysis (*22,* hypophyseal stalk) and the hemal milieu of the dura mater (*23*). *24–29,* Within the median eminence, axons are exposed (*24*) to the perivascular spaces or separated from them (*25*) by tanycytal endings (*2*), which may carry synaptoid contacts (*26*). *27,* Mutual influences between regulatory neuroendocrine axons. *28,* Basal laminae; *29,* pars tuberalis of the adenohypophysis. Modified from Krisch et al. (1983)

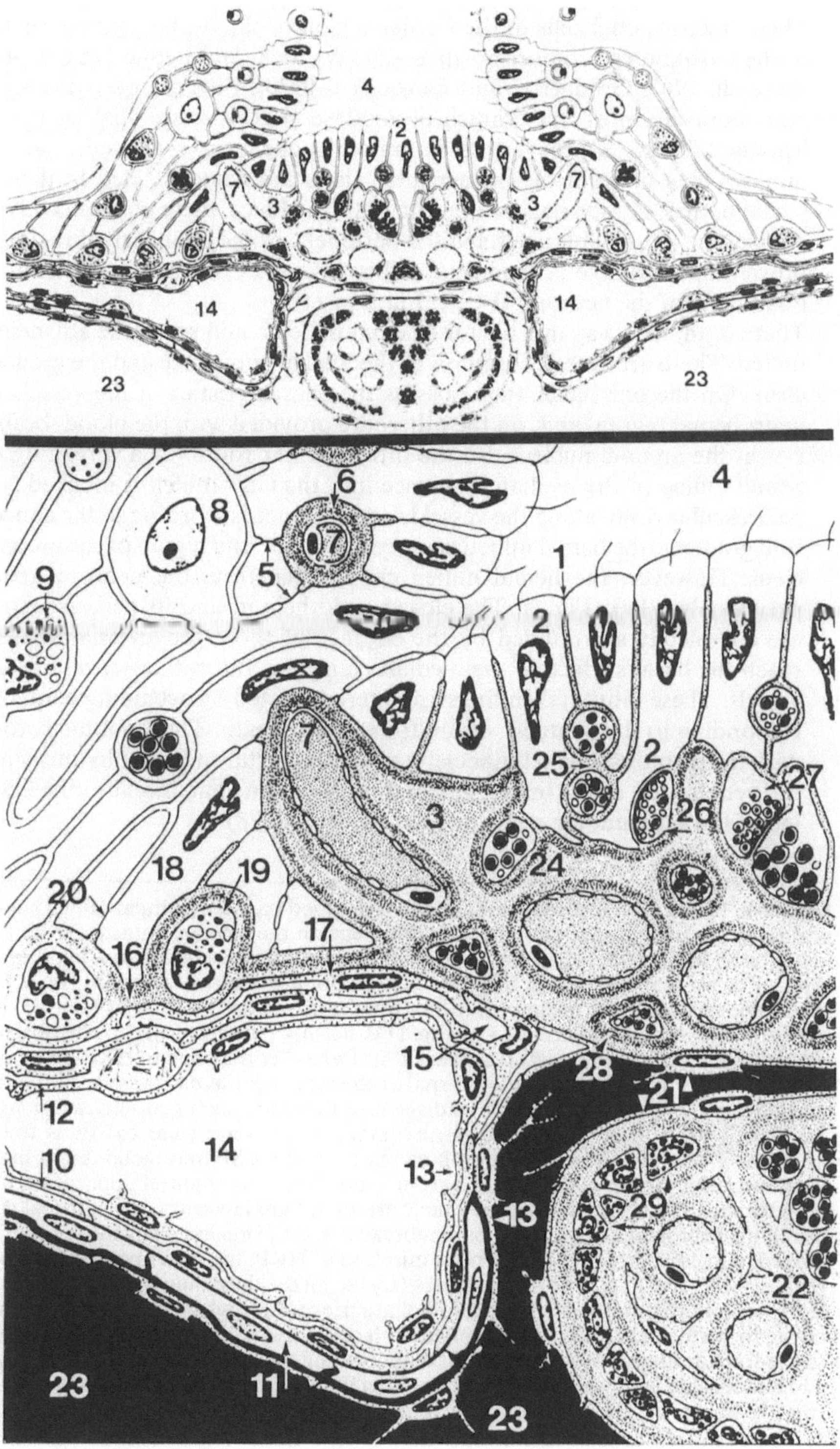
4
2
7
3
7
3
14
14
23
23
6
4
8
07
9
5
1
2
2
7
25
27
26
3
24
18
19
20
17
16
15
28
21
12
14
10
13
13
29
22
23
11
23

1. The numerous glial cells in the border region of the median eminence belong to the well-known Gomori-positive glia (Wislocki and Leduc 1954; Koritsansky et al. 1967; Goldgefter and Korochkin 1970). Five minutes after intravenous application of horseradish peroxidase (HRP), these glial elements are heavily labeled. Moreover, in immunocytochemical specimens, neurohormones taken in from the intracellular clefts are demonstrated in their lysosomal bodies (Krisch and Leonhardt 1980 b). Hence, these glial cells obviously control the intercellular clefts in the periphery of the neurohemal region, thus preventing a leakage of the hemal milieu into the CSF milieu via the intercellular clefts of the neuropil (Fig. 1; Fig. 2 inset).
2. There is another way in which the hemal milieu could reach the adjacent CSF milieu. There are vessels common to the median eminence and the arcuate nucleus. On the one hand, they possess the characteristics of the vessels in the neurohemal region, and, on the other, are provided with the blood–brain barrier in the arcuate nucleus. So the intravascular route for a spreading of the hemal milieu of the median eminence into the CSF milieu is blocked, yet the perivascular route along the vessel is open. Hence, according to the concentration gradient, the hemal milieu can penetrate into the region of the arcuate nucleus. However, the hemal milieu cannot penetrate the neuropil from the perivascular cleft (Fig. 1). The closed endothelium and its perivascular clefts are completely surrounded by the endings of those tanycytes which do not reach the basal surface of the median eminence, but rather terminate around vessels. These tanycytal endings are interconnected by occluding contacts corresponding to the contacts within the myelin sheath. These contacts, together with the tortuous intercellular clefts of the tanycytal endfeet, obviously impede the penetration of the hemal milieu of the median eminence into the CSF milieu of the surrounding neuropil (Krisch et al. 1978).

Fig. 2 a–d. Blood–brain, blood–CSF, and blood–blood barriers (compare Fig. 1). **a** Meningeal cells (*M*) separate the pars tuberalis (*pt*) beneath the median eminence from the dura mater, establishing the blood–blood barrier, and encompass small nerve bundles (*N*). *D*, layer of meningeal cells facing the dura mater. Double-contrasted section. *Inset*, Frontal vibratome section of the median eminence after intravenous injection of HRP. The hemal milieu of the median eminence (*3* in Fig. 1) is heavily labeled owing to the absence of a blood–brain barrier. At the *left side*, black dots (*arrow*) correspond to HRP-phagocytosing glial cells (*9* in Fig. 1), being partly blurred at the *right side* (*arrow*) because of the very low magnification. The neuropil directly adjacent to the median eminence (*clear triangle*) and the covering meninges (*black triangle*) is unlabeled. *V*, ventricular cavity, is free of the tracer, as is the remainder of neuropil, owing to blood–CSF and blood–brain barriers. **b** Demonstration of the blood–blood barrier. Subsequent to epidural application of HRP, only the meningeal layers (*D*) facing the dura mater are labeled; the meningeal cells (*M*) facing the neurohemal region (*pt*, pars tuberalis) and the median eminence proper are unlabeled. **c** In contrast, after intravenous injection of HRP, the tracer penetrates the meningeal cells (*M*) from both sides. The layers (*D*) facing the dura mater are labeled, as are the layers covering a portal vessel, the endothelial surface (*E*) of which is cleared from the tracer owing to perfusion fixation. *Broken lines* in **b** and **c**, reduction of space for montage. **d** A freeze-fracture replica of the meningeal cells covering the neuroadenohypophyseal complex and establishing the blood–blood barrier (*21* in Fig. 1) reveals extended tight junctions. (**a–c** from Krisch et al. 1983; **d** from Krisch and Buchheim 1984) **a** × 2800; *inset,* × 5; **b** × 12675; **c** × 10163; **d** × 44720

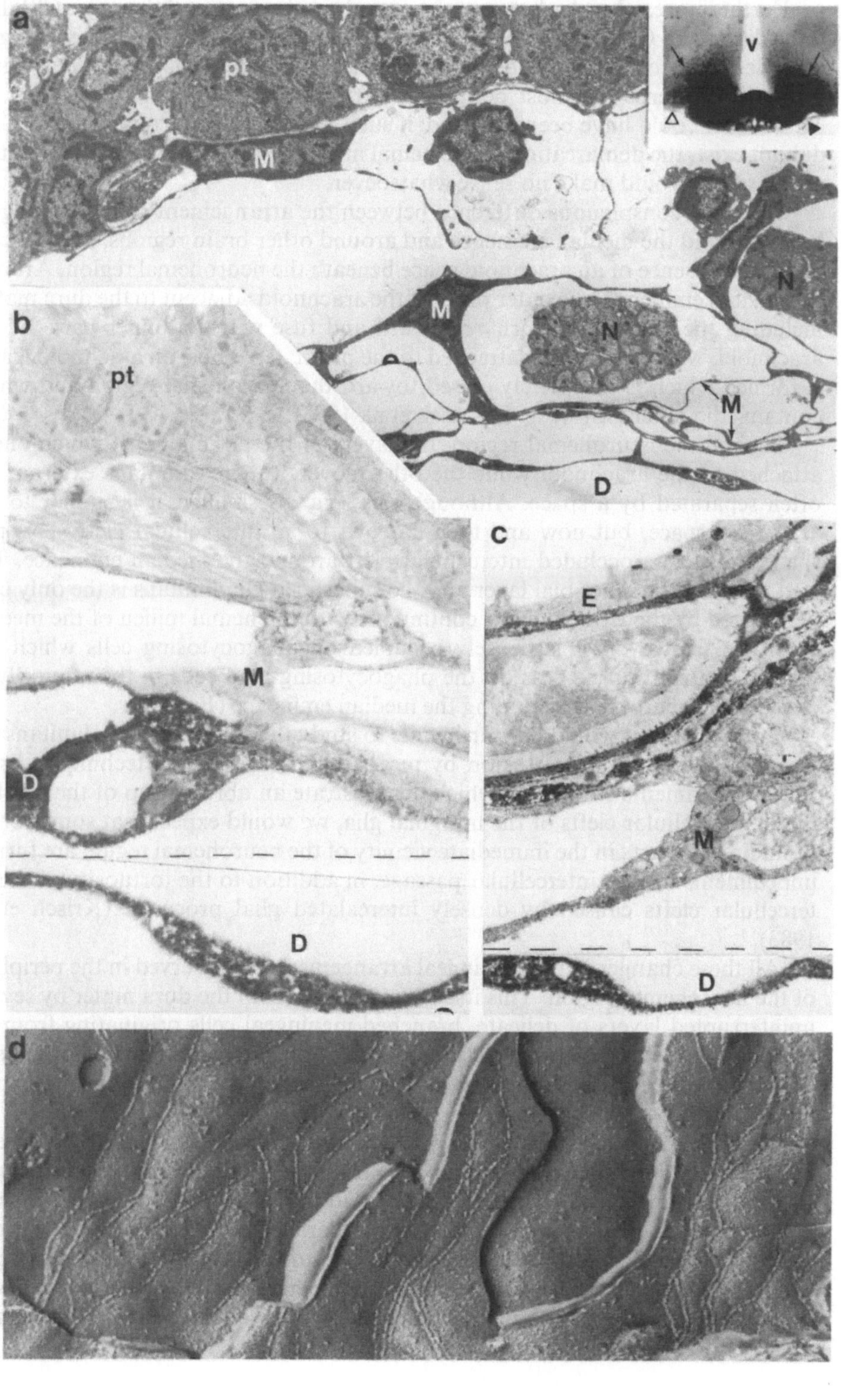
a
pt
M
M
e
N
N
M
D
V
b
pt
M
D
D
c
E
M
D
d

Furthermore, a border between the neurohemal region and the outer CSF has to be discussed. Corresponding to the multilayered architecture of the meninges, a border has been shown to be built up in a graduated fashion, a border which primarily concerns the most peripheral parts of the median eminence (Fig. 1). This border could have been expected: if such a demarcation from the outer CSF did not exist, the demarcation of the hemal milieu from the ventricular and intercellular CSF would make no sense whatsoever.

The most conspicuous difference between the arrangement of the meningeal layers around the median eminence and around other brain regions, i.e., the cortex, is the absence of an arachnoid space beneath the neurohemal region. Around the median eminence, the outer parts of the arachnoid adjacent to the dura mater, including the neurothele, turn brainward and fuse with the inner parts of the arachnoid, which are closely attached to the pia mater. Thus, an arachnoid frame is formed which is completely closed toward the neurohemal region and which contains the outer CSF (Fig. 1; Krisch et al. 1983).

Beyond the neurohemal region, the layers of the pia mater, of which one is attached to the arachnoid while the other covers the glia limitans externa, are often separated by a space. Although this space is smaller, it is similar to the arachnoid space, but now and then narrows to an intercellular cleft. This pial space ends as an occluded intercellular cleft around the median eminence. The cleft between the inner pial layer and the superficial glia limitans is the only cleft dominated by the CSF which is continuous with the hemal milieu of the median eminence. However, it is densely occupied by phagocytosing cells which obviously control this cleft, as do the phagocytosing glial cells in the intercellular clefts of the neuropil surrounding the median eminence (Fig. 1).

Furthermore,it would be worthwhile to study the superficial glia limitans adjacent to the neurohemal region by means of the freeze-etch technique. From tracer experiments with HRP which demonstrate an abrupt stop of the labeling in the intercellular clefts of the marginal glia, we would expect that some sort of occluding contacts in the immediate vicinity of the neurohemal region are further impediments for the intercellular passage, in addition to the tortuosity of the intercellular clefts caused by densely intercalated glial processes (Krisch et al. 1983).

All these changes in the meningeal arrangement are observed in the periphery of the neurohemal region. This itself is separated from the dura mater by several uninterrupted layers of delicate, branched meningeal cells originating from the terminating meningeal layers at the periphery of the median eminence (Fig. 1). The significance of these meningeal cells is illustrated by the following experiments (Fig. 2a–c).

If HRP is injected intravenously, the hemal milieu both of the median eminence and of the dura mater contains the tracer, i.e., the meningeal cells beneath the median eminence are labeled from both sides. However, if the tracer is injected epidurally in a distant region, e.g., in the parietal cortex, the meningeal cells beneath the median eminence are only labeled toward the dura mater (Krisch et al. 1983). Their surface toward the median eminence is unlabeled, as is the neurohemal region proper, thus suggesting occluding intercellular contacts between the meningeal cells as impediments for the penetration of HRP.

Hence, the question regarding the border between the hemal milieu and the outer CSF is not correctly put; it should rather focus on a blood–blood barrier between the hemal milieu of the entire neuro- and adenohypophysis and the hemal milieu of the general circulation in the dura mater, which possesses fenestrated capillaries. In terms of physiological conditions, this means that the meningeal cells beneath the median eminence impede an undirected drainage of neurohormones into the general circulation, or, vice versa, that they prevent a dilution of blood enriched with neurohormones by peripheral blood from the dura mater.

The sheath of meningeal cells continues uninterruptedly along the hypophyseal stalk down to the distal neuro- and adenohypophysis. Results on the meningeal sheath of the adenohypophysis gained by means of the freeze-etch technique demonstrate that the meningeal cells are indeed interconnected by occluding contacts. Their ridges are the more numerous the closer the meningeal cells are located to the dura mater (Fig. 2d).

Thus, the special hemal milieu of the neuro/adenohypophyseal complex is secluded, at least for short time intervals, from all the adjacent milieus – from the ventricular, the intercellular, and the outer CSF, and from the hemal milieu of the general circulation. Ultrashort feedback pathways are possible within the neuro/adenohypophyseal complex, i.e., from the adenohypophysis to the neurohypophysis. From the level of the latter to the brain, however, the idea of a passive diffusion of neurohormones into regions dominated by the CSF should be replaced by the concept of an actively controlled passage of substances. These substances use transcellular pathways (e.g., meningeal cells; Kaplan et al. 1981) or intercellular pathways through intercellular clefts, the intercellular contacts of which are subject to humoral or hormonal influences (Pitelka et al. 1973; Tice et al. 1975; Toshimori et al. 1983).

It is suggested that the neuroendocrine centers are predominantly regulated by two factors: neural inputs on the one hand, and differentiated properties of their endothelial cell membranes on the other. Owing to the selectivity of the blood–brain barrier, these membranes mediate the influences from the different levels peripheral target organs.

3 Neurons Producing Direct and Indirect Regulatory Neurohormones and Their Possible Target Sites

Mostly, the regulatory neuroendocrine neurons appear as spindle-shaped bipolar or triangular elements with a centrally located nucleus. Typically, they are smoothly outlined and exhibit a few, poorly branched processes. The perikarya are always smaller than those of the effectory neuroendocrine neurons. Now and then, collateral processes run from the regulatory neuroendocrine soma to the intercellular clefts of cuboid ependymal cells (Krisch 1981 b). In mammals, there is no convincing evidence that those fibers reach the ventricular surface. Generally, perikarya producing a distinct neurohormone all look alike, with the exception of LRF-producing cells (Fig. 3).

If the regulatory neuroendocrine perikarya are dispersed in the hypothalamic gray, they most frequently occupy the preoptic region or the caudally following subependymal neuropil; if they share hypothalamic nuclei, the paraventricular and arcuate nuclei are mostly concerned (see Tables 1 and 2).

3.1 Ultrastructure of Neurons Producing Regulatory Hormones

The ultrastructure of the various regulatory neuroendocrine neurons is rather uniform, differing at most in the size and electron density of granules in which the hormones are condensed (Fig. 4). Like any other neuron they project into very distant regions of the brain, mostly those belonging to structures of the limbic system, and it is conceivable that they are the site of afferent synapses attributable to the different transmitter systems and functional circuits, as described for the LRF system (synapses on LRF-immunoreactive perikarya: Hisano et al. 1981; Kiss and Halász 1985; Silverman and Witkin 1985; transmitters: dopamine and GABA, Jennes et al. 1983; dopamine, Hoffmann et al. 1982; GABA, Léránth et al. 1985; serotonin and norepinephrine, Jennes et al. 1982a). At least in the rat and in man, as already mentioned, two morphologically distinct forms of LRF-immunoreactive perikarya are observed (Krisch 1980b): one is bipolar with smooth contours, while the other is apparently unipolar with irregular outlines. According to Jennes et al. (1985), in the bipolar perikarya, smoothly outlined perikaryal synapses occur predominantly on the trunk of either process, whereas in the irregularly outlined cells the synaptic input is usually related to cytoplasmic protrusions present on the soma and on the initial axon segment.

3.1.1 Soma and Axon

Like any other active secretory cell, the somata of regulatory neuroendocrine neurons display nuclei with clear chromatin and one or two nucleoli, as well as one or more medium-sized Golgi areas per section, around the periphery of which neurosecretory "elementary" granules are found. The rough endoplasmic reticulum is rather randomly oriented and is rarely arranged in parallel stacks. Lysosomal bodies are always observed, and most of them display a spotty, positive im-

Fig. 3a–e. Light microscopic view of regulatory neuroendocrine neurons. **a–c** LRF-immunoreactive perikarya in the preoptic area of the rat occur **a** in two forms, **b** and *inset* as irregularly outlined cells, and **c** as smoothly contoured perikarya. *V*, preoptic recess of the third ventricle. From Krisch (1980b). **d** Somatostatin system. The parasagittal section through the mediobasal hypothalamus reveals the relationship between at least the most caudal group of somatostatin-immunoreactive perikarya (*asterisk*), the heavy projection into the neurohypophysis (*ME*, median eminence; *ST*, hypophyseal stalk), and the projection into the arcuate (*star*) and mamillary (*triangle*) nuclei. **e** Slightly rostral and dorsal from **d** in the same section, the main assembly of somatostatin-immunoreactive perikarya is located. **a** × 140; **b** *inset*, and **c** X 560; **d** × 40; **e** × 220

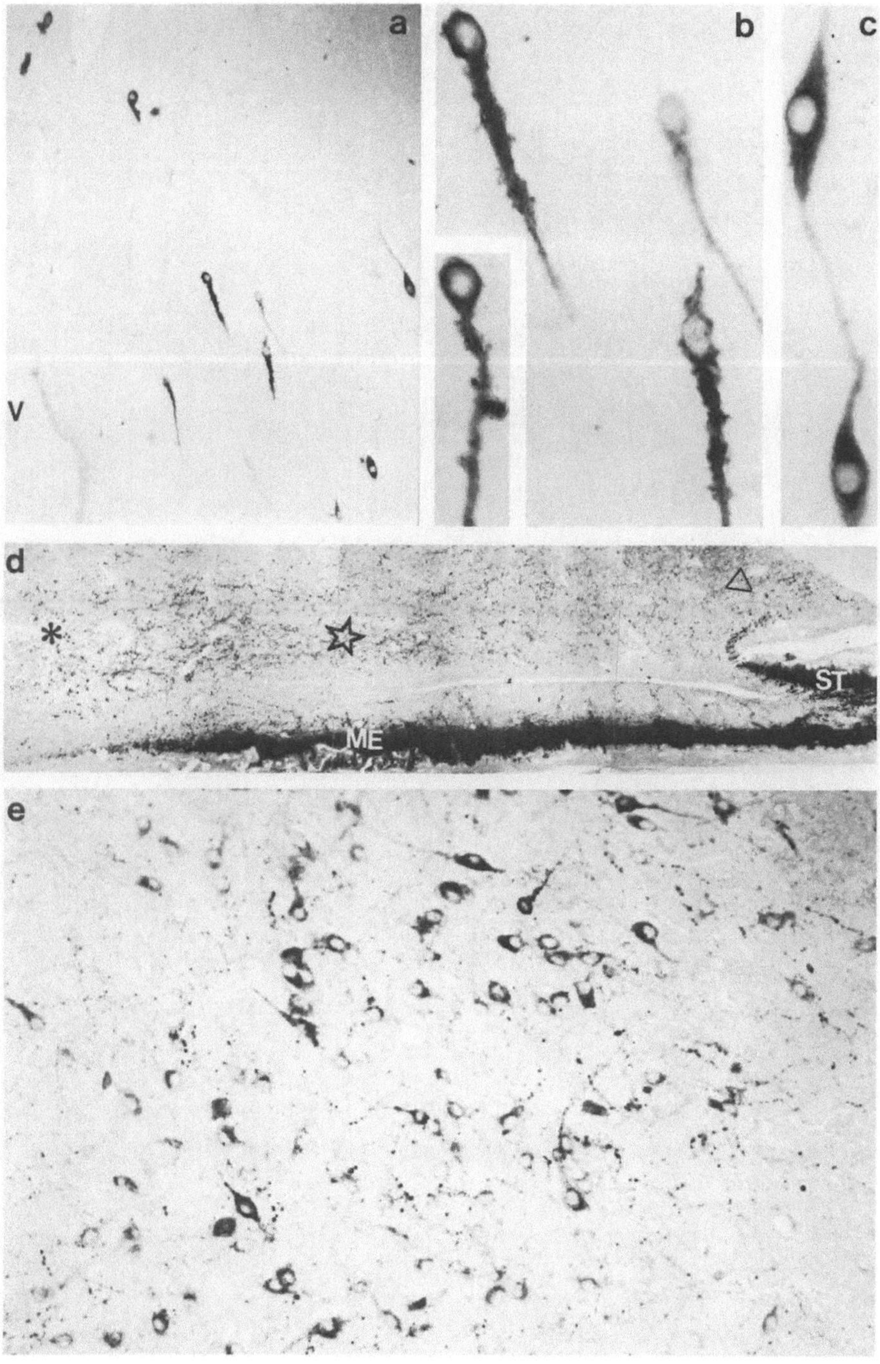
a
b
c
V
d
ST
ME
e

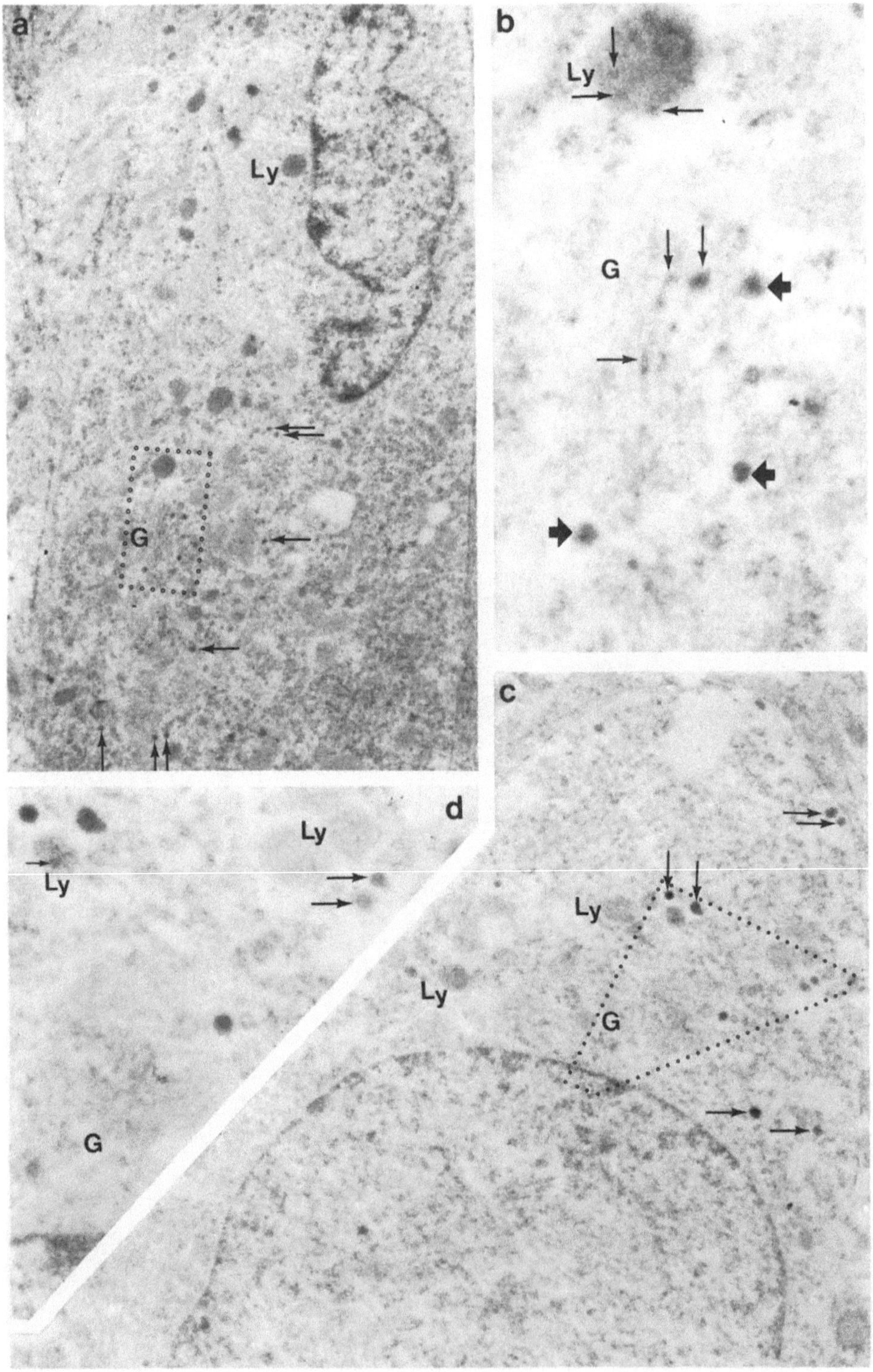
a
Ly
G
b
Ly
G
c
Ly
Ly
G
d
Ly
Ly
G

munoreaction to the respective regulatory neurohormone produced in the cell (Fig. 4). Compared with the rather detailed knowledge on the vasopressin neuron regarding, for example, its biosynthesis and biochemistry, genetic defects, increased or decreased function, and ultrastructural correlates in the vasopressin-producing cell, hardly any information on these topics is available for the regulatory neuroendocrine neurons. However, something can be learned from the comparison between magnocellular effectory neuroendocrine neurons and those smaller neurons synthesizing regulatory neurohormones. In the latter, the number of parallelized cisterns of rough endoplasmic reticulum, the extent of the Golgi areas, and in particular the number of hormone-containing granules, seem to be smaller even taking into account their smaller cell size as compared to vasopressin-containing neurons, although no systematic quantitative analysis has been done. In contrast to the effectory neuroendocrine neurons and in parallel to the low number of dense-core vesicles in the periphery of the regulatory neuroendocrine somata, the axon terminals of the smaller neurons synthesizing regulatory neurohormones lack storage capability and capacity.

However, like vasopressin and oxytocin the regulatory neurohormones are synthesized on the ribosomes of the rough endoplasmic reticulum as large precursor polypeptides (Brownstein et al. 1982; Brownstein 1985; Gainer et al. 1985; Richter 1985). By definition, preprohormones carry the signal peptide at their NH_2-terminus, which is essential only for the entry of the precursor into the cisterns of rough endoplasmic reticulum and which is then cleaved off, resulting in the prohormone. Further post-translational steps can include amidation at the N-terminal side, glycosylation, disulfide bridge formations and possibly phosphorylation, and sulfatation.

The prohormone is transported to the Golgi cisterns probably by a shuttle procedure in small vesicles observed in the spaces of the cisterns of the rough endoplasmic reticulum and the Golgi apparatus (Brownstein 1985). It mainly contains different biologically active peptides separated from one another by two

Fig. 4a–d. Ultrastructural characteristics of regulatory neuroendocrine perikarya in general, after postembedding immunocytochemical staining. **a** LRF-immunoreactive smoothly outlined perikaryon in the preoptic area of a male rat. The small immunoreactive granules are partly indicated by *arrows;* some lysosomes (*Ly*) can easily be seen owing to their dark coloration. *G*, Golgi area. The *dotted rectangle* is shown in higher magnification in **b**. **b** The higher magnification demonstrates pentagonal-shaped peroxidase–antiperoxidase (PAP) complexes (*arrows*) on the lysosome (*Ly*) and a Golgi area (*G*), the inner secretory cisterns of which are clearly labeled (*arrows*), as are the small elementary granules abutting from it (*arrowheads*). **c** In contrast, in this somatostatin-immunoreactive perikaryon, the few elementary granules of which are indicated by *arrows,* the Golgi area (*G*) can hardly be seen in the low magnification and the lysosomes (*Ly*) appear unstained. The *dotted triangle* is shown in higher magnification in **d**. **d** The higher magnification reveals that in this case neither is the Golgi area (*G*) immunoreactive, nor do the lysosomes (*Ly*) exhibit and positive staining beyond at most a positive inherent peroxidase reaction (*single arrow*). Moreover, obviously not all elementary granules are immunoreactive; two of them are negative (*arrows*), either owing to the loss of their antigen during tissue preparation or because they contain another neurohormone. **a** × 11400; **b** × 55440; **c** × 12300; **d** × 28500 basic amino acids which are cleaving sites for endopeptidases, and is packed in the Golgi apparatus into secretory vesicles budding off from the Golgi cisterns and migrating to the cell periphery (opioids, i.e., enkephalins, endorphins, dynorphins, Beaumont 1983; LRF, Curtis and Fink 1983; Seeburg and Adelman 1984; growth hormone-releasing factor/somatocrinin, Mayo et al. 1985; somatostatin, Lauber et al. 1979; Böhlen et al. 1980; Zingg and Patel 1980). Depending on the system, the earliest cleaving steps can occur within the Golgi apparatus or at least within the perikaryon directly after the formation of secretory vesicles. Thus Guy et al. (1985) could demonstrate in light and electron microscopy by means of specific antisera to somatostatin 28, to the C-terminus of somatostatin 28 (1–12), and to the eight N-terminal amino acids of somatostatin 14, that (subsequent to colchicine treatment) the somatostatin-producing perikarya in the wall of the third ventricle are not only immunoreactive to somatostatin 28, but also to somatostatin 28 (1–12) and, most probably, to somatostatin 14. This means that in this system post-translational cleaving partially occurs within the perikaryon.

In contrast, the study of King and Anthony (1983) illustrates and explains the well-known difficulties of immunohistochemically staining the LRF-producing perikarya in the preoptic region: as inferred from this study, neither the N-nor the C-terminus of the decapeptide LRF is free within the perikaryon, but the molecule is extended at both sides. Here, as in other systems, the secretory granule is the major site for the proteolytic processing of prohormones during axonal transport (Loh et al. 1984).

It is not yet definitely known whether the processing enzymes are segregated and packaged together with the prohormone or whether the enzymes and their substrates are first contained in different populations of granules which fuse in a second step. Proteolytic processing frequently leads to a spectrum of biologically active peptides which thus belong to a common peptide family when they are cleaving products of a single preprohormone. Owing to the fact that repetitive sequences or pairwise arrangements of closely related or identical sequences are frequently observed (Sundler et al. 1985), a common precursor of different (biologically active) peptides can be suspected if two peptides share significant sequence homologies. Even though the velocity of fast axonal transport is rather constant, the quantitative relationships of the members of such a family may still be altered because of afferent influences on the perikaryon, resulting in altered translational and post-translational steps and/or altered equipment of the granule membrane with cleaving enzymes. The significance of preprohormone derivatives released from the same axon terminals is illustrated by very recent preliminary results of Seeburg and colleagues (Nikolics et al. 1985; Phillips et al. 1985), which suggest that the carboxy-terminus of the LRF precursor, contained in, transported along, and released from, the LRF-immunoreactive neurons and their axon terminals, acts as a prolactin-inhibiting factor (Fink 1985).

The heterogeneity of peptide hormones in many regulatory neuroendocrine neurons results not only from the heterogeneity of cleavage products of a common precursor, but also from the coexistence of unrelated peptides deriving from different prohormones and of peptides and monoamines within a single neuroendocrine perikaryon [CRF and dynorphin, Roth et al. 1983; CRF and PHI (peptide which has histidine and isoleucine as terminal amino acids); vasoactive intestinal polypeptide (VIP) and met-enkephalin, Hökfelt et al. 1983; CRF and vasopressin, Tramu et al. 1983; Sawchenko et al. 1984; Wolfson et al. 1985; LRF and pro-opiomelanocortin derivatives, Tramu et al. 1977; Leonardelli and Tramu 1979; Beauvillain et al. 1981; neurotensin and dopamine, Hökfelt et al. 1984; growth hormone releasing factor and dopamine, Okamura et al. 1985]. In view of the plasticity (Wolfson et al. 1985) of neurohormonal perikarya, the coexistence of unrelated neurohormones seems to pose fewer problems for the synthesis than for the regulation of release and the function of the coexisting substances. However, the fact must not be disregarded that the coexistence of unrelated peptides is usually demonstrated by light microscopic – less frequently by electron microscopic – immunocytochemistry. Hence, only "immunoreactivities" can be described, and the results have to be regarded as preliminary as long as they are not corroborated by functional methods.

3.1.2 Axon Terminal and Release Site

The axon terminals of neuroendocrine neurons are the structures which determine the particular position these neurons have within the central nervous system. The axon terminals containing the different regulatory neurohormones in dense-core vesicles are densely packed in the floor of the third ventricle, i.e., the proximal neurohypophysis or median eminence, close to the portal vessels. Groups of axons are incompletely separated from one another by tanycytal processes, whose somata establish the blood–CSF barrier with the ventricular CSF, and whose endings separate the axon terminal more or less from the perivascular spaces of the fenestrated capillaries (Fig. 5). Nerve fibers rather frequently exhibit synaptic contacts with tanycytal processes in the median eminence, a finding which further illustrates experimental evidence according to which the extension of tanycytal endings separating the regulatory neuroendocrine terminals from the perivascular spaces is subject to and dependent on the endocrine status (Wittkowski and Scheuer 1974). However, it has to be emphasized that the basal lamina of the neuropil always remains preserved uninterruptedly. Even when neuroendocrine processes penetrate far between the submeningeal vessels, as somatostatin-immunoreactive fibers frequently do, they are overlaid by a complete basal lamina separating the neuropil from the perivascular spaces.

A role corresponding to that of tanycytes can be attributed to the pituicytes, in particular, as they receive synapselike contacts from enkephalin-immunoreactive fibers (Van Leeuwen 1982 a, b). Like tanycytes, the pituicytes seem to regulate the extent of the neurohemal contact zone: in the homozygote untreated Brattleboro rat they appear retracted, exposing the axon terminals to the perivascular spaces. After desmopressin (vasopressin analog) therapy, however, the axon terminals are largely encompassed by pituicyte processes which separate them from the perivascular spaces (R. Martin, Ulm, personal communication).

The obviously parallel organization of the median eminence and the neural lobe should remind one that both are parts of a single "organ," the neurohypophysis. On the basis of the history of neurosecretion, this has been subject to an artificial functional separation. Both parts of the neurohypophysis, the median eminence and the neural lobe, develop ontogenetically from a common anlage.

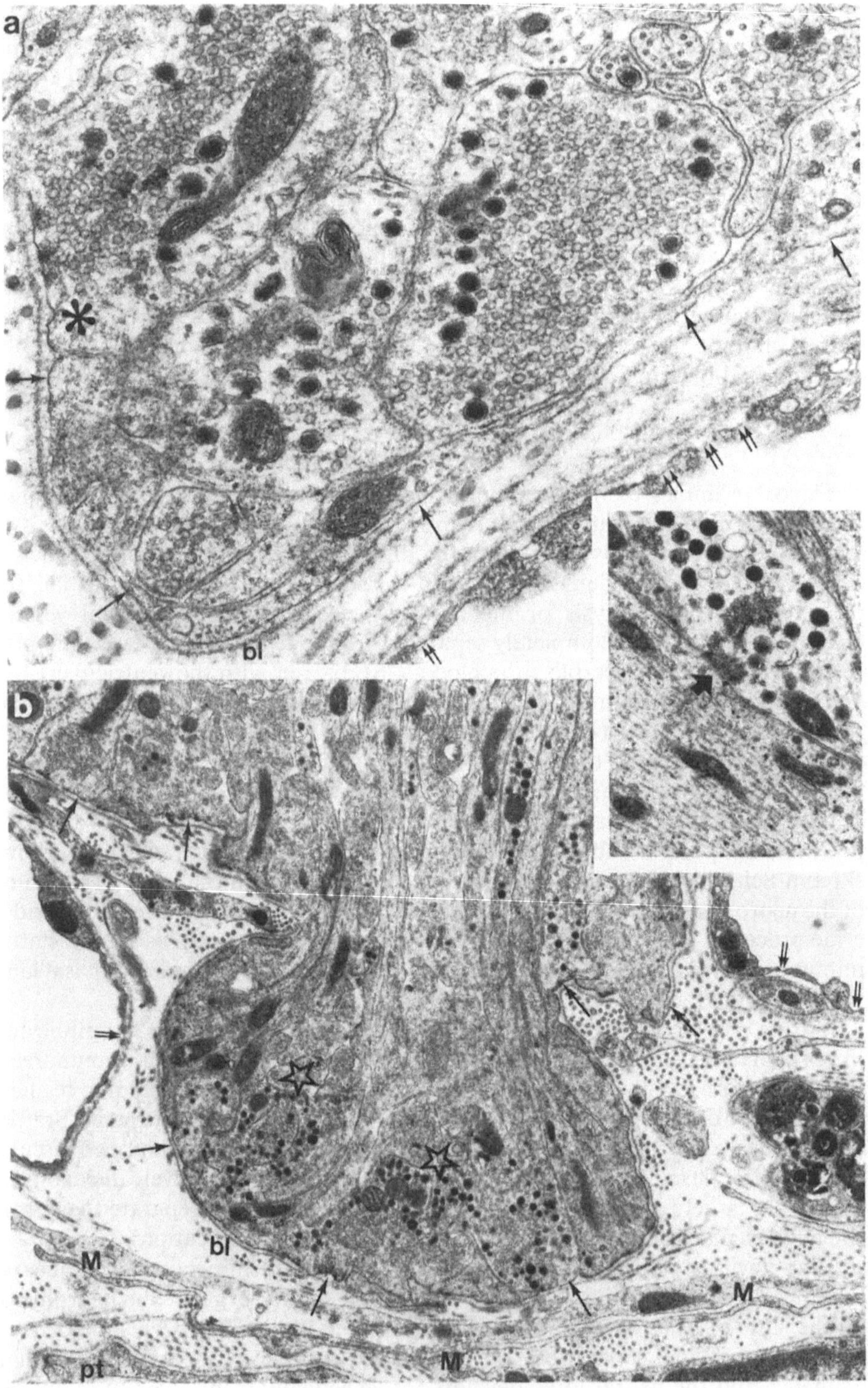
a
b
bl
bl
M
M
M
pt

Common to both is the basic histologic organization, both being structurally coupled to the adenohypophysis by vascular connections. Since regulatory neurohormones occur in the neural lobe [somatostatin, Pelletier 1980; met-enkephalin together with oxytocin and cholecystokinin (CCK), Martin and Voigt 1981; Martin et al. 1983; vasopressin and dynorphin precursors, Martin and Voigt 1982; Voigt and Martin 1985; dopamine, Moore and Demarest 1982], it should be considered a possible release site of neurohormones directed to the adenohypophysis.

The following examples support this proposal:

1. Results gained by Baertschi et al. (1980) strongly indicate that vasopressin of the neural lobe modulates ACTH release.
2. According to Froehlich and Ben-Jonathan (1984), in the rat a substance deriving from the neural lobe participates in the regulation of prolactin and LH secretion, but not FSH secretion, during the estrous cycle.
3. In the entire neurohypophysis and in the intermediate lobe of the adenohypophysis, dopamine-containing terminals have been demonstrated (Moore and Demarest 1982; Holzbauer et al. 1983). In the proximal neurohypophysis, the median eminence, their main function is that of a prolactin-inhibiting factor (Moore and Demarest 1982), whereas in the intermediate lobe they are thought to regulate the release of pro-opiomelanocortin derivatives (Tilders and Smelik 1977; Farah et al. 1982) by neuroglandular contacts. In the distal neurohypophysis, the neural lobe, dopamine terminals are found close to neurosecretory axons and pituicytes, but also close to the perivascular spaces. To judge from experiments with dopamine antagonists and agonists comparing their effects on the intermediate lobe and the striatum with those on the median eminence and the neural lobe, results obtained by Lookingland et al. (1985) indicate that the dopamine terminals in the neural lobe do not necessarily influence the effectory axon terminals or pituicytes, but that dopamine might be released into the bloodstream and exert its function as a regulatory neurohormone.
4. According to Frawley et al. (1985), oxytocin selectively attenuates TRF-induced thyroid-stimulating hormone (TSH) release from pituitary cells without affecting the prolactin response to TRF.

Some of the axon terminals of the different regulatory neurosecretory neurons occupy particular regions in the median eminence (Fig. 6), as do the LRF fibers,

Fig. 5 a, b. Release site of regulatory neurohormones: relationships between axon terminals, tanycytal endings, and perivascular spaces in the median eminence. **a** Most axon terminals are separated from the perivascular space of the fenestrated capillary (fenestrae, *double arrows*) by tanycytal endings (*arrows*). Only one fiber (*asterisk*) has free access to the perivascular space. *bl*, basal lamina. **b** The neuropil, including its basal lamina, protrudes up to the meningeal cells (*M*) separating the median eminence from the adenohypophyseal pars tuberalis (*pt*). Tanycytal endings encompass small groups of axon terminals (*stars*). As compared to **a**, it is clear that more axon terminals abut directly (*arrows*) on the perivascular spaces of the fenestrated capillaries (fenestrae, *double arrows*). *Inset* Synaptoid contact (*arrowhead*) of a peptide-containing axon with a tanycyte. **a** × 33 260; **b** × 15 200; *inset*, × 18 050

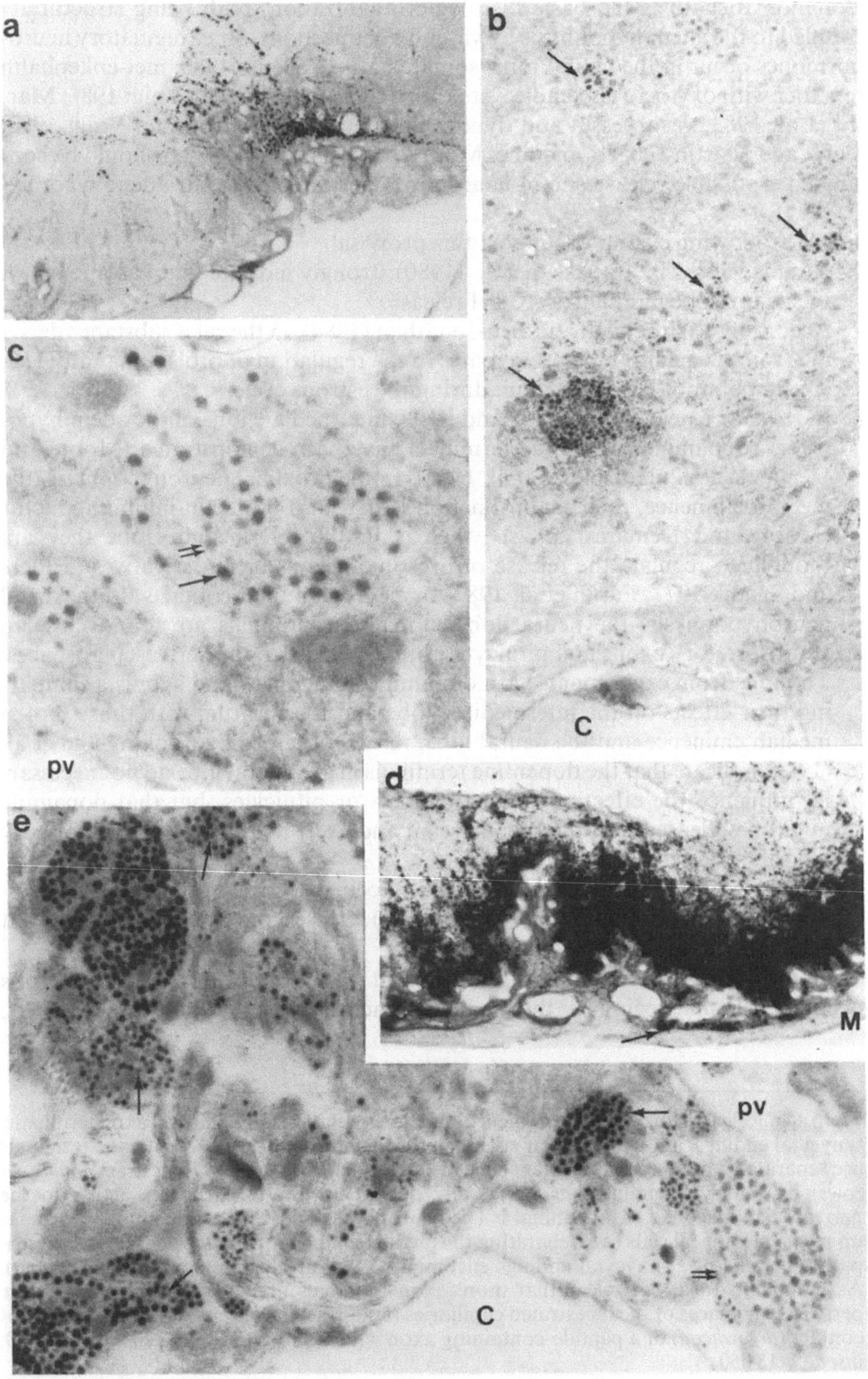
a
b
c
pv
C
d
M
e
pv
C

which are mainly restricted to its lateral parts; only in its most rostral pole LRF fibers are evenly distributed over the whole width of the organ. The fiber tracts of most of the other systems are rather evenly distributed over the bilateral extension of the ventricular floor. They are situated just beneath the fiber tracts of the effectory neurohormones vasopressin and oxytocin; for example, somatostatin (Krisch 1980a), TRF (Choy and Watkins 1977), CCK/gastrin (Vanderhaeghen et al. 1980), neurotensin (Jennes et al. 1982b), CRF (Pelletier et al. 1982), serotonin (Warembourg and Poulain 1985), etc. (Fig. 7). Somatocrinin-immunoreactive fibers are also found all over the bilateral extension of the median eminence, just adjacent to the perivascular spaces (Lechan et al. 1984). Pro-opiomelanocortin derivatives are mainly resricted to the subependymal and adjacent internal layers of the median eminence; immunoreactive fibers close to the perivascular spaces are rather sparse (Bloom et al. 1980; Kiss et al. 1985). This latter finding suggests that the perikarya containing pro-opiomelanocortin derivatives exert less influence as neuroendocrine neurons than as structures for possible neuronal feedback pathways to the arcuate nucleus (Fig. 7a, b).

Before the development of immunohistochemistry, neuroendocrinologists frequently described the presence of Gomori-positive material near the perivascular spaces of the median eminence (cf. Leonhardt 1980). The amount of this material increased after adrenalectomy. From these results, the hypothesis on the participation of vasopressin in the stimulatory regulation of ACTH cells developed (cf. Leonhardt 1980). Following immunostaining of the median eminence with antibodies against vasopressin, it becomes evident that from the main tract of effectory hormone-containing axons in the internal zone fibers project to the perivascular spaces. Ultrastructurally, some terminals are found here which abut on the perivascular spaces of fenestrated capillaries and which contain vasopressin-immunoreactive granules clearly smaller than those in the internal zone of the median eminence (Krisch 1980a). These early results are further explained by studies visualizing the onset of vasopressin synthesis in the parvocellular perikarya of the paraventricular nucleus following adrenalectomy (Tramu et al. 1983;

Fig. 6a–e. Direct regulatory neurohormones (**a–c**, LRF, **d–e**, somatostatin) in the median eminence. **a** By frontal light microscopic sectioning of the median eminence, the restriction of LRF-immunoreactive fibers to its lateral part is visualized. **b** Low electron microscopic magnification of the lateral pericapillary region (*C*, capillary) of the median eminence, demonstrating several rather small immunoreactive fiber profiles (*arrows*). **c** Higher magnification of an immunoreactive axon close to the perivascular space (*pv*), showing the differing immunoreactivity of the granules within the same fiber profile (*arrow* and *double arrow*), the result either of the loss of the antigen during tissue preparation or of the presence of another granule population. **d** Frontal light microscopic section of the median eminence, with incubation with antisomatostatin; it demonstrates the dense arrangement of immunoreactive fibers, some of which (*arrow*) protrude up to the meningeal layers (*M*). The fiber density in the middle part of the median eminence is somewhat less than in its lateral parts. **e** Ultrastructurally numerous immunoreactive fibers abut on the perivascular spaces (*pv; C*, capillary), obviously heterogeneous as regards granule diameter. However, a few granules within the same terminal are always less immunoreactive (*arrows*) than the rest. *Double arrow*, nonimmunoreactive axon profile. **a** × 140; **b** × 6640; **c** × 27500; **d** × 220; **e** × 11700

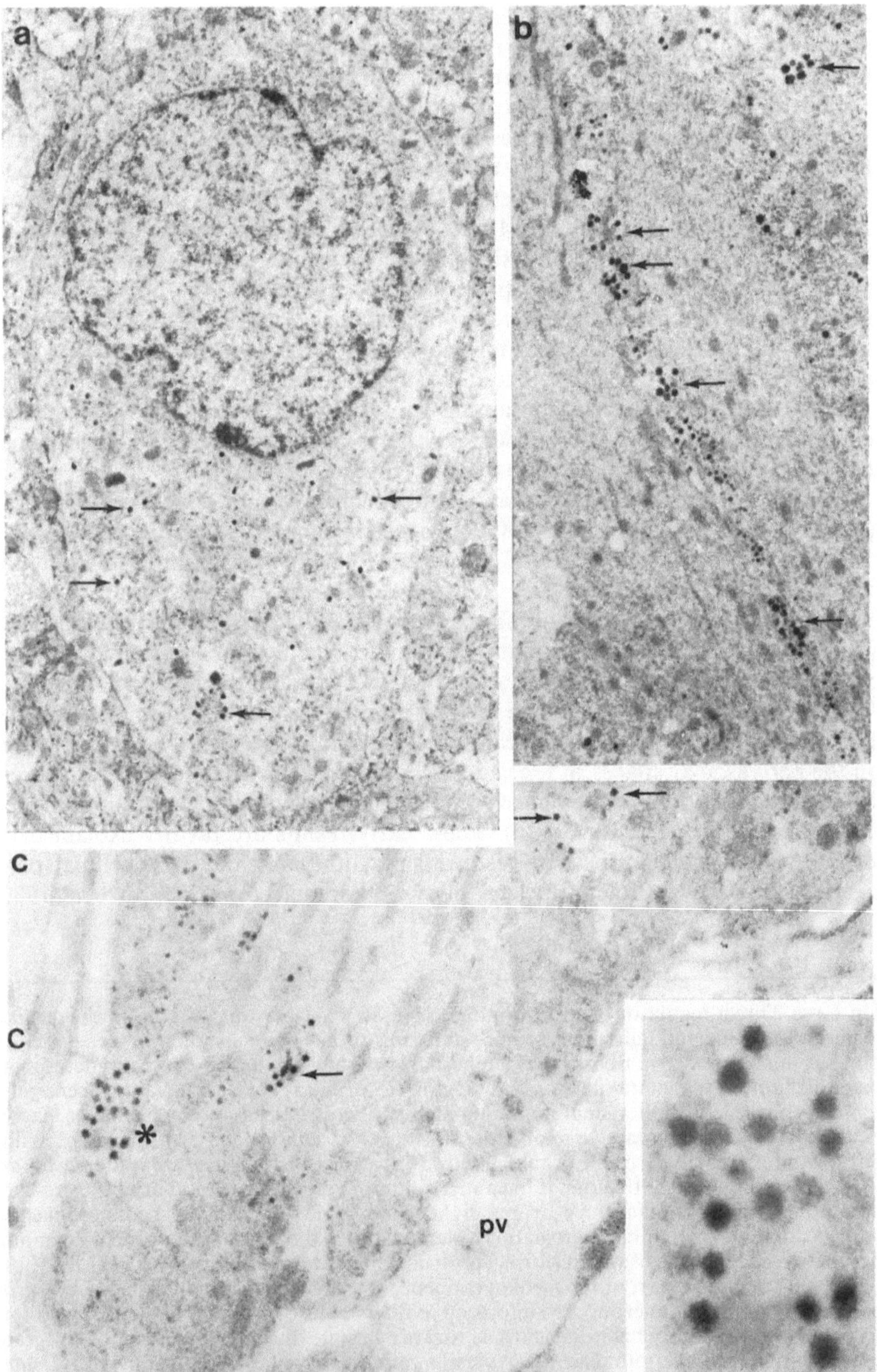
a
b
c
pv

Sawchenko et al. 1984), and are further underscored by the demonstration of vasopressin-mRNA in CRF-immunoreactive neurons following the same experiment (Wolfson et al. 1985).

These last-mentioned findings indicate a considerable plasticity of the regulatory neuroendocrine neuron in the adult animal (Purves and Hadley 1985), depending on the endocrine status. In this connection, results should be considered on the occurrence of several biologically active cleavage products of a single prohormone in axon terminals of the median eminence and in the portal blood – e.g., somatostatin 28 and 14 are seen by light (Lechan et al. 1983) and electron (Guy et al. 1985) microscopy to occur in the same granule. Moreover, somatostatin 28 binding sites have been localized by ultrastructural autoradiography on adenohypophyseal cells (Morel et al. 1985), illustrating biochemical results on its selective release (Bakhit et al. 1984; Charpenet and Patel 1985) into the portal blood vessels (Patel et al. 1981; Millar et al. 1983) and on the high potency and long-lasting effect of somatostatin 28 as compared to somatostatin 14 (Brown et al. 1981; Shaffer Tannenbaum et al. 1982). Thus, depending on the afferent influences on the regulatory neuroendocrine neuron, the pattern of substances released from its axon terminals in the median eminence can change, resulting in an enhanced function if the effect of the substances is additive, or in an altered function if the different substances have different biological effects.

When staining any regulatory neuroendocrine system by means of the postembedding immunocytochemical method, it becomes obvious that some equatorially sectioned granules remain unstained within one and the same terminal. This can be assumed to be an indirect indication of the presence of another substance (peptide/monoamine) within this terminal, packed in another population of granules. However, direct evidence of the occurrence of different peptides is given exclusively for their occurrence in the same granule, e.g., for somatostatin 14 and somatostatin 28 (Guy et al. 1985) or for LRF and a pro-opiomelanocortin derivative (Beauvillain et al. 1981). Certainly, ultrastructural double-staining methods will result in more numerous examples of the coexistence of substances in one or in different populations of granules within the same axon terminal.

The occurrence of more than one neurohormone in an axon terminal again leads to considerations and hypotheses about the possible function of biologically active substances which, in case of occurrence in the same granules, should necessarily be released simultaneously. Regarding the release of PHI, enkephalin,

Fig. 7 a–c. Indirect regulatory neurohormones in the mediobasal hypothalamus. **a** β-endorphin-immunoreactive perikaryon in the arcuate nucleus (antibody 19646-9/10/78, courtesy of Dr. Dubois, Nouzilly, France); immunoreactive granules are partly indicated by *arrows*. **b** β-endorphin-immunoreactive fiber profiles (*arrows*) in the internal zone of the median eminence. **c** Some serotonin-immunoreactive fiber profiles (*arrows*) abut on the perivascular spaces of the median eminence (*C*, capillary; *pv*, perivascular space; antibody courtesy of Dr. Steinbusch, Amsterdam, The Netherlands). The axon indicated by an *asterisk* is shown in higher magnification in the *inset*. As in other systems (cf. Figs. 4 and 6), the granules differ in immunoreactivity either as a result of the loss of their antigen during tissue preparation or of the presence of different granule populations within the same axon terminal. **a** × 8750; **b** × 17710; **c** × 14260; *inset*, × 50600 and CRF, Hökfelt et al. (1983) proposed on the basis of biochemical and physiological findings that enkephalin would inhibit dopamine release from adjacent axons, thus decreasing the prolactin-inhibiting factor dopamine, whereas PHI is reported to increase the release of prolactin in vitro. Simultaneous release would thus result in an additive effect on the prolactin cell. Thus, indirect regulatory neurohormones could modulate the release of direct regulatory neurohormones in a promoting/inhibiting way via axoaxonal influences.

A comparison between the regulatory neuroendocrine neuron and any other final efferent neuron in the central or peripheral nervous system shows that both receive multivarious inputs, being integrated in different neuronal circuits, and that the remainder of efferent neurons convey their signals – qualitatively uniform, but varying in frequency – to specific target cells in a more or less directed manner via the intercellular clefts. In contrast, the regulatory neuroendocrine neuron conveys its signals, which may change qualitatively, in an undirected manner via the bloodstream. It has not yet-been determined whether the adenohypophysis is the only direct target. As in other efferent neurons, the regulatory hormone/hormones in the neurosecretory neuron are released quantally in a calcium-dependent way (Mulder 1982), but as has been mentioned before, the pattern of quantally released substances may change. In addition, the lack of directedness may be compensated for by a number of mechanisms and involved structures which participate in regulatory functions, which are subject to neurohormonal influence, and which are thus possibly direct targets for regulatory neurohormones. On the one hand, these structures are situated on the level of the regulatory hormone-releasing axon terminal – the participation of tanycytal endings has already been mentioned as one of several examples. On the other hand, these structures are situated on the level of the target region, in the adenohypophysis.

In other words, the influence of regulatory neurohormones on the adenohypophysis may be mediated by structures either inherent in the architecture of the release site of the regulatory hormones or inherent in their final target region. Hence, when describing the ultrastructure of regulatory neuroendocrine neurons in order to understand their concerted function, a description of the structures participating in the regulation of adenohypophyseal activity has to be included.

3.2 Ultrastructure of Possible Targets Influenced by Indirect Regulatory Neurohormones

In this section, possible targets for indirect regulatory neurohormones will be discussed. Although they are mere hypotheses supported by a few indirect indications, these considerations can perhaps stimulate further research on the confusing functional multifariousness of indirect regulatory neurohormones (cf. Table 2), particularly as the structures discussed as targets for indirect regulatory neurohormones are integral elements of the neuro/adenohypophyseal complex.

These structures include (a) the glial elements in the entire neurohypophysis (i.e., tanycytes which are glial derivatives in the median eminence, and pituicytes in the neural lobe); (b) the transport routes of regulatory neurohormones, i.e., mainly the portal vessels and the capillaries of the adenohypophysis; and (c) the structures which finally determine the extent of distribution of regulatory neurohormones in the adenohypophysis, i.e., stellate cells and intercellular contacts.

The possible role of tanycytes and pituicytes as targets of regulatory neurohormones has alredy been mentioned in Sect. 3.1.2.

The fenestrated capillaries of the entire neurohypophysis are connected via long and short portal vessels with the adenohypophysis, which does not possess its own arterial supply but is completely dependent on these long and short portal vessels (Murakami 1975; Murakami et al. 1985). Earlier results obtained from vascular casts of the adenohypophysis (Adams et al. 1964, 1969; Daniel 1966; Porter et al. 1972) indicate that the complete vascular bed of the gland is evidently not always perfused to the same degree. In addition, short-term experiments with intravenously injected HRP demonstrate that some adenohypophyseal capillaries are not perfused within the 5 min following injection of the tracer, whereas the neighboring ones do contain the tracer (Fig. 8; Krisch and Buchheim 1984). It can be assumed that the portal vessels differ in their caliber owing to innervation and possibly to vasoactive substances. These can reach the smooth-muscle cells from the vessel lumen, being secreted into capillaries of the neurohypophysis. Possible candidates as mediators of vasoconstriction or -dilatation are serotonin (Léránth et al. 1983; Holzbauer et al. 1985), abutting on numerous terminals at the perivascular spaces of the median eminence (Fig. 7c); VIP and substance P (Stoeckel et al. 1982; Tsuruo et al. 1983; Rønnekleiv et al. 1984; Edvinsson 1985); and epinephrine of hypothalamic origin released into the portal vessels (Gibbs 1985). Hence, as regards the vascular connection between neuro- and adenohypophysis, the smooth-muscle cells of the portal vessels might be a target for indirect regulatory neurohormones (Page and Dovey-Hartman 1984).

Moreover, the endothelia of the adenohypophyseal capillaries may indirectly participate in the regulation of the activity of the gland. It is generally accepted that a fenestrated endothelium is not necessarily permeable to every substance at all times. In all capillaries, the interendothelial contacts are more or less leaky tight junctions (Simionescu et al. 1975), and the degree of tightness seems to be subject to hormonal influences (Pitelka et al. 1973; Tice et al. 1975; Toshimori et al. 1983). In fenestrated capillaries, substance permeation is thought to occur via the diaphragms in the fenestrae of the endothelial cell. However, it can be shown that 5 min after HRP injection the luminal surface of most of the adenohypophyseal capillaries is labeled by the tracer, but that the latter does *not* penetrate every fenestrated endothelial layer, and thus does not spread into all the perivascular spaces of the gland (Fig. 8a, b). Hence, it is obviously the luminal endothelial cell membrane of the adenohypophyseal capillary which participates in the regulation of transcapillary substance permeation, and it might be speculated that the endothelial cell membrane could in some way be subject to neurohormonal influences from the neurohypophysis.

The stellate cells are structures in the adenohypophysis that roughly correspond to tanycytes and pituicytes in the neurohypophysis; they encompass adenohypophyseal cell groups to a varying extent and partly separate them from the perivascular spaces. In addition, they are the elements which mainly determine the tortuosity of the intercellular clefts (Nicholson et al. 1979) in the gland. Earlier results obtained by Bogdanove (1963) demonstrated that the substances of hor-

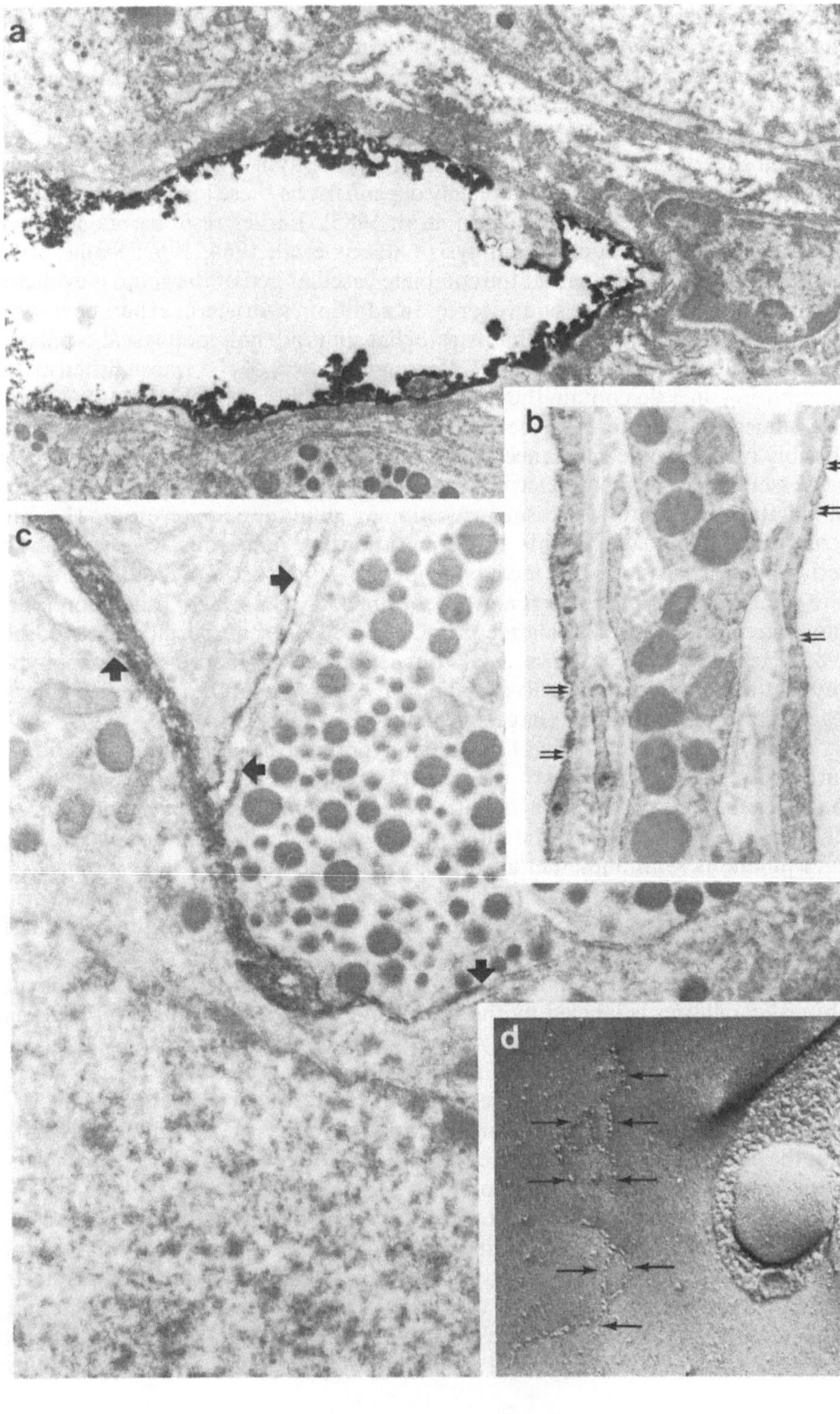
a
b
c
d

mone pellets implanted into the adenohypophysis spread in an irregular way, results which can be mimicked in short-term experiments with HRP: the marker spreading from a labeled perivascular space obviously meets impediments at varying distances from the vessel (Fig. 8c). Freeze-etch preparations of the adenohypophysis show maculae occludentes between hormone-producing cells, between these and stellate cells, or between stellate cells (Fig. 8d). These maculae occludentes belong to the category of "leaky" tight junctions, and as such are not thought to totally impede substance penetration through the intercellular clefts. However, together with the stellate cell processes which form extremely tortuous intercellular clefts, they may lead to the formation of short-term microcompartments within the adenohypophysis which either seclude the endocrine cells from the influence of regulatory neurohormones or expose them to it. Certainly, further factors participate in the varying accessibility of endocrine cells to neurohormones, e.g., the charge of the extracellular matrix, the diffusion coefficients of the regulatory neurohormones (Rice et al. 1985), and the number and distribution of receptor molecules on the adenohypophyseal cells. But in addition to the last-mentioned factor, the degree of extension of the stellate cells and, in particular, the maculae occludentes as intercellular contacts, are thought to be subject to endocrine influences (Pitelka et al. 1973; Tice et al. 1975; Toshimori et al. 1983; Krisch and Buchheim 1984), in this case possibly from the neurohypophysis (besides feedback influences from the peripheral glands).

To summarize the above hypotheses: targets of indirect regulatory neurohormones might be tanycytes and pituicytes at the level of the neurohypophysis, smooth-muscle and endothelial cells in the connecting transport routes, and stellate cells and their intercellular contacts at the level of the adenohypophysis. The structures in the neurohypophysis would influence the degree of accessibility of regulatory neurohormones to the bloodstream; those in the adenohypophysis would regulate the accessibility of the final target cells to the regulatory neurohormones.

4 Summary

Rather than merely describing the apparently valid facts about regulatory neuroendocrine neurons, the present – necessarily incomplete – review was intended to

Fig. 8a–d. Possible targets of indirect regulatory neurohormones in the adenohypophysis: **a** and **b** capillary endothelia; **c** and **d** intercellular clefts and contacts. **a** Heavy luminal deposition of HRP on a capillary endothelium following intravenous injection of the tracer which, however, leaves the surrounding parenchyma completely free. **b** Two fenestrated capillaries (fenestrae are indicated by *double arrows*), the *right* one of which obviously was not perfused during the 5-min survival time after HRP injection. The *left* capillary is labeled on its luminal surface; the perivascular space, however, is unlabeled. **c** Originating from a labeled perivascular space (not shown), the labeling of the more or less tortuous intercellular clefts extends to differing degrees (compare the distance of the *arrowheads* from the heavily labeled intercellular cleft). **d** Between the different cellular elements of the adenohypophysis, focal maculae occludentes (*arrows*) can be evidenced in freeze-fracture replicas; the contacts might participate in the regulation of the accessibility of the intercellular clefts. (From Krisch and Buchheim 1984) **a** × 8753; **b** × 5400; **c** × 31850; **d** × 59000 outline the increasing number of problems and questions that arise from the attempt to integrate the rapidly multiplying insights into the details of regulatory neuroendocrine neurons to form a comprehensive concept. The open questions concern possible functions of related and unrelated neurohormones coexisting within one axon terminal and, in particular, their possible targets in the broadest sense.

In the light of the recent immunohistochemical results briefly mentioned in this review.

1. It is proposed to reject the distinction between two functionally separate parts of the neurohypophysis, i.e., the distinction between the proximal neurohypophysis, the median eminence, from which regulatory neurohormones are released, and the distal neurohypophysis, the neural lobe, as the release site of effectory neurohormones. It seems reasonable to treat the neurohypophysis as a functional entirety: the close functional association of its two parts with regulatory or effectory neurohormones, respectively, seems to be an oversimplification.
2. It is proposed to include in further research the possibility that each of the elements inherent in the neuroadenohypophyseal complex might participate in the modulation of adenohypophyseal activity as possible targets of indirect regulatory neurohormones.

Acknowledgement. This study is dedicated to Professor Dr. Drs. h.c. Berta Scharrer on the occasion of her 80th birthday.

References

Abe H, Engler D, Molitch ME, Bollinger-Gruber J, Reichlin S (1985) Vasoactive intestinal peptide is a physiological mediator of prolactin release in the rat. Endocrinology 116:1383–1390

Adams JH, Daniel PM, Prichard MML (1964) Distribution of hypophysial blood in the anterior lobe of the pituitary gland. Endocrinology 75:120–126

Adams JH, Daniel PM, Prichard MML (1969) The blood supply of the pituitary gland of the ferret with special reference to infarction after stalk section. J Anat 104:209–225

Ambach G, Palkovits M, Szentágothai J (1976) Blood supply of the rat hypothalamus. IV. Retrochiasmatic area, median eminence, arcuate nucleus. Acta Morphol Hung 24:93–119

Ambach G, Kivovics P, Palkovits M (1978) The arterial and venous blood supply of the preoptic region in the rat. Acta Morphol Hung 26:21–41

Anhut H, Knepel W, Holland A, Meyer DK (1982) β-Endorphin release by angiotensin II: studies on the mechanism of action. Regul Pept 4:83–90

Baertschi AJ, Vallet P, Baumann JB, Girard J (1980) Neural lobe of pituitary modulates corticotropin release in the rat. Endocrinology 106:878–882

Bai FL, Yamano M, Inagaki S, Shiosaka S, Yamazoe M, Shibasaki T, Ling N, Tachibana S, Hamaoka T, Tohyama M, Shiotani Y (1984) Distribution of neuropeptides in the hypothalamo-hypophyseal system in the rat: an immunohistochemical observation. Cell Mol Biol 30:437–452

Bakhit C, Koda L, Benoit R, Morrison JH, Bloom FE (1984) Evidence for selective release of somatostatin-14 and somatostatin-28(1–12) from rat hypothalamus. J Neurosci 4:411–419

Bargmann W (1949) Über die neurosekretorische Verknüpfung von Hypothalamus und Neurohypophyse. Z Zellforsch Mikroskop Anat 34:610–634

Bargmann W (1966) Neurosecretion. Int Rev Cytol 19:183–201
Barry J (1979) Immunohistochemistry of luteinizing hormone-releasing hormone-producing neurons of the vertebrates. Int Rev Cytol 60:179–221
Beaumont A (1983) Putative peptide neurotransmitters: the opioid peptides. Int Rev Exp Pathol 25:279–298
Beauvillain JC, Tramu G, Dubois MP (1981) Ultrastructural immunocytochemical evidence of the presence of a peptide related to ACTH in granules of LHRH nerve terminals in the median eminence of the guinea pig. Cell Tissue Res 218:1–6
Beauvillain JC, Tramu G, Garaud JC (1984) Coexistence of substances related to enkephalin and somatostatin in granules of the guinea-pig median eminence: demonstration by use of colloidal gold immunocytochemical methods. Brain Res 301:389–393
Bloom FE, Battenberg ELF, Shibasaki T, Benoit R, Ling N, Guillemin R (1980) Localization of γ-melanocyte stimulating hormone (γMSH) immunoreactivity in rat brain and pituitary. Regul Pept 1:205–222
Bloom FE, Battenberg ELF, Rivier J, Vale W (1982) Corticotropin releasing factor (CRF): immunoreactive neurones and fibres in rat. Regul Pept 4:43–48
Bogdanove EM (1963) Direct gonad-pituitary feedback: an analysis of effects of intracranial estrogenic depots on gonadotrophin secretion. Endocrinology 73:696–712
Böhlen P, Brazeau P, Benoit R, Ling N, Esch F, Guillemin R (1980) Isolation and amino acid composition of two somatostatin-like peptides from ovine hypothalamus: somatostatin-28 and somatostatin-25. Biochem Biophys Res Commun 96:725–734
Brightman MW, Prescott L, Reese TS (1975) Intercellular junctions of special ependyma. In: Knigge KM, Scott DE, Kobayashi H, Ishii S (eds) Brain-endocrine interaction II. The ventricular system in neuroendocrine mechanisms. Karger, Basel, pp 146–165
Brown M, Rivier J, Vale W (1981) Somatostatin-28: selective action on the pancreatic β-cell and brain. Endocrinology 108:2391–2393
Brownstein MJ (1985) Peptide processing: an overview. In: Håkanson R, Thorell J (eds) Biogenetics of neurohormonal peptides. Academic, New York, pp 105–112
Brownstein MJ, Russell JT, Gainer H (1982) Biosynthesis of posterior pituitary hormones. In: Ganong WF, Martini L (eds) Frontiers in neuroendocrinology 7. Raven, New York, pp 31–43
Carrillo AJ, Pool TB, Sharp ZD (1985) Vasoactive intestinal peptide increases prolactin messenger ribonucleuc acid content in GH_3 cells. Endocrinology 116:202–206
Charpenet G, Patel YC (1985) Characterization of tissue and releasable molecular forms of somatostatin-$28_{(1-12)}$-like immunoreactivity in rat median eminence. Endocrinology 116:1863–1868
Cheung CY (1984) Does β-endorphin modulate basal and dopamine-inhibited prolactin release by an action at the anterior pituitary? Neuroendocrinology 39:489–495
Chihara K, Arimura A, Chihara M, Schally AV (1978) Studies on the mechanism of growth hormone and thyrotropin responses to somatostatin antiserum in anesthetized rats. Endocrinology 103:1916–1923
Choy VJ, Watkins WB (1977) Immunohistochemical localization of thyrotropin-releasing factor in the rat median eminence. Cell Tissue Tes 177:371–374
Curtis A, Fink G (1983) A high molecular weight precursor of luteinizing hormone releasing hormone from rat hypothalamus. Endocrinology 112:390–392
Daniel PM (1966) The anatomy of the hypothalamus and pituitary gland. In: Martini L, Ganong WF (eds) Neuroendocrinology 1. Academic, New York, pp 15–80
Denef C, Schramme C, Baes M (1985) Stimulation of growth hormone release by vasoactive intestinal peptide and peptide PHI in rat anterior pituitary reaggregates. Neuroendocrinology 40:88–91
De Souza EB, Perrin MH, Whitehouse PJ, Rivier J, Vale W, Kuhar MJ (1985) Corticotropin-releasing factor receptors in human pituitary gland: autoradiographic localization. Neuroendocrinology 40:419–422
Duvernoy H (1972) The vascular architecture of the median eminence. In: Knigge KM, Scott DE, Weindl A (eds) Brain-endocrine interaction. Median eminence: structure and function. Karger, Basel, pp 79–108
Edvinsson L (1985) Functional role of perivascular peptides in the control of cerebral circulation. Trends Neuro Sci 8:126–131

Emson PC, Goeder M, Benton H, St.-Pierre S, Rioux F (1982) The regional distribution and chromatographic characterization of neurotensin-like immunoreactivity in the rat. In: Costa E, Trabucchi M (eds) Regulatory peptides: from molecular biology to function. Raven, New York, pp 477–486

Enjalbert A, Epelbaum J, Arancibia S, Tapia-Arancibia L, Bluet-Pajot MT, Kordon C (1982) Reciprocal interactions of somatostatin with thyrotropin-releasing hormone and vasoactive intestinal peptide on prolactin and growth hormone secretion in vitro. Endocrinology 111:42–47

Ermisch A, Rühle HJ, Klauschenz E, Kretzschmar R (1984) On the blood–brain barrier to peptides: (^{3}H) gonadotropin-releasing hormone accumulation by eighteen regions of the rat brain and by anterior pituitary. Exp Clin Endocrinol 84:112–116

Ermisch A, Barth T, Rühle HJ, Škopková J, Hrbas P, Landgraf R (1985) On the blood–brain barrier to peptides: accumulation of labelled vasopressin, DesGlyNH_2-vasopressin and oxytocin by brain regions. Endocrinol Exp 19:29–37

Farah JM jr, Malcolm DS, Mueller GP (1982) Dopaminergic inhibition of β-endorphin-like immunoreactivity secretion in the rat. Endocrinology 110:657–659

Fink G (1985) Has the prolactin inhibiting peptide at last been found? Nature 316:487–488

Fink G, Koch Y, Ben Aroya N (1982) Release of thyrotropin releasing hormone into hypophysial portal blood is high relative to other neuropeptides and may be related to prolactin secretion. Brain Res 243:186–189

Frawley LS, Leong DA, Neill JD (1985) Oxytocin attenuates TRH-induced TSH release from rat pituitary cells. Neuroendocrinology 40:201–204

Froehlich JC, Ben-Jonathan N (1984) Posterior pituitary involvement in the control of luteinizing hormone and prolactin secretion during the estrous cycle. Endocrinology 14:1059–1064

Gainer H, Russell JT, Loh YP (1985) The enzymology and intracellular organization of peptide precursor processing: the secretory vesicle hypothesis. Neuroendocrinology 40:171–184

Gibbs DM (1985) Hypothalamic epinephrine is released into hypophysial portal blood during stress. Brain Res 335:360–364

Goldgefter L, Korochkin L (1970) Periventricular Gomori-positive glial cells in rat hypothalamus. Arch Anat Histol Embryol (Strasb) 59:9–16

Grandison L, Guidotti A (1977) Regulation of prolactin release by endogenous opiates. Nature 270:357–359

Guillemin R, Brazeau P, Böhlen P, Esch F, Ling N, Wehrenberg WB (1982) Growth hormone-releasing factor from a human pancreatic tumor that caused acromegaly. Science 218:585–587

Guy J, Benoit R, Pelletier G (1985) Immunocytochemical localization of somatostatin-28_{1-12} in the rat hypothalamus. Brain Res 330:283–289

Hisano S, Kawano H, Maki Y, Daikoku S (1981) Electron microscopic study of immunoreactive LHRH perikarya with special reference to neuronal regulation. Cell Tissue Res 220:511–518

Hoffman GE, Wray S, Goldstein M (1982) Relationship of catecholamines and LHRH: light microscopic study. Brain Res Bull 9:417–430

Hökfelt T, Fahrenkrug J, Tatemoto K, Mutt V, Werner S, Hulting AL, Terenius L, Chang KJ (1983) The PHI (PHI-27)/corticotropin-releasing factor/enkephalin immunoreactive hypothalamic neuron: possible morphological basis for integrated control of prolactin, corticotropin, and growth hormone secretion. Proc Natl Acad Sci USA 80:895–898

Hökfelt T, Barry JE, Theodorsson-Nordheim E, Goldstein M (1984) Occurrence of neurotensin-like immunoreactivity in subpopulations of hypothalamic, mesencephalic, and medullary catecholamine neurons. J Comp Neurol 222:543–559

Holzbauer M, Muscholl E, Rache K, Sharman DF (1983) Evidence that dopamine is a neurotransmitter in the neurointermediate lobe of the hypophysis. In: The neurohypophysis. Structure, function and control. Prog Brain Res 60:357–364

Holzbauer M, Racké K, Sharman DF (1985) Release of endogenous 6-hydroxytryptamine from the neural and the intermediate lobe of the rat pituitary gland evoked by electrical stimulation of the pituitary stalk. Neuroscience 15:723–728
Ibata Y, Kawakami F, Fukui K, Okamura H, Obata-Tsutu HL, Tsuto T, Terubayashi H (1984) Morphological survey of neurotensin-like immunoreactive neurons in the hypothalamus. Peptides 5 [Suppl] 109–120
Jennes L, Beckmann WC, Stumpf WE, Grzanna R (1982a) Anatomical relationships of serotoninergic and noradrenalinergic projections with the GnRH system in septum and hypothalamus. Exp Brain Res 46:331–338
Jennes L, Stumpf WE, Kalivas PW (1982b) Neurotensin: topographical distribution in rat brain by immunohistochemistry. J Comp Neurol 210:211–224
Jennes L, Stumpf WE, Tappaz ML (1983) Anatomical relationships of dopaminergic and GABAergic systems with the GnRH-systems in the septo-hypothalamic area. Immunohistochemical studies. Exp Brain Res 50:91–99
Jennes L, Stumpf WE, Sheedy ME (1985) Ultrastructural characterization of gonadotropin-releasing hormone (GnRH)-producing neurons. J Comp Neurol 232:534–547
Johannsson O, Hökfelt T (1980) Immunohistochemical distribution of thyrotropin-releasing hormone, somatostatin and enkephalin with special reference to the hypothalamus. In: Wuttke W, Weindl A, Voigt KH, Driess RR (eds) Brain and pituitary peptides. Karger, Basel, pp 202–212
Johannsson O, Hökfelt T, Jeffcoate SL, White N, Sternberger LA (1980) Ultrastructural localization of TRH-like immunoreactivity. Exp Brain Res 38:1–10
Johannsson O, Hökfelt T, Elde RP (1984) Immunohistochemical distribution of somatostatin-like immunoreactivity in the central nervous system of the adult rat. Neuroscience 13:265–339
Kaplan GP, Hartman KB, Creveling CR (1981) Localization of catechol-*O*-methyltransferase in the leptomeninges, choroid plexus and ciliary epithelium: implications for the separation of central and peripheral catechols. Brain Res 204:353–360
Kawata M, Hashimoto K, Takahara J, Sano Y (1982a) Immunohistochemical demonstration of the localization of corticotropin releasing factor-containing neurons in the hypothalamus of mammals including primates. Anat Embryol 165:303–313
Kawata M, Hashimoto K, Takahara J, Sano Y (1982b) Immunohistochemical demonstration of corticotropin releasing factor containing nerve fibers in the median eminence of the rat and monkey. Histochemistry 76:15–19
Kawata M, Hashimoto K, Takahara J, Sano Y (1983) Immunohistochemical identification of neurons containing corticotropin-releasing factor in the rat hypothalamus. Cell Tissue Res 230:239–246
Kerdelhué B, Tartar A, Lenoir V, El Abed A, Hublau P, Millar RP (1985) Binding studies of substance P anterior pituitary binding sites: changes in substance P binding sites during the rat estrous cycle. Regul Pept 10:133–143
Khachaturian H, Watson SJ, Lewis ME, Coy D, Goldstein A, Akil H (1982) Dynorphin immunocytochemistry in the rat central nervous system. Peptides 3:941–954
King JC, Anthony ELP (1983) Biosynthesis of LHRH: inferences from immunocytochemical studies. Peptides 4:963–970
Kiss J, Halász B (1985) Demonstration of serotoninergic axons terminating on luteinizing hormone-releasing hormone neurons in the preoptic area of the rat using a combination of immunocytochemistry and high resolution autoradiography. Neuroscience 14:69–78
Kiss JZ, Mezey E, Cassell MD, Williams TH, Mueller GP, O'Donohue TL, Palkovits M (1985) Topographical distribution of pro-opiomelanocortin-derived peptides (ACTH/β-END/α-MSH) in the rat median eminence. Brain Res 329:169–176
Knigge KM, Morris M, Scott DE, Joseph SA, Notter M, Schock D, Krobisch-Dudley G (1975) Distribution of hormones by cerebrospinal fluid. In: Cserr HF, Fenstermacher JD, Fencl V (eds) Fluid environment of the brain. Academic, New York, pp 237–253
Knowles F, Bern HA (1966) The function of neurosecretion in endocrine regulation. Nature 210:271–272

Kobayashi H, Matsui T, Ishii S (1970) Functional microscopy of the hypothalamic median eminence. Int Rev Cytol 29:282–381

Koritsánsky S, Vigh B, Aros B (1967) Studies on the Gomori-positive glial cells. I. The changes in the periventricular Gomori-positive glial cells in rats of various ages. Acta Biol Hung 18:9–19

Krieger DT (1981) Circadian rhythms. In: Martin JB, Reichlin S, Bick KL (eds) Neurosecretion and brain peptides. Raven, New York, pp 447–448

Krisch B (1980 a) Immunocytochemistry of neuroendocrine systems (vasopressin, somatostatin, luliberin). Prog Histochem Cytochem 13(2):1–166

Krisch B (1980 b) Two types of luliberin-immunoreactive perikarya in the preoptic area of the rat. Cell Tissue Res 212:443–455

Krisch B (1980 c) Differing immunoreactivities of somatostatin in the cortex and the hypothalamus of the rat. A light and electron microscopic study. Cell Tissue Res 212:457–464

Krisch B (1981 a) Somatostatin-immunoreactive fiber projections into the brain stem and the spinal cord of the rat. Cell Tissue Res 217:431–552

Krisch B (1981 b) Distribution of somatostatin and luliberin in the rat brain. In: Farner DS, Lederis K (eds) Neurosecretion. Molecules, cells, systems. Plenum, New York, pp 49–59

Krisch B, Buchheim W (1984) Access and distribution of exogenous substances in the intercellular clefts of the rat adenohypophysis. Cell Tissue Res 236:439–452

Krisch B, Leonhardt H (1980 a) Luliberin and somatostatin fiber-terminals in the subfornical organ of the rat. Cell Tissue Res 210:33–45

Krisch B, Leonhardt H (1980 b) Neurohormones in the intercellular clefts and in glia-like cells of the rat brain. Cell Tissue Res 211:251–268

Krisch B, Leonhardt H, Buchheim W (1978) The functional and structural border of the neurohemal region of the median eminence. Cell Tissue Res 192:327–339

Krisch B, Leonhardt H, Oksche A (1983) The meningeal compartments of the median eminence and the cortex. A comparative analysis in the rat. Cell Tissue Res 228:597–640

Lauber M, Camier M, Cohen P (1979) Higher molecular weight forms of immunoreactive somatostatin in mouse hypothalamic extracts: evidence of processing in vitro. Proc Natl Acad Sci USA 76:6004–6008

Lechan RM, Goodman RH, Rosenblatt M, Reichlin S, Haberer JF (1983) Prosomatostatin-specific antigen in rat brain: localization by immunocytochemical staining with an antiserum to a synthetic sequence of preprosomatostatin. Proc Natl Acad Sci USA 80:2780–2784

Lechan RM, Lin HD, Ling N, Jackson IMD, Jacobson S, Reichlin S (1984) Distribution of immunoreactive growth hormone releasing factor (1-44)NH_2 in the tuberoinfundibular system of the rhesus monkey. Brain Res 309:55–61

Leonardelli J, Tramu G (1979) Immunoreactivity for β-endorphin in LH-RH neurons of the fetal human hypothalamus. Cell Tissue Res 203:201–207

Leonhardt H (1980) Ependym und circumventrikuläre Organe. In: Oksche A, Vollrath L (eds) Neuroglia I. Springer, Berlin Heidelberg New York, S 177–666 (Handbuch der mikroskopischen Anatomie des Menschen, vol 4/10)

Leonhardt H, Krisch B, Hartwig HG (1983) Circumventricular organs as targets and release sites for peptide hormones and monoamines. In: Sano Y, Ibata Y, Zimmerman EA (eds) Structure and function of peptidergic and aminergic neurons. Taniguchi Symposia on brain science No. 6. Japan Scientific Societies, Tokyo, pp 55–71

Léránth C, Palkovits M, Krieger DT (1983) Serotonin immunoreactive nerve fibres and terminals in the rat pituitary. Light- and electron-microscopic studies. Neuroscience 9:289–296

Léránth C, MacLusky NJ, Sakamoto H, Shanabrough M, Naftolin F (1985) Glutamic acid decarboxylase-containing axons synapse on LHRH neurons in the rat medial preoptic area. Neuroendocrinology 40:536–539

Lind WR, Swanson LW, Bruhn TO, Ganten D (1985) The distribution of angiotensin II-immunoreactive cells and fibers in the paraventriculo-hypophysial system of the rat. Brain Res 338:81–89

Liotta AS, Advis JP, Krause JE, McKelvy JF, Krieger DT (1984) Demonstration of in vivo synthesis of proopiomelanocortin-, β-endorphin-, and α-melanotropin-like species in the adult rat brain. J Neurosci 4:956–965

Liposits Z, Léngvari J, Vigh S, Schally A, Flerkó B (1983) Immunohistological detection of degenerating CRF-immunoreactive nerve fibers in the median eminence after lesion of paraventricular nucleus of the rat. A light and electron microscopic study. Peptides 4:941–953

Loh YP, Brownstein MJ, Gainer H (1984) Proteolysis in neuropeptide processing and other neural functions. Ann Rev Neurosci 7:189–222

Lookingland KJ, Farah JM Jr, Lovell KL, Moore KE (1985) Differential regulation of tuberohypophysial dopaminergic neurons terminating in the intermediate lobe and in the neural lobe of the rat pituitary gland. Neuroendocrinology 40:145–151

Maeda K, Frohman LA (1980) Release of somatostatin- and thyrotropin-releasing hormone from rat hypothalamic fragments in vitro. Endocrinology 106:1837–1842

Mann DR, Evans D, Edoimioya F, Kamel F, Butterstein GM (1985) A detailed examination of the in vivo and in vitro effects of ACTH on gonadotropin secretion in the adult rat. Neuroendocrinology 40:297–302

Martin R, Voigt KH (1981) Enkephalins co-exist with oxytocin and vasopressin in nerve terminals of rat neurohypophysis. Nature 289:502–504

Martin R, Voigt KH (1982) Leucine-enkephalin-like immunoreactivity in vasopressin terminals is enhanced by treatment with peptidases. Life Sci 31:1729–1732

Martin R, Geis R, Holl R, Schäfer M, Voigt KH (1983) Co-existence of unrelated peptides in oxytocin and vasopressin terminals of rat neurohypophyses: immunoreactive methionine-enkephalin-, leucine-enkephalin- and cholecystokinin-like substances. Neuroscience 8:213–227

Martini L (1974) Recent advances in the study of the hypothalamic releasing factors. In: Knowles F, Vollrath L (eds) Neurosecretion – the final neuroendocrine pathway. Springer, Berlin Heidelberg New York, pp 135–147

Matsumura M, Yamanoi A, Yamemoto S, Saito S (1983) In vivo and in vitro effects of cholecystokinin octapeptide on the release of β-endorphin-like immunoreactivity. Neuroendocrinology 36:443–448

Matsumura M, Yamanoi A, Yamamoto S, More H, Saito S (1984) In vivo and in vitro effects of cholecystokinin octapeptide on the release of growth hormone in rats. Horm Metabol Res 16:626–630

Matsumura M, Yamanoi A, Yamamoto S, Mori H, Saito S (1985) In vivo and in vitro effects of cholecystokinin octapeptide on the release of prolactin in rats. Horm Metabol Res 17:293–297

Mayo KE, Cerelli GM, Rosenfeld MG, Evans RM (1985) Characterization of cDNA and genomic clones encoding the precursor to rat hypothalamic growth hormone-releasing factor. Nature 314:464–467

McCann SM, Vijayan E, Samson WK, Koenig J, Krulich L (1980) Role of brain peptides in the control of pituitary hormone release. In: Wuttke W, Weindl A, Voigt KH, Driess RR (eds) Brain and pituitary peptides. Karger, Basel, pp 223–233

Merchenthaler I, Vigh S, Petrusz P, Schally AV (1983) The paraventriculo-infundibular corticotropin releasing factor (CRF) pathway as revealed by immunocytochemistry in long-term hypophysectomized or adrenalectomized rats. Regul Pept 5:295–305

Mezey É, Kiss JZ (1985) Vasoactive intestinal peptide-containing neurons in the paraventricular nucleus may participate in regulating prolactin secretion. Proc Natl Acad Sci USA 82:245–247

Mezey É, Palkovits M (1982a) Two-way transport in the hypothalamo-hypophyseal system. In: Ganong WF, Martini L (eds) Frontiers in neuroendocrinology 6. Raven, New York, pp 1–29

Mezey É, Palkovits M (1982b) Meningeal relations of the rat hypothalamo-hypophysial system. Extravascular fluid spaces in and around the median eminence. Brain Res 250:21–30

Mezey É, Palkovits M, de Kloet ER, Verhoef J, de Wied D (1978) Evidence for pituitary-brain transport of a behaviorally potent ACTH analog. Life Sci 22:831–838

Millan MJ, Herz A (1985) The endocrinology of the opioids. Int Rev Neurobiol 26:1–81
Millar RP, Sheward WJ, Wegener I, Fink G (1983) Somatostatin-28 is an hormonally active peptide secreted into hypophysial portal vessel blood. Brain Res 260:334–337
Moore KE, Demarest KT (1982) Tuberoinfundibular and tuberohypophysial dopaminergic neurons. In: Ganong WF, Martini L (eds) Frontiers in neuroendocrinology 7. Raven, New York, pp 161–190
Moore RY (1981) The suprachiasmatic nucleus, circadian rhythms, and regulation of brain peptides. In: Martin JB, Reichlin S, Bick KL (eds) Neurosecretion and brain peptides. Raven, New York, pp 449–459
Moore RY, Bloom FE (1978) Central catecholamine neuron systems: anatomy and physiology of the dopamine systems. Annu Rev Neurosci 1:129–169
Morel G, Leroux P, Pelletier G (1985) Ultrastructural autoradiographic localization of somatostatin-28 in the rat pituitary gland. Endocrinology 116:1615–1620
Morris JF (1983) Organization of neural inputs to the supraoptic and paraventricular nuclei: anatomical aspects. In: The neurohypophysis: structure, function and control. Prog Brain Res 60:3–18
Motta M, Martini L (1972) Hypothalamic releasing factors: a new class of "neurotransmitters." Arch Int Pharmacodyn [Suppl] 196:191–204
Mulder AH (1982) An overview of subcellular localization, release and termination of action of amine, amino acid and peptide neurotransmitters in the central nervous system. In: Chemical transmission in the brain. The role of amines, amino acids and peptides. Prog Brain Res 55:135–156
Murakami T (1975) Pliable methacrylate casts of blood vessels: use in a scanning electron microscope study of the microcirculation in rat hypophysis. Arch Histol Jpn 38:151–168
Murakami T, Ohtsuka A, Taguchi T, Kikuta A, Ohtani O (1985) Blood vascular bed of the rat pituitary intermediate lobe, with special reference to its development and portal drainage into the anterior lobe. A scanning electron microscope study of vascular casts. Arch Histol Jpn 48:69–87
Nicholson C, Phillips JM, Gardner-Medwin AR (1979) Diffusion from a iontophoretic point source in the brain: role of tortuosity and volume fraction. Brain Res 169:580–584
Nikolics K, Mason AJ, Szönyi E, Ramachandran J, Seeburg PH (1985) A prolactin-inhibiting factor within the precursor for human gonadotropin-releasing hormone. Nature 316–511–517
O'Donohue TL, Dorsa D (1982) The opiomelanotropinergic neuronal and endocrine systems. Peptides 3:353–395
Okamura H, Murakami S, Chihara K, Nagatsu I, Ibata Y (1985) Coexistence of growth hormone releasing factor-like and tyrosine hydroxylase-like immunoreactivities in neurons of the rat arcuate nucleus. Neuroendocrinology 41:177–179
Olschowka JA, O'Donohue TL, Mueller GP, Jacobowitz DM (1982) Hypothalamic and extrahypothalamic distribution of CRF-like immunoreactive neurons in the rat brain. Neuroendocrinology 35:305–308
Page RB, Dovey-Hartman BJ (1984) Resistance vessels supplying the median eminence of the rabbit, rat and cat. Anat Rec 210:647–655
Pardridge WM, Frank HJL, Cornford WM, Braun LD, Crane PD, Oldendorf WH (1981) Neuropeptides and the blood-brain barrier. In: Martin JB, Reichlin S, Bick KL (eds) Neurosecretion and brain peptides. Raven, New York, pp 321–328
Patel GC, Wheatley T, Ning Ch (1981) Multiple forms of immunoreactive somatostatin: comparison of distribution in neural and nonneural tissues and portal plasma of the rat. Endocrinology 190:1943–1949
Paull WK, Gibbs FP (1983) The corticotropin releasing factor (CRF) neurosecretory system in intact, adrenalectomized, and adrenalectomized-dexamethasone treated rats. Histochemistry 78:303–316
Pelletier G (1980) Immunohistochemical localization of somatostatin. Prog Histochem Cytochem 12(3):1–41

Pelletier G, Désy L, Lôté J, Lefèvre G, Vaudry H, Labrie F (1982) Immunoelectron microscopic localization of corticotropin-releasing factor in the rat hypothalamus. Neuroendocrinology 35:402–404

Phillips HS, Nikolics K, Branton D, Seeburg PH (1985) Immunocytochemical localization in rat brain of a prolactin release-inhibiting sequence of gonadotropin-releasing hormone prohormone. Nature 316:542–545

Pitelka DR, Hamamoto ST, Duafala JG, Nemanic MK (1973) Cell contacts in the mouse mammary gland. I. Normal gland in postnatal development and the secretory cycle. J Cell Biol 56:797–818

Porter JC, Mical RS, Ondo JG, Kamberi IA (1972) Perfusion of the rat anterior pituitary via a cannulated portal vessel. Acta Endocrinol (Copenh) [Suppl] 158:249–269

Purves D, Hadley RD (1985) Changes in the dendritic branching of adult mammalian neurones revealed by repeated imaging in situ. Nature 315:404–406

Reymond MJ, Kaur Ch, Porter JC (1983) An inhibitory role for morphine on the release of dopamine into hypophysial portal blood and on the synthesis of dopamine in tuberoinfundibular neurons. Brain Res 262:253–258

Rice ME, Gerhardt GA, Hierl PM, Nagy G, Adams RN (1985) Diffusion coefficients of neurotransmitters and their metabolites in brain extracellular fluid space. Neuroscience 15:891–902

Richardson I, Schönbrunn A (1981) Inhibition of adrenocorticotropin secretion by somatostatin in pituitary cells in culture. Endocrinology 108:281–290

Richter D (1985) Biosynthesis of vasopressin. In: Ganten D, Pfaff D (eds) Neurobiology of vasopressin. Springer, Berlin Heidelberg New York Tokyo, pp 1–16 (Current topics in neuroendocrinology 4)

Rivier C, Vale W (1983 a) Interaction of corticotropin-releasing factor and arginine vasopressin on adrenocorticotropin secretion in vivo. Endocrinology 113:939–942

Rivier C, Vale W (1983 b) Effect of angiotensin II on ACTH release in vivo: role of corticotropin-releasing factor. Regul Pept 7:253–258

Rivier J, Spiess J, Thorner MJ, Vale W (1982) Characterization of a growth hormone-releasing factor from a human pancreatic tumor. Nature 300:276–278

Rivier C, Rivier J, Lederis K, Vale W (1983 a) In vitro and in vivo ACTH-releasing activity of ovine CRF, sauvagine and urotensin I. Regul Pept 5:139–143

Rivier J, Spiess J, Vale W (1983 b) Characterization of rat hypothalamic corticotropin-releasing factor. Proc Natl Acad Sci USA 80:4851–4855

Rivier C, Rivier J, Mormede P, Vale W (1984) Studies of the nature of the interaction between vasopressin and corticotropin-releasing factor on adrenocorticotropin release in the rat. Endocrinology 115:882–886

Rønnekleiv OK, Kelly MJ, Eskay RL (1984) Distribution of immunoreactive substance P neurons in the hypothalamus and pituitary of the rhesus monkey. J Comp Neurol 224:51–59

Rostène WH (1984) Neurobiological and neuroendocrine functions of the vasoactive intestinal peptide (VIP). Prog Neurobiol 22:103–129

Roth K, Weber E, Barchas JD, Chang D, Chang JK (1983) Immunoreactive dynorphin-(1-8) and corticotropin-releasing factor in subpopulation of hypothalamic neurons. Science 219:189–191

Samson WK, Burton KP, Reeves JP, McCann SM (1981) Vasoactive intestinal peptide stimulates luteinizing hormone-releasing hormone release from median eminence synaptosomes. Regul Pept 2:253–264

Samson WK, Lumpkin MD, McDonald JK, McCann SM (1983) Prolactin-releasing activity of porcine intestinal peptide (PHI-27). Peptides 4:817–819

Sawchenko PE, Swanson LW, Vale WW (1984) Co-expression of corticotropin-releasing factor and vasopressin immunoreactivity in parvocellular neurosecretory neurons of the adrenalectomized rat. Proc Natl Acad Sci USA 81:1883–1887

Sawchenko PE, Swanson LW, Rivier J, Vale WW (1985) The distribution of growth-hormone-releasing factor (GFR) immunoreactivity in the central nervous system of the rat: an immunohistochemical study using antisera directed against rat hypothalamic GFR. J Comp Neurol 237:100–115

Scharrer E (1936) Vergleichende Untersuchungen über die zentralen Anteile des vegetativen Systems. Z Anat Entwickl Gesch 106:169–192
Scharrer E, Scharrer B (1954) Neurosekretion. In: Barmann W (Hrsg) Handbuch der mikroskopischen Anatomie des Menschen 5. Springer, Berlin Göttingen Heidelberg, S 953–1066
Seeburg PH, Adelman JP (1984) Characterization of cDNA for precursor of human luteinizing hormone releasing hormone. Nature 311:666–668
Shaffer Tannenbaum G, Ling N, Brazeau P (1982) Somatostatin-28 is longer acting and more selective than somatostatin-14 on pituitary and pancreatic hormone release. Endocrinology 111:101–107
Shimatsu A, Kato Y, Matsushita N, Katakami H, Yanaihara N, Imura H (1982) Effects of glucagon, neurotensin, and vasoactive intestinal polypeptide on somatostatin release from perfused rat hypothalamus. Endocrinology 110:2113–2117
Silverman AJ, Witkin JW (1985) Synaptic interactions of luteinizing hormone-releasing hormone (LHRH) neurons in the guinea pig preoptic area. J Histochem Cytochem 33:69–72
Simionescu M, Simionescu N, Palade GE (1975) Segmental differentiations of cell junctions in the vascular endothelium. The microvasculature. J Cell Biol 67:863–885
Spiess J, Rivier J, Vale W (1983) Characterization of rat hypothalamic growth hormone-releasing factor. Nature 303:532–535
Spinedi E, Negro-Vilar A (1983) Angiotensin II and ACTH release: site of action and potency relative to corticotropin releasing factor and vasopressin. Neuroendocrinology 37:446–453
Spinedi E, Negro-Vilar A (1984) Arginine vasopressin and adrenocorticotropin release: correlation between binding characteristics and biological activity in anterior pituitary dispersed cells. Endocrinology 114:2247–2251
Stoeckel ME, Porte A, Klein MJ, Cuello AC (1982) Immunocytochemical localization of substance P in the neurohypophysis and hypothalamus of the mouse compared with the distribution of other neuropeptides. Cell Tissue Res 223:533–544
Sundler F, Alumets J, Ekblad E, Böttcher G, Håkanson R (1985) Coexistence of peptides in the neuroendocrine system. In: Håkanson R, Thorell J (eds) Biogenetics of neurohormonal peptides. Academic, New York, pp 213–243
Swaab DF (1982) Neuropeptides. Their distribution and function in the brain. In: Chemical transmission in the brain. The role of amines, amino acids and peptides. Prog Brain Res 55:97–122
Swanson LW, Sawchenko PE, Rivier J, Vale WW (1983) Organization of ovine corticotropin-releasing factor immunoreactive cells and fibers in the rat brain: an immunohistochemical study. Neuroendocrinology 36:165–186
Tice LW, Wollman SH, Carter RC (1975) Changes in tight junctions of thyroid epithelium with changes in thyroid activity. J Cell Biol 66:657–663
Tilders FJH, Smelik PG (1977) Direct neural control of MSH secretion in mammals: involvement of dopaminergic tuberohypophysial neurons. In: Tilders FJH, Swaab DF, van Wimersma Greidanus TB (eds) Frontiers in Hormon Research, vol 4. Karger, Basel, pp 80–93
Tilders F, Tatemoto K, Berkenbosch F (1984) The intestinal peptide PHI-27 potentiates the action of corticotropin-releasing factor on ACTH release from rat pituitary fragments in vitro. Endocrinology 115:1633–1635
Tizabi Y, Skofitsch G, Jacobowitz DM (1985) Effect of chronic reserpine and desmethylimipramine treatment on CRF-like immunoreactivity of discrete brain areas of rat. Brain Res 335:389–391
Toshimori K, Higashi R, Oura C (1983) Quantitative analysis of zonulae occludentes between oviductal epithelial cells at diestrous and estrous stages in the mouse: freeze-fracture study. Anat Rec 206:257–266
Tramu G, Leonardelli J (1979) Immunohistochemical localization of enkephalins in median eminence and adenohypophysis. Brain Res 168:457–471
Tramu G, Leonardelli J, Dubois MP (1977) Immunohistochemical evidence for an ACTH-like substance in hypothalamic LH-RH neurons. Neurosci Lett 6:305–309

Tramu G, Beauvillain JC, Croix D, Leonardelli J (1981) Comparative immunocytochemical localization of enkephalin and somatostatin in the median eminence, hypothalamus and adjacent areas of the guinea-pig brain. Brain Res 215:235–255

Tramu G, Croix C, Pillez A (1983) Ability of the CRF immunoreactive neurons of the paraventricular nucleus to produce a vasopressin-like material. Neuroendocrinology 37:467–469

Tsuruo Y, Kawano H, Nishiyama T, Hisano S, Daikoku S (1983) Substance P-like immunoreactive neurons in the tuberoinfundibular area of rat hypothalamus. Light and electron microscopy. Brain Res 289:1–9

Vale W, Spiess J, Rivier C, Rivier J (1981) Characterization of a 41-residue ovine hypothalamic peptide that stimulates secretion of corticotropin and β-endorphin. Science 213:1394–1397

Vale W, Rivier C, Brown MR, Spiess J, Koob G, Swanson L, Bilezikjian L, Bloom F, Rivier J (1983a) Chemical and biological characterization of corticotropin releasing factor. Rec Prog Horm Res 39:245–270

Vale W, Vaughan J, Smith M, Yamamoto G, Rivier J, Rivier C (1983b) Effects of synthetic ovine corticotropin-releasing factor, glucocorticoids, catecholamines, neurohypophysial peptides, and other substances on cultured corticotropic cells. Endocrinology 113:1121–1131

Vanderhaeghen JJ, Lotstra F, De Mey J, Gilles C (1980) Immunohistochemical localization of cholecystokinin- and gastrinlike peptides in the brain and hypophysis of the rat. Proc Natl Acad Sci USA 77:1190–1194

Vanderhaeghen JJ, Lotstra F, Vandesande F, Dierickx K (1981) Co-existence of cholecystokinin and oxytocin-neurophysin in some magnocellular hypothalamo-hypophysial neurons. Cell Tissue Res 221:227–231

Van Houten M, Khan MN, Walsh RJ, Baquiran GB, Renaud L, Bourque C, Sgro S, Gauthier S, Chretien M, Posner BI (1985) NH_2-terminal specificity and axonal localization of adrenocorticotropin binding sites in rat median eminence. Proc Natl Acad Sci USA 82:1271–1275

Van Leeuwen F (1982a) Enkephalin in the rat neural lobe: immunocytochemical evidence for its presence within synaptic elements on pituicytes. In: Chemical transmission in the brain. The role of amines, amino acids and peptides. Prog Brain Res 55:135–156

Van Leeuwen FW (1982b) Enkephalin immunoreactivity in fibers terminating in a synaptoid fashion on pituicytes in the rat neural lobe. In: Costa E, Trabucchi M (eds) Regulatory peptides. From molecular biology to function. Raven, New York, pp 203–208

Voight KH, Martin R (1985) Coexistence of unrelated neuropeptides in nerve terminals. In: Håkanson R, Thorell J (eds) Biogenetics of neurohormonal peptides. Academic, New York, pp 245–272

Vollrath L (1974) New trends in vertebrate neurosecretion. In: Knowles F, Vollrath L (eds) Neurosecretion – The final neuroendocrine pathway. Springer, Berlin Heidelberg New York, pp 276–284

Warembourg M, Poulain P (1985) Localization of serotonin in the hypothalamus and the mesencephalon of the guinea-pig. An immunohistochemical study using monoclonal antibodies. Cell Tissue Res 240:711–721

Watson SJ, Akil H, Fischli W, Goldstein A, Zimmerman EA, Nilaver G, van Wimersma Greidanus TB (1982) Dynorphin and vasopressin: common localization in magnocellular neurons. Science 216:85–87

Watson SJ, Khachaturian H, Taylor L, Fischli W, Goldstein A, Akil H (1983) Pro-dynorphin peptides are found in the same neurons throughout rat brain: immunocytochemical study. Proc Natl Acad Sci USA 80:891–894

Weitzman ED, Czeisler ChA, Zimmerman JC, Moore-Ede MC (1981) Biological rhythms in man: relationship of sleep-wake, cortisol, growth hormone and temperature during temporal isolation. In: Martin JB, Reichlin S, Bick KL (eds) Neurosecretion in brain peptides. Raven, New York, pp 475–499

Wislocki GB, Leduc EH (1954) The cytology of the subcommissural organ, Reissner's fiber, periventricular glial cells and posterior collicular recess of the rat's brain. J Comp Neurol 101:283–309

Wittkowski W, Scheuer A (1974) Functional changes of the neuronal and glial elements at the surface of the external layer of the median eminence. Z Anat Entwickl Gesch 143:255–262

Wolfson B, Manning RW, Davis LG, Arentzen R, Baldino FJr (1985) Co-localization of corticotropin releasing factor and vasopressin mRNA in neurones after adrenalectomy. Nature 315:59–61

Zingg HH, Patel YC (1980) Processing of somatostatin precursors: Evidence for enzymatic cleavage by hypothalamic extract. Biochem Biophys Res Commun 93:1274–1279

Zlokovic BV, Begley DJ, Chain-Eliash DG (1985) Blood–brain barrier permeability to leucine-enkephalin, D-alanine^2-D-leucine^5-enkephalin and their N-terminal amino acid (tyrosine). Brain Res 336:125–132

Synaptogenesis and Neuronal Plasticity to Gonadal Steroids: Implications for the Development of Sexual Dimorphism in the Neuroendocrine Brain

Y. Arai, A. Matsumoto, and M. Nishizuka

Contents

1 Introduction

Sex differences in adult neuroendocrine and behavioral responses largely depend on differences in the organization of the brain. Many of these are brought about by the exposure of undifferentiated brain to perinatal gonadal hormones, especially aromatizable androgen or estrogen (Goy and McEwen 1980; MacLusky and Naftolin 1981; Arai et al. 1983). Recent studies indicate that sex steroid hormones modulate and promote neuronal maturation and neurite growth in certain brain regions (Toran-Allerand 1984). Synaptogenesis can also be facilitated by estrogenic action in the hypothalamic arcuate nucleus (ARCN) and medial amygdaloid nucleus (MAN) during early postnatal development (Arai 1981). These organizational effects of gonadal steroids appear to be regionally specific and correlated with the presence and topographic localization of the sex steroid receptor-containing neurons (Goy and McEwen 1980).

For the expression of sexually dimorphic neuroendocrine and behavioral functions, some quantitative and/or qualitative difference between males and females, not only in the metabolism of neurotransmitters but also in neural circuitry, may be needed. Recently the presence of sex differences in brain morphology has been documented in various parameters, such as nuclear volume, dendritic pattern, and synaptic organization (Arai 1981; Arnold and Gorski 1984). In this article, morphological correlates of sexually dimorphic neuroendocrine brain are discussed.

Department of Anatomy, Juntendo University School of Medicine, 2-1-1 Hongo, Bunkyo-ku, Tokyo 113, Japan

Current Topics in Neuroendocrinology, Vol. 7

2 Cytoarchitectonic Sex Differences in the Neuroendocrine Brain

In the rat, Gorski and his coworkers (1978) have described a region with a striking sex difference which they have called the sexually dimorphic nucleus of the preoptic area (SDN-POA). The SDN-POA occupies a greater volume and contains more neurons in males than in females. These characteristics are dependent on perinatal gonadal steroid conditions (Döhler et al. 1982). The SDN-POA is embedded within the medial preoptic nucleus of Gurdjian (1927) and corresponds to the central part of the medial preoptic nucleus (Simerly et al. 1984). Analogues of the rat SDN-POA have been also identified in gerbils (Yahr and Commins 1984), ferrets (Tobet et al. 1983), and guinea pigs (Hines et al. 1985). More recently, the presence of an analogue of the SDN-POA in the human brain has been demonstrated (Swaab anf Fliers 1985). Morphometric analysis showed that the SDN-POA is 2.5 times as large in men as in women and contains 2.5 times as many neurons. The functional significance of the SDN-POA is still not clear. There is a report that the absolute volume of the rat SDN-POA does not correlate with ovulatory function or with lordosis responsiveness (Gorski 1984).

Sexual dimorphism has also been observed in other areas of the neuroendocrine brain. On the basis of a comparative neuroanatomical study, Bleier et al. (1982) have postulated the concept of the sexually dimorphic nuclear complex of the medial preoptic–anterior hypothalamic area. Other than the SDN-POA, another major cell group comprising the sexually dimorphic nuclear complex is identified in the rostroventral periventricular gray of the POA just caudal to the organum vasculosum lamina terminalis. At present, there is a discrepancy in its terminology. According to the nomenclature of Bleier et al. (1982), this structure is designated as the medial preoptic nucleus (this is not the medial preoptic nucleus of Gurdjian), but so far its ventral portion has often been called the nucleus preopticus pars suprachiasmatica and the dorsal portion corresponds to the nucleus preopticus periventricularis (König and Klippel 1963). Reevaluation of the terminology for these neural structures is needed. However, the significant finding here by Bleier et al. (1982) is that the rostroventral periventricular gray of the POA and its adjacent area (their medial preoptic nucleus) is larger and more densely cellular in the female rat than that of the male. Furthermore, this structure has been found to play a critical role in regulating the cyclic release of gonadotropins in the female rat (Wiegand et al. 1978; Terasawa et al. 1980). Small lesions confined to this part block spontaneous ovulation, inducing anovulatory persistent estrous syndrome. However, whether the reduction of the volume of this neuronal group in the male rat is dependent on perinatal androgen remains to be shown.

Sex differences in nuclear volume have also been found in the ventromedial nucleus (VMN) (Matsumoto and Arai 1983 a), the bed nucleus of the stria terminalis (Hines et al. 1985), and the MAN (Mizukami et al. 1983). For instance, in the VMN, which is considered to play a key role in regulating various behavioral and neuroendocrine functions, the nuclear volume is significantly greater in males than in females (Fig. 1). Castration of neonatal male rats on the day of birth to produce functionally "feminine males" decreases this volume to a level comparable to that of normal females. However, the VMN volume is no longer influenced

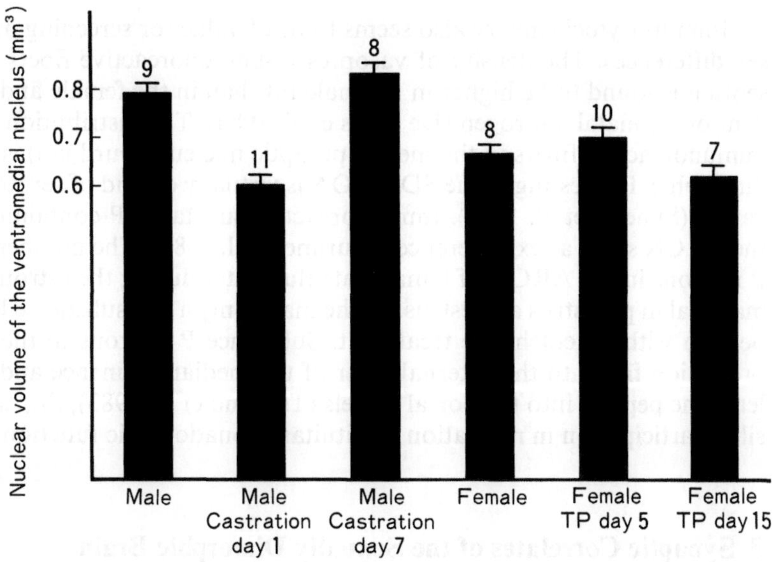

Fig. 1. Sexual dimorphism in volume of the rat VMN and the effect of the manipulation of the neonatal hormonal environment

by castration at 7 days of age. This suggests that the volume of this nucleus is not determined entirely by the genetic factor, but rather the postnatal growth of the VMN is modified by the organizational action of testicular hormones. The critical period for this neonatal imprinting by testicular androgen in the developing VMN seems to be before day 7. A single injection of testosterone propionate (TP) to 5-day-old females does not have any significant influence on the volume of the VMN. However, failure of a single dose of TP to modify the volume of the VMN in females may be due to the time and/or manner of androgen injection. Probably a cumulative effect of androgen during the critical period is needed. The similar sex difference in nuclear volume that has been detected in the MAN can also be modified by manipulation of the neonatal hormone environment (Mizukami et al. 1983).

A sex difference in dendritic branching has been studied in the POA neurons using the Golgi impregnation method. In male hamsters the dendritic field is significantly larger than in females (Greenough et al. 1977). A similar dimorphism has been found in the POA of the monkey brain (Ayoub et al. 1983); neurons of males show more dendritic branching and a higher frequency of spines. A sex difference in the density of the dendritic spines in one type of the neurons, which usually have two to three principal dendrites arranged in a cylindrical pattern, has been found in the MAN in rats; the number of spines is significantly greater in males than in females (Nishizuka et al. unpublished). Since spines are thought to be a postsynaptic element, and the width of the dendritic field correlates with the neuronal receptivity, the presence of a sex difference in these parameters may suggest that the capacity for processing neural information is different between the two sexes in these brain areas.

Immunocytochemistry also seems to be of value for screening morphological sex differences. The density of vasopressin-immunoreactive fibers in the lateral septum is found to be higher in the male rat than in the female and to be dependent on neonatal androgen (De Vries et al. 1984). The distribution of serotonin-immunoreactive fibers in the medial preoptic nucleus (Gurdjian) is also sexually dimorphic. Interestingly, the SDN-POA is virtually devoid of serotonergic innervation (Simerly et al. 1984). Immunoreactive substance P-containing neurons in the ARCN show a sex difference (Tsuruno et al. 1984). The number of substance P neurons in the ARCN of female rats fluctuates during the estrous cycle, being maximal in proestrus and estrus. In the male, only a few substance P neurons can be seen without colchicine treatment. Substance P neurons in the ARCN send projection fibers to the external layer of the median eminence and probably release the peptide into the portal vessels (Tsuruno et al. 1983); this suggests a possible participation in regulation of pituitary gonadotropic function.

3 Synaptic Correlates of the Sexually Dimorphic Brain

Raisman and Field (1973) provided the first electron microscopic evidence for a sex difference in the types of dendritic synapses found in the rat dorsomedial POA. The number of dendritic spine synapses of nonamygdaloid origin in females is greater than in males. This difference can be reversed appropriately by neonatal manipulation of androgen level – removal of neonatal testes or injection of androgen. A sexually dimorphic synaptic pattern has also been demonstrated in four other regions: the ARCN (Matsumoto and Arai 1980), the VMN (Matsumoto and Arai 1986), the suprachiasmatic (SCN) (Guldner 1982; Le Blond et al. 1982), and the MAN (Nishizuka and Arai 1981 a).

In those VMN that show a regional difference in distribution of sex steroid receptors (Pfaff and Keiner 1973), a sexually dimorphic synaptic pattern is found in the ventrolateral part of the VMN (VL-VMN) where sex steroid receptors are abundant. In contrast no sex difference is seen in the dorsomedial part of the VMN (DM-VMN), which contains only a few sex steroid receptors (Matsumoto and Arai 1986). As shown in Fig. 2, the number of shaft and spine synapses in the VL-VMN of normal male rats are significantly greater than those in the VL-VMN of normal females. Neonatal castration of the males on day 1 significantly reduces the numbers of shaft and spine synapses in the VL-VMN to levels comparable to those of normal females. On the other hand neonatal treatment of females with TP significantly increases the number of shaft and spine synapses in the VL-VMN to a level comparable to that of normal males. The numbers of shaft and spine synapses in the DM-VMN are not significantly different among the groups examined. The synaptic pattern of the VL-VMN of normal females and neonatally castrated males, which were not exposed to the organizational action of androgen, is not significantly different from that of the DM-VMN. These results confirm the importance of the neonatal sex steroid environment for the development of the sexually dimorphic synaptic pattern of the VMN. Furthermore, a high correlation between the sexually dimorphic synaptic pattern and the pres-

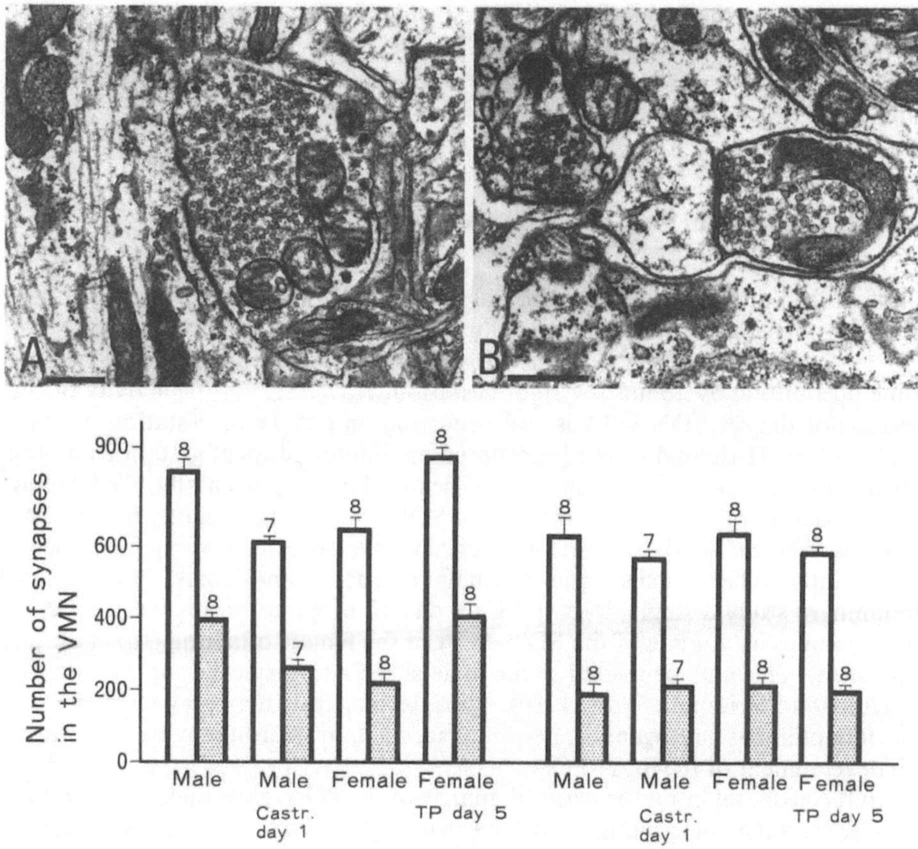

Fig. 2 A, B. Synapse in the neuropil of the VMN. **A** Shaft synapse, × 20800. **B** Spine synapse, × 28300. *Scale,* 0.5 μm. The *lower figure* indicates the number of shaft and spine synapses in the VL-VMN (*left*) and DM-VMN (*right*) per 10000 μm^2 in normal males and females, males castrated on the day of birth, and females treated with 1.25 mg TP. *White column* and *dotted column* indicate shaft and spine synapses, respectively

ence of neurons containing sex steroid receptors may indicate that the occurrence of synaptic sexual differentiation is rather specific to the sex steroid-concentrating neuronal system. It is of interest to note that neural outputs of the VMN projecting to the midbrain central gray are sexually dimorphic in neural connectivity and sensitivity to estrogen, both of which are critically dependent on the neonatal sex steroid environment (Sakuma and Pfaff 1981; Sakuma 1984). The sex difference in both inputs and outputs of the VMN may reflect the different behavioral functions – its role in regulating sexual, feeding, and aggressive behavior (Pfaff 1980; Yamanouchi et al. 1985).

Other regions which show a sexually dimorphic synaptic pattern, the dorsomedial POA, the ARCN, the SCN, and the MAN, are all (except the SCN) known to be abundant in sex steroid receptors (Pfaff and Keiner 1973). For the SCN the possible factors inducing a sexually dimorphic synaptic pattern may partly arise from the facts that the afferent side converges from the sex steroid-concentrating neuronal system to the SCN. One supportive piece of evidence for this is that sex steroids promote axonal growth in neonatal preoptic and hypothalamic tissue in vitro (Toran-Allerand 1976).

4 Synaptogenesis in the Developing Neuroendocrine Brain

The final cell divisions of the neuroblasts which give rise to preoptic, hypothalamic, or amygdaloid neurons occur during days 11 to 18 of gestation in the rat (Ifft 1972; Altman and Bayer 1978). Most of the POA neurons, for example, become postmitotic by about day 16 of gestation. However, mitotic activity of the neurons of the rat SDN-POA is still occurring on day 18 of gestation (Gorski 1984). When ^{3}H-thymidine is administered on different days of gestation and the labeling index is examined at the time of death (day 30) postnatally, the labeling index at day 14 of gestation in the female SDN-POA is significantly higher than in males, whereas on day 17 this difference is reversed, the labeling index being significantly higher in males than in females (Jacobson and Gorski 1981). In our preliminary study a single dose of TP on day 17 of gestation was very effective in increasing the volume of the SDN-POA in the female offspring (Ito et al. unpublished). This may be related to the male-specific testosterone surge on day 18 of gestation (Weisz and Ward 1980). Thus, testosterone may play an important role in modifying neurogenesis, neuronal survival, or neuronal migration during the development of the SDN-POA.

In the neonatal brain the neuropil matrix of the POA, hypothalamic, and limbic areas is still in an immature state characterized by the presence of extracellular space and growth cones and a paucity of synapses (Fig. 3). The major neural cir-

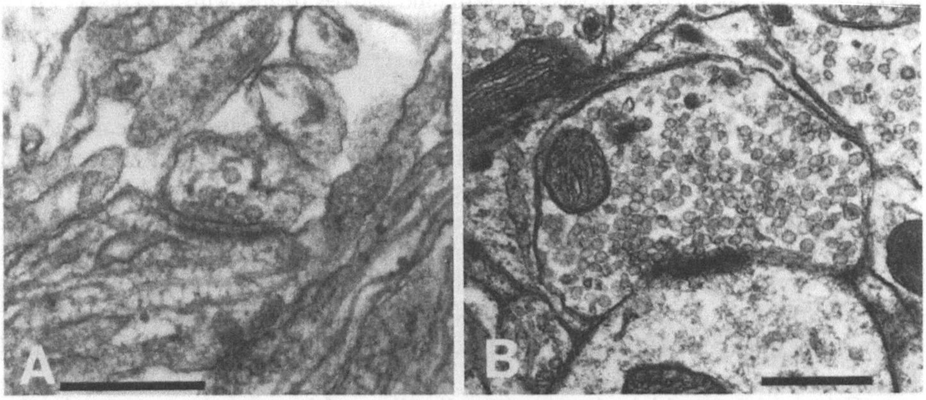

Fig. 3. A Axodendritic shaft synapse in the VMN of a 5-day-old female rat. Note the presence of extracellular space, that the axon terminal contains only a small number of synaptic vesicles, and that pre- and postsynaptic membrane specialization are poorly developed. × 38000. **B** Shaft synapse in the VMN of a 100-day-old female rat, × 29800. *Scale,* 0.5 μm

cuits driving postpubertal neuroendocrine and/or behavioral functions are not yet established. Thus, the neuropil environment of these areas is easily accessible to the organizational action of sex steroids (Arai 1981).

Using in vitro culture methods, stimulation of axonal and dendritic growth in the preoptic and hypothalamic explants in response to sex steroid treatment has been reported (Toran-Allerand 1976; Toran-Allerand et al. 1983; Uchibori and Kawashima 1985). In the brain transplantation experiment the growth of the newborn rat POA tissues grafted into the third ventricle is also effectively enhanced by the treatment with estrogen via the host (Arai et al. 1986). The volume of the POA grafts exposed to estrogen via the host are significantly greater than the volume of those not exposed. However, the volume of the cortical graft is not influenced by estrogen (Fig. 4).

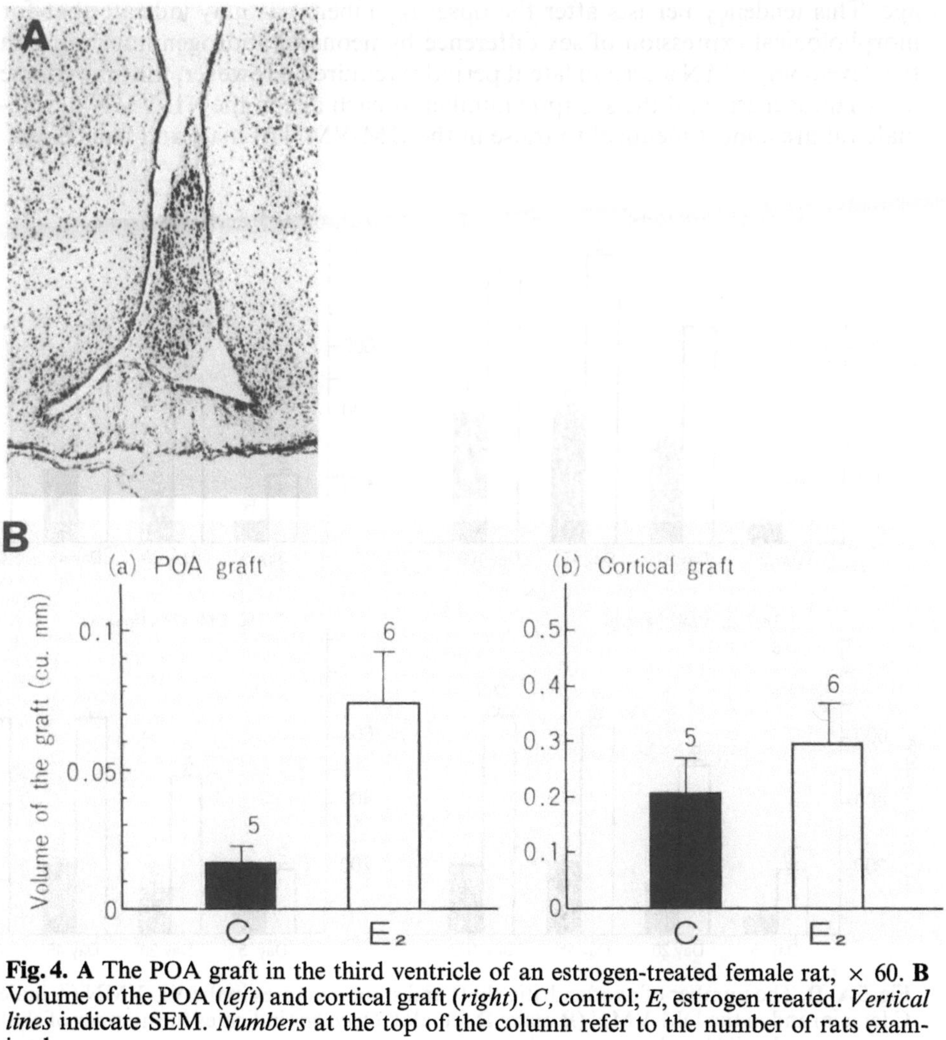

Fig. 4. **A** The POA graft in the third ventricle of an estrogen-treated female rat, × 60. **B** Volume of the POA (*left*) and cortical graft (*right*). *C*, control; *E*, estrogen treated. *Vertical lines* indicate SEM. *Numbers* at the top of the column refer to the number of rats examined

Maturation of the neuropil may vary among the neuronal groups. As mentioned before, the synaptic number at neonatal ages is quite small in the neuroendocrine brain, compared to that after the onset of puberty. In the MAN the synaptic density reaches the adult level before day 21 (Nishizuka and Arai 1981 b). The SCN also shows a similar tendency (Le Blond et al. 1982). The synaptic density of the POA reaches a plateau between days 20 and 30 (Lawrence and Raisman 1980). In the ARCN the completion of synaptogenesis is delayed to around the onset of puberty (Matsumoto and Arai 1976a, 1981a; Arai and Matsumoto 1978). Figure 5 shows the time course of synaptogenesis and the development of a sexually dimorphic synaptic pattern in the VMN. Synaptic number in the VMN is increased to 70%–80% of that at postpubertal age during the first 20 days. Although a sex difference in synaptic density is not detected at 5 days of age, a synaptic sex difference in the VL-VMN develops by 20 days of age. This tendency persists after the onset of puberty. It may indicate that for morphological expression of sex difference by neonatal androgen imprinting in the developing VMN a certain latent period is required. However, the time course of synaptogenesis and the synaptic number at each age in the VL-VMN of a female rat are almost identical to those in the DM-VMN of male and female rats.

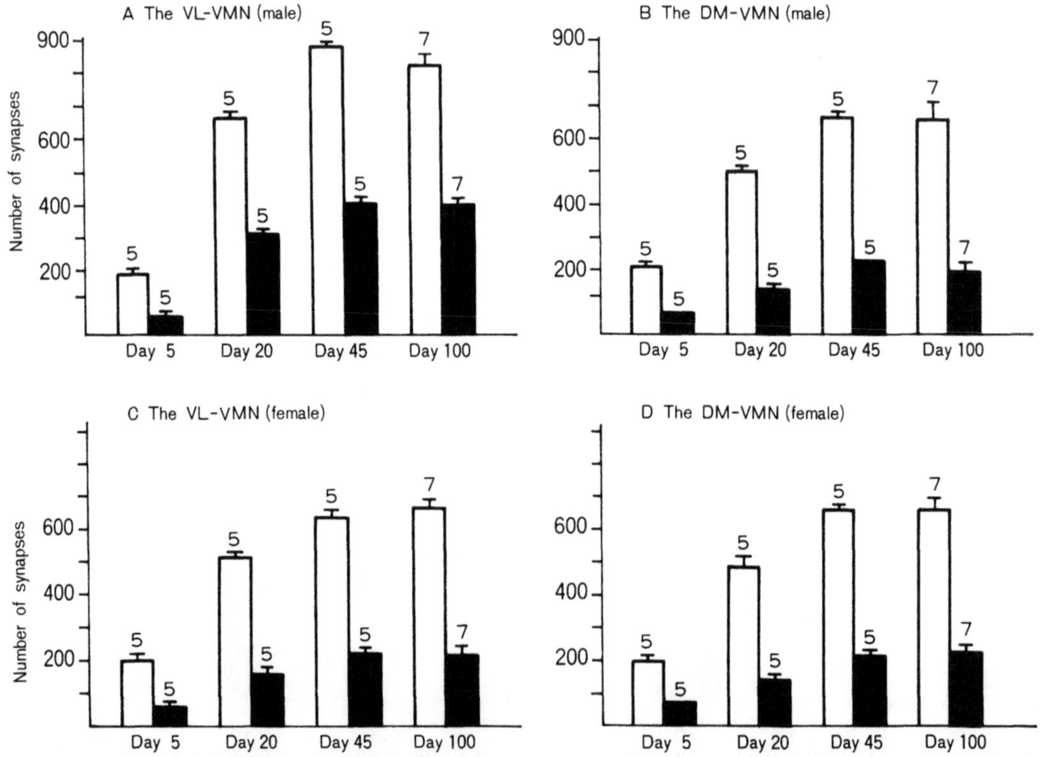

Fig. 5 A–D. The number of axodendritic shaft and spine synapses in the VL-VMN (**A** male; **C** female) and in the DM-VMN (**B** male; **D** female). *Open bar* indicates the number of shaft synapses, and *solid bar* the number of spine synapses

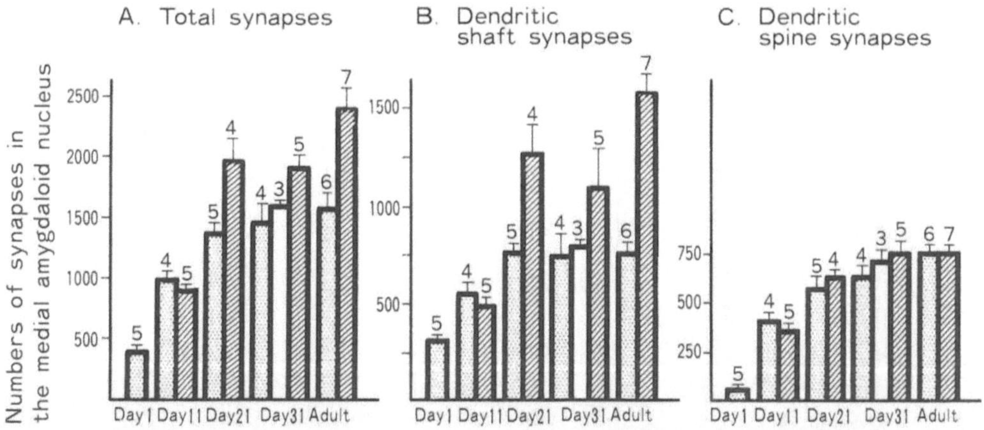

Fig. 6 A–C. Number of the total (**A**), dendritic shaft (**B**), and dendritic spine (**C**) synapses per 10 000 μm^2 in the MAN of untreated (*dotted bar*), oil (*ofen bar*), and neonatally-estrogenized (*hatched bar*) female rats at 1, 11, 21, 31 and 90–100 days of age

Again, this suggests that the basic synaptogenic program of the VMN can be modified by the presence of sex steroids and their receptors in neonatal days.

Estrogen has been found to facilitate synaptogenesis markedly during the postnatal development of the ARCN (Matsumoto and Arai 1976b; Arai and Matsumoto 1978) and MAN (Nishizuka and Arai 1981b). In the MAN estrogen given in early postnatal days specifically promotes shaft synaptogenesis (Fig. 6). In this nucleus, the total number of synapses is actually larger in males and androgenized females than in females and neonatally castrated males (Nishizuka and Arai 1981a). This sex difference can be attributed to a significant increase in the number of shaft synapses in response to the organizational action of testicular steroids in males. In addition, shaft synaptogenesis in the newborn medial amygdaloid tissue grafted into the anterior eye chamber of adult ovariectomized rats is also markedly facilitated by estrogen given via the host animals (Nishizuka and Arai 1982). The predominance of shaft synaptogenesis in response to the organizational action of estrogen seems to be a characteristic feature of the neurons of the medial amygdala. Estrogen also stimulates synaptogenesis in newborn rat POA tissue grafted into the third ventricle of adult ovariectomized rats (Arai et al. 1986). The fact that the numbers of shaft and spine synapses in the POA grafts exposed to estrogen through the host are significantly increased compared to those without estrogen treatment (Fig. 7) may indicate that axonal growth, terminal arborization, dendritic branching, and spine formation are stimulated in the grafted POA by estrogen (Fig. 8).

The mechanism of estrogen-stimulated synaptogenesis is still not fully understood. According to Toran-Allerand (1984) gonadal steroids exert organizational effects on these processes in a cascading manner. Stimulation of axonal and dendritic growth by steroids could lead to sex differences in neuron size, number, and synaptic pattern. Compared to steroid-insensitive neurons, those stimulated by sex steroids may make more effective synaptic contacts which can promote their

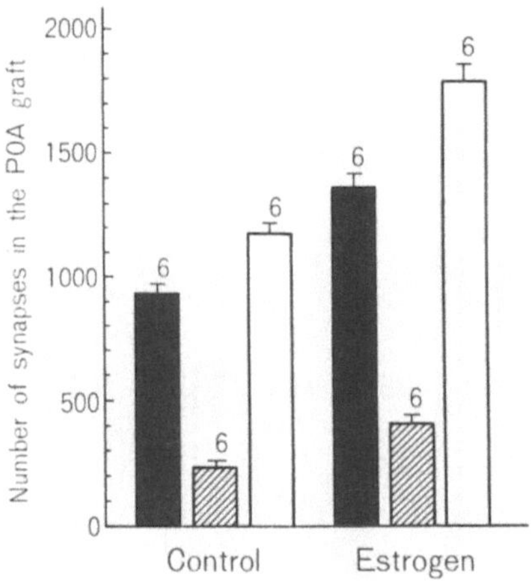

Fig. 7. The number of shaft, spine, and total synapses per 10 000 μm^2 of the POA grafts in the control and estrogen-treated females. *Solid bar,* shaft synapse; *hatched bar,* spine synapse; *open bar*, total synapses

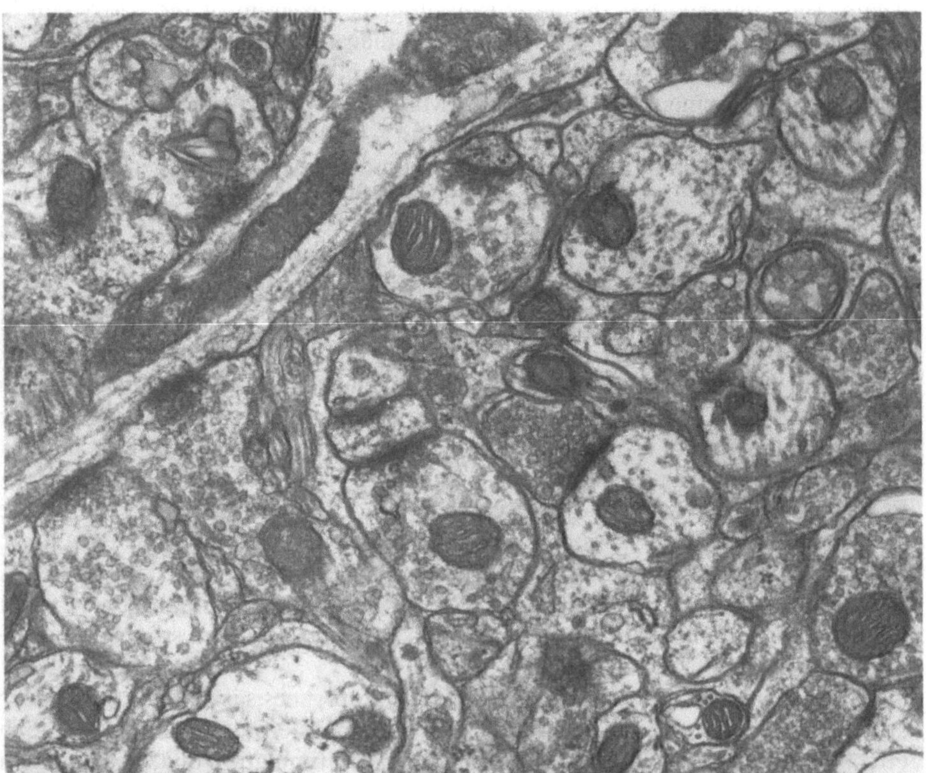

Fig. 8. Neuropil of the POA graft in an estrogen-treated host female rat, × 32 000; *scale,* 0.5 μm

subsequent survival during a period of cell death. The fact that estrogen facilitates synaptogenesis in medial amygdala grafted into the eye (Nishizuka and Arai 1982) suggests that estrogen would stimulate not only axonal growth of the neurons in the graft but also of the postsynaptic elements of the neurons. There is evidence that α-bungarotoxin binding capacity in the medial amygdala is sexually dimorphic and can be permanently modified by neonatal sex steroid manipulation (Arimatsu 1983; Arimatsu et al. 1981). Furthermore, an addition of estrogen to a culture medium of amygdala cell aggregates causes a significant increase in the toxin-binding capacity (Arimatsu et al. 1985). Since a spurt of estrogen-induced synaptogenesis in the MAN in situ is seen between days 11 and 21 after estrogen exposure (Nishizuka and Arai 1981 b), neonatal activation of neurotransmitter receptors by estrogen may be correlated with the onset of synaptogenesis.

5 Estrogen and Synaptogenesis in the Peripubertal and Adult Neuroendocrine Brain

As well as its organizational effect in the neonate brain, estrogen has a facilitatory effect on synapse formation in peripubertal and adult rats. As mentioned before, the synaptic density of the ARCN reaches a plateau around the onset of puberty. In the experimental situation in which precocious puberty is induced by pregnant mare serum gonadotropin (PMSG) injection into prepubertal rats, synaptogenesis in the ARCN can be accelerated (Matsumoto and Arai 1977). When a single dose of 20 IU PMSG is given to 28-day-old female rats, precocious ovulation occurs 3 days later. Concurrently the synaptic number reaches an adult level within 3 days. Since PMSG fails to increase the number of synapses in the ARCN in ovariectomized rats, the ovarian estrogen would appear to play a facilitatory role in the maturation of synaptic organization in the ARCN in PMSG-induced precocious puberty. Similar precocious synaptogenesis has been reported in immature rats in which precocious ovulation can be induced by the administration of estrogen (Clough and Rodriguez-Sierra 1983) or by the placement of an electrolytic lesion in the basal hypothalamus (Ruf 1982).

Estrogen also facilitates synaptogenesis in the ARCN of the adult brain. In this case, however, an injury to the neural substrates is needed for the subsequent expression of the facilitatory effect of estrogen, probably because the vacated synaptic sites are made available by the lesions (Matsumoto and Arai 1979, 1981 b). Three weeks after complete deafferentation of the medial basal hypothalamus (MBH-island), the number of shaft and spine synapses in the ARCN decreases to two-fifths and three-fifths of the intact controls respectively, because of transection of the inputs to the MBH (Table 1). However, estrogen treatment for 3 weeks produces a marked increase in the number of both shaft and spine synapses in the ovariectomized MBH-island females, with the number of shaft synapses being restored to almost 75% of the intact level. The incidence of spine synapses is also significantly greater than that in the intact control animals. In these estrogen-treated MBH-island rats, double synapses (spine–spine and spine–

Table 1. Effect of estrogen on numbers of axodendritic shaft and spine synapses in the arcuate nucleus with or without complete deafferentation in the medial basal hypothalamus

Group	Treatment	No. of rats	Axodendritic shaft synapses	Axodendritic spine synapses
1	Intact control	8	1544 ± 91[a] (22.6 ± 2.6)[b]	215 ± 23[a] (0.62 ± 0.62)[b]
2	OVX	6	1539 ± 87 (18.2 ± 3.4)	207 ± 23 (0.33 ± 0.21)
3	OVX-EB	6	1636 ± 153 (18.2 ± 3.1)	241 ± 37 (0.33 ± 0.33)
4	MBH island-OVX	9	696 ± 46 (6.75 ± 1.10)	156 ± 20 (0.50 ± 0.50)
5	MBH island-OVX-EB	8	1159 ± 70 (17.0 ± 2.0)	380 ± 24 (2.00 ± 0.71)

[a] Mean ± SEM
[b] The number of synapses containing 5-hydroxydopamine labeled small granular vesicles is given in parenthesis. In order to differentiate catecholaminergic terminals from non-catecholaminergic terminals, intraventricular infusion of 5-hydroxydopamine was made before death. Animals were killed 21 day following the operation.
OVX, ovariectomy; EB, treatment with 2 µg of estradiol benzoate for 20 days

shaft double synapses) are frequently observed. The occurrence of double synapses has been discussed in relation to neuronal plasticity during the process of recovery from brain lesions (Raisman 1969). In ovariectomized females without MBH deafferentation, however, estrogen fails to increase the numbers of shaft, spine, and double synapses, which are almost comparable to the intact and ovariectomized controls. These results suggest that estrogen enhances neuronal plasticity of adult ARCN neurons and stimulates axonal sprouting and dendritic spine formation in the intact ARCN neurons in the MBH-island.

In the VL-VMN, terminals and synaptic contacts have been found to be increased after estrogen treatment (Carrer and Aoki 1982), associated with an increase in the number of cored vesicles in perikarya in adult animals (Cohen and Pfaff 1981). In this region synaptic circuits seem to be plastic and subject to spontaneous remodelling in the adult brain. Degenerating terminals coexisted with growth cones and possible signs of axonal or dendritic regeneration (Van Houten and Brawer 1978). Availability of vacated synaptic sites during the process of spontaneous remodelling may enable estrogen to stimulate synaptogenesis in this neuropil environment. The response of the neuropil matrix and the neuronal perikarya of the VL-VMN of adult females to estrogen is activational in nature. However, the possibility that the response is one of the organizational events continuing into adult life cannot be excluded. In canaries, for example, certain neural structures of the striatum controlling song exhibit seasonal changes in synaptic number that parallel seasonal changes in androgen and singing (De Voogd et al. 1982), suggesting that this type of morphological response in the adult brain is activational. Thus, the stimulation of synaptogenesis by sex steroids in the adult brain may be dependent on the potential for axonal or synaptic reformation in the neuropil environment.

On the other hand, administration of a large dose of a long-acting estrogen (estradiol valerate) to young cycling female rats induces neuronal degeneration with astrocytic and microglial hypertrophy in the ARCN (Brawer et al. 1978,

1980). The ARCN degeneration appears to be associated with a reduction of the LH surge after stimulation of the POA, suggesting the occurrence of a functional disconnection between the POA and the MBH. Furthermore, a marked decrease in synaptic number, especially that of somatic and shaft synapses, is found in the ARCN following the injection of a long-acting estrogen. In this case, however, degenerative changes in the cell bodies apparently preceded the estrogen-induced decrease in synaptic number (Garcia-Segura et al. 1986). These pathological changes may be due to a neurotoxic effect of estrogen when specific target neurons are continuously exposed to large doses of it.

6 Summary and Conclusion

Sex steroids exert very complex effects on the neuroendocrine brain. Whether the effect is organizational or activational, and whether neurotropic or neurotoxic, the morphological responses to sex steroids are primarily specific to the steroid receptor-containing neuronal system. The circuitries involved in neuroendocrine and behavioral functions, especially regulation of gonadotropin secretion and sexual behaviors, are under the direct control of gonadal steroids throughout life. In perinatal animals aromatizable androgen or estrogen affects brain sexual differentiation by regulating the number of target neurons, stimulating axonal and dendritic growth, and modifying synaptic connectivity (Arai 1981; Gorski 1984; Toran-Allerand 1984). The exact mechanisms for these organizational actions of gonadal steroids remain to be discovered.

The onset of puberty is dependent on a complex process of neuroendocrinological events which involve the maturation of the hypothalamic pituitary gonadal axis. This may also be correlated with a sex steroid facilitation of synapse formation in the neuroendocrine brain to complete the neural circuits that operate postpubertal feedback regulation of gonadotropin secretion (Matsumoto and Arai 1977; Ruf 1982; Clough and Rodriguez-Sierra 1983). In adult animals sex steroids can affect synaptic remodelling in certain neural substrates which possess considerable plasticity and where vacated synaptic sites are made available by spontaneous degeneration (Carrer and Aoki 1982) or surgical deafferentation (Matsumoto and Arai 1979). However, a long-lasting estrogen such as estradiol valerate appears to exert a dystrophic effect, resulting in degeneration of the ARCN neurons associated with glial hypertrophy (Brawer et al. 1978, 1980). In the process of aging, gliosis is found to occur spontaneously in the ARCN (Schippers et al. 1981; Matsumoto and Arai 1983 b). The development of the age-related gliosis in the ARCN appears to be dependent on estrogen, since ovariectomy at 2 months of age significantly suppresses gliosis later (Schippers et al. 1981). The continuous exposure of the brain to estrogen might be at least one of the factors that cause hypothalamic aging. At the synaptic level, however, there is evidence suggesting that estrogen may prevent certain neuronal groups from age-dependent decrease of synaptic density, presumably by facilitating neuronal plasticity. A striking decrease in synaptic number has been reported to occur in the ARCN of aged male rats, whereas in aged females the synaptic number is not markedly decreased, suggesting the possible existence of a sex difference in synaptic degeneration here (Matsumoto and Arai 1983 b). The observed reduction of synaptic number might be partly due to the loss of neuronal populations in the ARCN of aged males and females. When island isolation of the MBH is performed in aged female rats, however, estrogen is capable of restoring the synaptic population in the ARCN to approximately 60% of that in intact aged females (Matsumoto and Arai 1985). This value is somewhat less than the 90% recovery obtained in young adult females subjected to the same experimental procedure (Matsumoto and Arai 1979). Thus, the ARCN neurons of aged females still retain the plasticity to react under the influence of estrogen after deafferentation.

Morphological response of the neuroendocrine brain to gonadal steroids is detectable at the different sequences of the developmental process and after aging. Whether the response is permanent or not seems to depend on the time at which the hormones act on the neural substrates. A dose-dependent response to perinatal gonadal steroids is of special interest. In the MAN, for example, the synaptic number of females treated with large doses of estrogen early postnatally is significantly greater than that of normal males (Nishizuka and Arai 1981 b). The excess or reduction of available gonadal steroids during neonatal days may cause overproduction or underproduction of synapses, consequently bringing about a less standard neural circuit ("wiring") pattern. A possible outcome would be some defective morphological or functional development of the neuroendocrine brain. This evidence provides an anatomical basis for neuronal differentiation in the neuroendocrine brain. The next problem is to discover how these morphological parameters correlate with neuroendocrine and behavioral functions.

Acknowledgment. This work from the author's laboratory was supported by research grants from the Japanese Ministry of Education, Culture, and Science, the Ministry of Health and Welfare, and Sasagawa Foundation for Medical Research.

References

Altman J, Bayer SA (1978) Development of the diencephalon in the rat. I. Autoradiographic study of the time of origin and settling patterns of neurons of the hypothalamus. J Comp Neurol 182:945–978

Arai Y (1981) Synaptic correlates of sexual differentiation. Trends Neurosci 4:291–293

Arai Y, Matsumoto A (1978) Synapse formation of the hypothalamic arcuate nucleus during post-natal development in the female rat and its modification by neonatal estrogen treatment. Psychoneuroendocrinology 3:31–45

Arai Y, Mori T, Suzuki Y, Bern HA (1983) Long-term effects of perinatal exposure to sex steroids and diethylstilbestrol on the reproductive system of male mammals. Int Rev Cytol 84:235–268

Arai Y, Matsumoto A, Nishizuka M, Murakami S (1986) Neurotropic action of estrogen on the developing brain: effect on newborn preoptic and amygdaloid tissues grafted into the brain or into the eye. In: Dörner G (ed) Systemic hormones, neurotransmitters and brain development. Karger, Basel (to be published)

Arimatsu Y (1983) Short- and long-lasting influences of neonatal sex steroids on α-bungarotoxin binding capacity in the mouse amygdala. Neurosci 9:873–877

Arimatsu Y, Seto A, Amano T (1981) Sexual dimorphism in α-bungarotoxin binding capacity in the mouse amygdala. Brain Res 213:432–437

Arimatsu Y, Kondo S, Kojima M (1985) Enhancement by estrogen treatment of α-bungarotoxin binding in fetal mouse amygdala cell reaggregated in vitro. Neurosci Res 2:211–220
Arnold AP, Gorski RA (1984) Gonadal steroid induction of structural sex differences in the central nervous system. Annu Rev Neurosci 7:423–442
Ayuob DM, Greenough WT, Juraska JM (1983) Sex differences in dendritic structure in the preoptic area of the juvenile macaque monkey brain. Science 219:197–198
Bleier R, Byne W, Siggelkow I (1982) Cytoarchitectonic sexual dimorphisms of the medial preoptic and anterior hypothalamic areas in guinea pig, rat, hamster, and mouse.J Comp Neurol 212:118–130
Brawer JR, Naftolin F, Martin J, Sonnenschein C (1978) Effects of a single injection of estradiol valerate on the hypothalamic arcuate nucleus and on reproductive function in the female rat. Endocrinology 103:501–512
Brawer JR, Ruf KB, Naftolin F (1980) Effects of estradiol-induced lesions of the arcuate nucleus on gonadotropin release in response to preoptic stimulation in the rat. Neuroendocrinology 30:144–149
Carrer HA, Aoki A (1982) Ultrastructural changes in the hypothalamic ventromedial nucleus of ovariectomized rats after estrogen treatment. Brain Res 240:221–233
Clough RW, Rodriguez-Sierra JF (1983) Synaptic changes in the hypothalamus of the prepubertal female rat administered estrogen. Am J Anat 167:205–214
Cohen R, Pfaff DW (1981) Ultrastructure of neurons of the ventromedial nucleus of the hypothalamus in ovariectomized rats with or without estrogen treatment. Cell Tissue Res 217:451–470
Commins D, Yahr P (1984) Adult testosterone levels influence the morphology of a sexually dimorphic area in the mongolian gerbil brain. J Comp Neurol 224:132–140
De Voogd TJ, Nixdorf B, Nottebohm F (1982) Recruitment of additional synapses into a brain network takes extra brain space. Soc Neurosci Abstr 8:140
De Vries GJ, Buijs RM, Van Leeuwen FW (1984) Sex differences in vasopressin and other neurotransmitter system in the brain. In: De Vries GJ, De Bruin JPC, Uyling HBM, Corner MA (eds) Sex differences in the brain. The relation between structure and function. Prog Brain Res 61:185–203
Döhler KD, Coquelin A, Davis F, Hines M, Shryne JE, Gorski RA (1982) Differentiation of the sexually dimorphic nucleus in the preoptic area of the rat brain is determined by the perinatal hormone environment. Neurosci Lett 33:295–298
Garcia-Segura LM, Baetens D, Naftolin F (1986) Synaptic remodelling in arcuate nucleus after injection of estradiol valerate in adult female rats. Brain Res 366:131–136
Gorski RA (1984) Critical role for the medial preoptic area in the sexual differentiation of the brain. In: De Vries GJ, De Bruin JPC, Uyling HBM, Corner MA (eds) Sex differences in the brain. The relation between structure and function. Prog Brain Res 61:129–146
Gorski RA, Gordon JH, Shryne JE, Southam AM (1978) Evidence for a morphological sex difference within the medial preoptic area of the rat brain. Brain Res 148:333–346
Goy RW, McEwen BS (1980) Sexual differentiation of the brain. MIT Press, Cambridge
Greenough WT, Carter CS, Steerman C, DeVoogd TJ (1977) Sex differences in dendritic patterns in hamster preoptic area. Brain Res 126:63–72
Guldner FH (1982) Sex dimorphism of axo-spine synapses and postsynaptic density material in the suprachiasmatic nucleus of the rat. Neurosci Lett 28:145–150
Gürdjian ES (1927) The diencephalon of the albino rat. J Comp Neurol 43:1–114
Hines M, Davies FC, Coquelin A, Goy RW, Gorski RA (1985) Sexually dimorphic regions in the medial preoptic area and the bed nucleus of the stria terminals of the guinea pig brain: a description and an investigation of their relationship to gonadal steroids in adulthood. J Comp Neurol 144:193–204
Ifft JD (1972) An autoradiographic study of the time of final division of neurons in rat hypothalamic nuclei. J Comp Neurol 144:193–204
Jacobson CD, Gorski RA (1981) Neurogenesis of the sexually dimorphic nucleus of the preoptic area of the rat. J Comp Neurol 196:519–524
König JFR, Klippel RA (1963) The rat brain: A stereotaxic atlas of the forebrain and lower parts of the brain stem: Williams and Wilkins, Baltimore

Lawrence JM, Raisman G (1980) Ontogeny of synapses in a sexually dimorphic part of the preoptic area in the rat. Brain Res 183:466–471
Le Blond CB, Morris S, Karakiulakis G, Powell R, Thomas PJ (1982) Development of sexual dimorphism in the suprachiasmatic nucleus of the rat. J Endocrinol 95:137–145
Mac Lusky NJ, Naftolin F (1981) Sexual differentiation of the central nervous system. Science 211:1294–1303
Matsumoto A, Arai Y (1976a) Developmental changes in synaptic formation in the hypothalamic arcuate nucleus of female rats. Cell Tissue Res 169:143–156
Matsumoto A, Arai Y (1976b) Effect of estrogen on early postnatal development of synaptic formation in the hypothalamic arcuate nucleus of female rats. Neurosci Lett 2:79–82
Matsumoto A, Arai Y (1977) Precocious puberty and synaptogenesis in the hypothalamic arcuate nucleus in pregnant mare serum gonadotropin (PMSG) treated immature female rats. Brain Res 129:275–278
Matsumoto A, Arai Y (1979) Synaptogenic effect of estrogen on the hypothalamic arcuate nucleus of the adult female rats. Cell Tissue Res 198:427–433
Matsumoto A, Arai Y (1980) Sexual dimorphism in "wiring pattern" in the hypothalamic arcuate nucleus and its modification by neonatal hormonal environment. Brain Res 190:238–242
Matsumoto A, Arai Y (1981a) Effect of androgen on sexual differentiation of synaptic organization in the hypothalamic arcuate nucleus: an ontogenetic study. Neuroendocrinology 33:166–169
Matsumoto A, Arai Y (1981b) Neuronal plasticity in the deafferented hypothalamic arcuate nucleus of adult female rats and its enhancement by treatment with estrogen. J Comp Neurol 197:197–205
Matsumoto A, Arai Y (1983a) Sex difference in volume of the ventromedial nucleus of the hypothalamus in the rat. Endocrinol Jpn 30:277–280
Matsumoto A, Arai Y (1983b) Synaptic changes in the hypothalamus of old rats. In: Endroczi E, de Wied D, Angelucci L, Scapagnini U (eds) Integrative neurohumoral mechanisms. Elsevier, Amsterdam, pp 401–407
Matsumoto A, Arai Y (1985) Estrogen stimulates neuronal plasticity in the deafferented hypothalamic arcuate nucleus in aged female rats. Neurosci Res 2:412–418
Matsumoto A, Arai Y (1986) Male–female difference in synaptic organization of the ventromedial nucleus of the hypothalamus in the rat. Neuroendocrinology 42:232–236
Mizukami S, Nishizuka M, Arai Y (1983) Sexual difference in nuclear volume and its ontogeny in the rat amygdala. Exp Neurol 79:569–575
Nishizuka M, Arai Y (1981a) Sexual dimorphism in synaptic organization in the amygdala and its dependence on neonatal hormone environment. Brain Res 212:31–38
Nishizuka M, Arai Y (1981b) Organizational action of estrogen on synaptic pattern in the amygdala: implications for sexual differentiation of the brain. Brain Res 213:422–426
Nishizuka M, Arai Y (1982) Synapse formation in response to estrogen in the medial amygdala developing in the eye. Proc Natl Acad Sci USA 79:7024–7026
Pfaff DW (1980) Estrogens and brain function. Neural analysis of a hormone-controlled mammalian reproductive behavior. Springer, New York Heidelberg Berlin
Pfaff DW, Keiner M (1973) Atlas of estradiol-concentrating cells in the central nervous system of the female rats. J Comp Neurol 151:121–158
Raisman G (1969) Neuronal plasticity in the septal nuclei of the adult rat. Brain Res 14:25–48
Raisman G, Field PM (1973) Sexual dimorphism in the neuropil of the preoptic area of the rat and its dependence on neonatal androgen. Brain Res 54:1–29
Ruf KB (1982) Synaptogenesis and puberty. In: Ruf KB, Tolis G (eds) Front Horm Res 10. Advances in neuroendocrine physiology. Karger, Basel, pp 65–72
Sakuma Y (1984) Influences of neonatal gonadectomy or androgen exposure on the sexual differentiation of the rat ventromedial hypothalamus. J Physiol (Lond) 349:273–286
Sakuma Y, Pfaff DW (1981) Electrophysiological determination of projections from ventromedial hypothalamus to midbrain central gray: differences between female and male rats. Brain Res 225:184–188

Schippers H, Brawer JR, Nelson JF, Finch CE (1981) Role of the gonads in the histologic aging of the hypothalamic arcuate nucleus. Biol Reprod 25:413–414

Simerly RB, Swanson LW, Gorski RA (1984) Demonstration of a sexual dimorphism in the distribution of serotonin-immunoreactive fibers in the medial preoptic nucleus of the rat. J Comp Neurol 225:151–166

Swaab DF, Fliers E (1985) A sexually dimorphic nucleus in the human brain. Science 228:1112–1115

Terasawa E, Wiegand SJ, Bridson WE (1980) A role for the medial preoptic nucleus on afternoon of proestrus in female rats. Am J Physiol 238:533–539

Tobet SA, Gallagher CA, Zahniser DJ, Cohen MH, Baum MJ (1983) Sexual dimorphism in the preoptic/anteriorhypothalamic area of adult ferrets. Endocrinology [Suppl] 112:240

Toran-Allerad CD (1976) Sex steroids and the development of the newborn mouse hypothalamus and preoptic area in vitro: implications for sexual differentiation. Brain Res 106:407–412

Toran-Allerand CD (1984) On the genesis of sexual differentiation of the central nervous system: morphogenic consequences of steroid exposure and possible role of α-fetoprotein. In: De Vries GJ, De Bruin JPC, Uyling HBM, Corner MA (eds) Sex differences in the brain. The relation between structure and function. Prog Brain Res 61:63–98

Toran-Allerand CD, Hashimoto K, William T, Greenough WT, Saltarelli M (1983) Sex steroids and the development of the newborn mouse hypothalamus and preoptic area in vitro: effects of estrogen on dendritic differentiation. Brain Res 7:97–101

Tsuruno Y, Kawano H, Nishiyama T, Hisano S, Daikoku S (1983) Substance P-like immunoreactive neurons in the tuberoinfundibular area of rat hypothalamus. Light and electron microscopy. Brain Res 289:1–9

Tsuruno Y, Hisano S, Okamura Y, Tsukamoto N, Daikoku S (1984) Hypothalamic substance P-containing neurons. Sex-dependent topographical differences and ultrastructural transformations associated with stages of the estrous cycle. Brain Res 305:331–341

Uchibori M, Kawashima S (1985) Effects of sex steroids on the growth of neuronal processes in neonatal rat hypothalamus-preoptic area and cerebral cortex in primary culture. Int J Dev Neurosci 3:169–176

Van Houten M, Brawer JR (1978) Regional variations in glia and neuropil within the hypothalamic nucleus. J Comp Neurol 179:719–738

Weisz J, Ward IL (1980) Plasma testosterone and progesterone titers of pregnant rats, their male and female fetuses, and neonatal offspring. Endocrinology 106:306–316

Wiegand SJ, Terasawa E, Bridson WE (1978) Persistent estrus and blockade of progesterone-induced LH release follows lesions which do not damage the suprachiasmatic nucleus. Neuroendocrinology 102:1645–1648

Yamanouchi K, Matsumoto A, Arai Y (1985) Neural and hormonal control of lordosis behavior in the rat. Zool Sci 2:617–627

Subject Index

MIX
Papier aus verantwortungsvollen Quellen
Paper from responsible sources
FSC® C105338

If you have any concerns about our products,
you can contact us on
ProductSafety@springernature.com

In case Publisher is established outside the EU,
the EU authorized representative is:
Springer Nature Customer Service Center GmbH
Europaplatz 3, 69115 Heidelberg, Germany

Printed by Libri Plureos GmbH
in Hamburg, Germany